Engineering with Fibre–Polymer Laminates

Peter C. Powell

Chair of Engineering Design with Plastics
Department of Mechanical Engineering
University of Twente
Enschede
The Netherlands

CHAPMAN & HALL

London · Glasgow · New York · Tokyo · Melbourne · Madras

Published by Chapman & Hall, 2–6 Boundary Row, London SE1 8HN, UK

Chapman & Hall, 2–6 Boundary Row, London SE1 8HN, UK

Blackie Academic & Professional, Wester Cleddens Road, Bishopbriggs, Glasgow G64 2NZ, UK

Chapman & Hall Inc., One Penn Plaza, 41st Floor, New York NY 10119, USA

Chapman & Hall Japan, Thomson Publishing Japan, Hirakawacho Nemoto Building, 6F, 1-7-11 Hirakawa-cho, Chiyoda-ku, Tokyo 102, Japan

Chapman & Hall Australia, Thomas Nelson Australia, 102 Dodds Street, South Melbourne, Victoria 3205, Australia

Chapman & Hall India, R. Seshadri, 32 Second Main Road, CIT East, Madras 600 035, India

First edition 1994

© 1994 Peter C. Powell

Typeset in Times 10/12 pt by Thomson Press (India) Ltd., New Delhi
Printed in England by Clays Ltd, St Ives plc

ISBN 0 412 49610 0 (HB)   0 412 49620 8 (PB)

A catalogue record for this is available from the British Library

Library of Congress Cataloging-in-Publication data
Powell, Peter C.
    Engineering with fibre–polymer laminates / Peter C. Powell. – 1st ed.
        p.   cm.
    Includes bibliographical references and index.
    ISBN   0–412–49610–0 (HB)   0–412–49620–8 (PB)
    1. Laminated plastics – Mechanical properties.   2. Polymeric composites – Mechanical properties.   3. Fibrous composites – Mechanical properties.   I. Title.
TA418.9.L3P69   1993
620.1'923 – dc20
                                                          93–16825
                                                          CIP

∞ Printed on permanent acid-free text paper, manufactured in accordance with the proposed ANSI/NISO Z 39.48-199X and ANSI Z 39.48-1984

Contents

Preface

This book has its recent origins in a Master's course in Polymer Engineering at Manchester. It is a rather extended version of composite mechanics covered in about twenty five hours within a two-week intensive programme on Fibre Polymer Composites which also formed part of the UK Government and Industry-sponsored Integrated Graduate Development Scheme in Polymer Engineering. The material has also been used in other courses, and in teaching to students of engineering and of polymer technology both in the UK and in mainland Europe.

There are already many books describing the analysis of and mechanical behaviour of polymer/fibre composites, so why write another? Most of these excellent books appear to be aimed at readers who already have a substantial understanding of stress analysis for linear elastic isotropic materials, who are thoroughly at home with mathematical analysis, and who seem often not to need much of the reassurance which numerical examples and illustrated applications can offer.

In teaching the mechanics of composites to many groups of scientists, technologists and engineers, I have found that most of them need and seek an introduction before consulting the advanced texts. This book is intended to fill the gap. Throughout this text is interspersed a substantial range of examples to bring out the practical implications of the basic principles, and a wide range of problems (with outline solutions) to test the reader and extend understanding. The reader is encouraged to develop this understanding of the basic principles at the level of physical intuition as well as by the use of detailed analysis. Non-numerical examples from a wide range of technologies are included to suggest that the principles have wide application beyond traditional engineering; the composites theme is indeed all pervasive. There are references to source books for more detailed discussions.

There are some modest prerequisites for this text; it is assumed that the reader can perform elementary integration and differentiation of simple functions, and can handle the simple matrix algebra operations of addition, subtraction, multiplication and inversion.

WHAT DOES THIS BOOK COVER?

This book first states briefly the main themes of polymer/fibre composites to provide a context for the more detailed discussion of mechanical behaviour. Second, this text addresses the philosophy and elementary concepts of solid body mechanics for a linear elastic isotropic material: this provides an introduction to the framework of understanding assumed in the more advanced texts on composite mechanics. Third, the language of plate theory is introduced and applied to a single ply of an isotropic material and then to a unidirectional ply. Fourth, the vocabulary of laminates is introduced and a simple analysis is applied for laminates where constituent plies are isotropic or unidirectional. Fifth the strength of plies and laminates is discussed. Finally the effect of change of temperature on mechanical performance of laminates is introduced, together with some comments about the stiffness of thin-walled structures.

Chapter 1 provides an introduction to composites. It discusses the basic building block of the unidirectional ply, and the extent to which it is possible to predict some of the main stiffness and strength properties of the ply from the bulk properties of the fibres and the matrix material. The need for laminate constructions becomes obvious, and the main classes of simple laminates are described. There is also a brief description of some of the major methods of making real products from those polymer fibre composites discussed in later chapters.

For those without a detailed knowledge of solid body mechanics, Chapter 2 provides an introduction for structural elements based on linear elastic isotropic materials. The emphasis is on what is needed as a basis for composites, and is therefore slightly different from normal texts. There is hardly a mention of principal stresses and strains, yield phenomena, and plane strain, because these do not usually feature much in composites within the scope of this book, if at all. On the other hand prominence is given to plane stress, to stresses and strains in the principal directions, and to bending and twisting curvatures. Some attention is given to the variation of stress and strain through the thickness under in-plane loads or bending moments, because these are so important in design to avoid failure in composites.

Chapter 2 provides an introduction to the three fundamental features of solid body mechanics: the principles of conservation of equilibrium, comptibility, and relations between stress and strain. To solve problems, one additionally needs to identify boundary conditions, and these are introduced in Chapter 2 and the ideas are applied throughout the book. In particular, later chapters show that there are some additional practical complications for composites, especially for single plies loaded off-axis, and most loading patterns for non-symmetric laminates: these have profound significance in design, testing and performance in service.

Chapter 3 provides the oppostunity to apply the elementary solid body mechanics to a single ply, using the ideas of composite mechanics to study the isotropic ply, the unidirectional ply loaded in the principal directions, and then the same ply loaded in-the-plane at some angle to the principal directions. The concept of coupling between applied in-plane loads and in-plane shear is explored.

Chapter 4 examines the behaviour of laminates based on isotropic plies: the concept of symmetry is explored, and the effects of lack of symmetry are studied without the complications of using unidirectional plies. The concept of coupling between bending and in-plane bahviour is explored.

Chapter 5 extends the concepts of laminate behaviour to those made from unidirectional materials arranged in a variety of ways commonly encountered in real products. Only laminates consisting of a few plies are discussed, but the principles also apply to more complicated lay-ups or to hybrid laminates. The full range of coupling between bending, twisting, shear and in-plane deformations is discussed, and raises some interesting challenges both for design and for manufacture.

Chapter 6 explores some of the factors which cause failure because the stresses in the principal directions in plies or within the laminate exceed the failure values. The main emphasis is on the use of the Tsai–Hill interactive failure criterion, which essentially says that when stresses in more than one principal direction are applied, the failure occurs at a rather lower stress than if the load were solely applied in the direction in which failure occurred. The Tsai–Hill criterion is useful and provides a good appreciation of the basic principles for strength under static load, and it would not be difficult to develop the understanding so gained by using a more advanced criterion where the need arose.

Chapter 7 provides an introduction to the behaviour of simple plies and laminates where a change of temperature occurs. The emphasis is on three situations: a change of uniform temperature, a temperature gradient, and (for laminates only) the effect of curing or solidification of the matrix at an elevated temperature and then cooling down to ambient temperature. Each of these situations leads to the development of internal stress and strain profiles which cause deformation and internal residual stresses, and are therefore of considerable practical importance. The emphasis is on stiffness but an indication is given about the effect of temperature on strength.

Chapter 8 is a simple view of how to adapt the principles in earlier chapters to the design of thin-walled sections for stiffness, and is intended to indicate some of the important aspects.

The analysis presented in this book is quite straightforward, but there are many small steps. Such a situation leads itself to the use of microcomputers, for which much good software is now available. This book is intended to provide a sound basis for using that software. The writing of this book was greatly facilitated by access to an early version of Laminate Analysis Programme (LAP) now available from the Centre for Composite Materials, Imperial College, London SW1X 2BY.

It is my experience that many students learn a great deal from studying examples and working through a range of problems from first principles: this develops and confirms understanding, adds confidence, and helps the student to develop problem-solving strategies for composites. An integral part of this book is the inclusion of a wide range of problems from which the reader can make a selection to try. Outline answers are at the back of the book. With this experience, the reader should be well-placed to explore other effects and the responses of more complicated laminates using computer software.

I will be pleased to learn of any errors, so that the next edition can benefit from corrections.

WHAT DOES THIS BOOK NOT COVER?

On the mechanics side, the scope of composites is vast, and in the interests of containing this book to manageable proportions some important topics have necessarily had to be omitted. In particular these include discussion of stresses and deflections of panels under out-of-plane loads, buckling and vibration of plates, resistance of composites to environmental attack, the principles of crack growth and fatigue in composite structures, design of joints, and testing of composites to gain data and prove performance. These topics are un-doubtedly important, and they are well covered in other texts. For these topics, and further development of what this book introduces, the reader is invited to dip into the more specialist books in the recommended further reading at the end of each chapter.

On the materials side, although the main emphasis is on the behaviour of plastics reinforced with fibres, nevertheless some attention is paid to applications involving the use of fibres to reinforce rubber. Although there is a large market for short fibre reinforced thermoplastics, these are not discussed in detail, not least because in most applications it is still quite difficult to predict accurately the alignment of fibres within the plastic as a function of position through the thickness: this is the subject of current research. Once the alignment has been determined, then some form of modelling with a basis in (local) unidirectional plies will be informative – but there is not the space to pursue this here.

Nevertheless, the scope of this book is wide, ambitious and challenging. The reader who masters its contents will not, of course, be an expert able to cope readily with the more advanced literature or with all aspects of composites behaviour. I would, however, hope that the diligent and patient reader would know enough about the behaviour of composites to be able to talk with the experts who have mastered the advanced texts; if the reader were able to engage in a two-way dialogue to advance his professional competence in the performance of composites, then this book will have achieved its major objective.

I am indebted to authors of texts quoted in the bibliographies, from whom I have learned a great deal. I would like to express my appreciation to friends and colleagues who have encouraged me to write this book, and who have given me valuable advice, guidance and material: Mr Frank Matthews of Imperial College London, Dr George Jeronimidis of the University of Reading, Dr C. Paul Buckley of Oxford University, Ir A. Jan Ingen Housz of the University of Twente, Professor Roy Crawford of the Queen's University Belfast and Professor A. Geoff Gibson of the University of Newcastle-upon-Tyne. I am grateful to Mr R. Kuilboer who prepared the illustrations. The responsibility for the published text and the solutions to the problems is mine.

Peter Powell
Enschede

Introduction to fibre–polymer composites

1

SUMMARY

This chapter sets the scene for the rest of the book. It introduces the main types of behaviour of fibre–polymer composites, with most emphasis on mechanical behaviour, and describes briefly the main methods for making products based on polymer–fibre composites.

One of the basic building blocks for composites is the unidirectional array of stiff strong fibres held together by a polymeric matrix. We shall call this a 'unidirectional ply' (or 'lamina'). We describe the need for a bond between polymer and fibre, the effect of using fibres of different lengths, the properties of representative polymers and fibres, and the influence of fibre volume fraction. In addition, the stiffnesses and strengths of such a lamina are described in terms of the properties of fibre and matrix and the volume fraction of fibres. Hooke's law, and some simple concepts of strength, are stated for the basic lamina loaded in-plane along and transverse to the fibre direction. These concepts are extended to a sheet which is loaded in-plane but at some angle to the fibre direction (the off-axis loading problem).

It becomes clear that the usefulness of the unidirectional lamina is severely limited by the poor transverse and shear properties, which are usually little better than those of the unreinforced polymer, and some strengths may be worse.

Stacks of bonded plies or laminae, called laminates, are used to compensate for the poor transverse properties of a single lamina.

A description is given of the stiffnesses and strengths of some simple laminates which are widely used in commerce, particularly the crossply laminate, the angle-ply laminate, and the random mat laminate. The importance of symmetry in laminate construction is emphasized. The discussion concludes with a note that a check should always be made of the interlaminar shear behaviour of a laminate subject to bending because of the lack of reinforcement through the thickness of the laminate.

Interspersed with the discussion about the mechanics of composites are descriptions of the processes commonly used to make products from (mainly long) fibre polymer composites.

To make the single unidirectional ply, or versions of it, techniques such as calendering, pultrusion and the winding of prepreg are used. Plies containing random in-plane arrangements of fibres wetted out by resin are also available as Sheet Moulding Compound (SMC).

For laminates based on unidirectional plies, various precision lay-up techniques are used and also filament winding. The use of woven or knitted cloth lends itself to hand lay-up (contact moulding) operations. Random fabrics can be processed by contact moulding, or by various resin-injection techniques together with closed moulds. Many laminates based on SMC are processed by compression moulding. Dough (Bulk) moulding compounds may be processed using either compression or (more commonly) the faster injection moulding processes.

Topics outside the scope of this introductory chapter include the buckling and fracture of fibre polymer composites, edge effects, and environmental effects. For detailed discussion of these topics and those covered in the chapter, the reader should refer to the further reading section.

1.1 INTRODUCTION

1.1.1 Design

It may seem strange to begin a book about composite materials by introducing the rather controversial topic of 'design'.

When we see something, or buy something, we see or buy a shape together with the defined volume of material used to make that product. That shape might be a dinner plate, a tennis racquet, a jacket, a sewing machine, a car or a house. One of the designer's primary jobs is to specify that shape (or the shape of each of the parts making the overall product), and the thickness and the material chosen to achieve it. To do this job well demands great skill. We might well argue strongly that an even more important task of the designer is to meet a specified need, or to fulfil a functional requirement: for example, is it better to ask a designer to design a bicycle, or to design a personal transport system for one person? It might be too blinkered to put the task in a very closed form.

Paintings and drawings can have great aesthetic and artistic appeal. But the shape which the designer determines is not practically usable until it has been made from a material into the desired product. (In this book we shall normally use the word 'material' in its general sense, rather than the special sense of a fabric for making garments.) The choice of material and the choice of method of manufacture can have an enormous influence on design and on each other.

Efficient design involves constantly switching between such questions as 'what is the best material?' to 'how would this processing method affect the shape I want to achieve?' 'can I make the shape I want using this material?' 'how would using this process affect the properties of the product?' and so on. The jargon shorthand for this process is 'concurrent engineering'.

We do this instinctively at the do-it-yourself level. For example, if we wish to decorate a wall in the house, we may either paint it, or apply wallpaper which may be patterned or textured, or both; or we may even apply a textured covering and then paint it.

Let us now focus on the materials aspect of design.

1.1.2 Uniform materials

Many materials from which commonplace items are made can be regarded as uniform in their composition and properties. Examples include glass for windows, steel for the housing of a washing machine, aluminium cooking foil for use in the kitchen, thick plastics sheet for illuminated advertising or motorway signs, or thick plastics pipes for gas or water supply.

Each of these materials has different properties, but they share some general and familar responses: under modest stretching forces they become uniformly longer and thinner (Figure 1.1(a)), and they expand uniformly on heating them so that a square becomes a slightly bigger square, and a circle becomes a slightly bigger circle (Figure 1.1(b)).

This assumption of uniformity of behaviour is at the core of almost all physics and engineering taught in schools, colleges, polytechnics and universities. On this basis, we choose materials and design many structures (or parts of structures) to be able to carry applied loads without causing distortion, excessive changes of shape or dimension, or breaking or undergoing permanent change of shape. Examples of such structures pervade our surroundings: there are complicated items or assemblies such as bridges, boats, cars, aeroplanes, trains; more humble but still useful and essential are things like cutlery, cups and saucers.

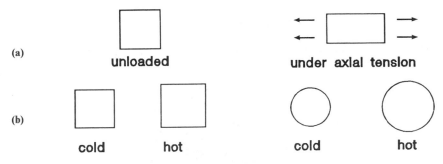

Figure 1.1 Behaviour of uniform materials.

Even using uniform materials presents the designer with more than enough complicated challenges in the quest for the most satisfactory and economical solution to the design problem. Chapter 2 introduces the mechanics of a special class of uniform materials, where the change of dimension or shape is directly proportional to the load applied. Chapter 2 is not about design, but the ideas are an essential (partial) input to design.

We do not want to become bogged down with detailed definitions and concepts at this stage. But it is helpful to remind ourselves that we cannot use the word 'uniform' without specifying the scale of dimension we have in mind. A house wall may seem to have a uniform composition from afar, yet looking closer we see bricks and mortar (Figure 1.2). A fabric for curtains or a cotton shirt may look uniform at a distance, yet on closer inspection can be seen as a woven structure (Figure 1.3), and under a hand lens we are likely to see that the yarn of the weave is twisted from many fibres. (The extremes of this approach may not help us – from astronomical distances all looks uniform, at the atomic scale nothing looks continuous: we choose a sensible scale and stick to it.)

1.1.3 Composite materials

But increasingly it becomes necessary to use (as one material) combinations of materials to solve problems because any one material alone cannot do so at an

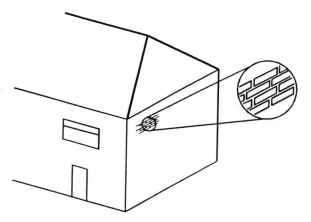

Figure 1.2 The brick wall of a house: homogeneous from afar, inhomogeneous close-up.

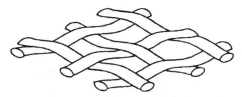

Figure 1.3 Detail of plain weave.

acceptable cost or performance. Indeed we have already quoted one example (the house wall). These materials are called 'composites'.

Many, if not most, materials, natural or man-made, are 'composites', or can be conceptually modelled as such. They do not have a uniform structure, and many (but not all) have properties which can vary considerably according to the directions in which they are measured. This can have a dramatic effect on the behaviour of products made from them.

A composite may be defined as a physical mixture of two or more different materials. The mixture has properties which are generally 'better' (to defined criteria) than those of any constituent on its own.

There are many forms in which materials can be combined to achieve composites. Our interest is in materials where there is usually a (more-or-less) continuous and often uniform phase, called a matrix, which usually surrounds a (more-or-less) discontinuous phase. Three types of discontinuous phase are of particular interest for solid materials: particles, plates, or rods (Figure 1.4).

We shall see that each type gives a different range of types of behaviour to the composite; for example it is not difficult to appreciate that particulate composites can have properties which do not normally depend much on the direction in which they are measured, whereas the behaviour of the rod composites depends on the direction of the rods in relation to the direction of the applied loads. The rods in rod-composites can be aligned in the plane, randomly directed in the plane, or random in all possible directions (Figure 1.5).

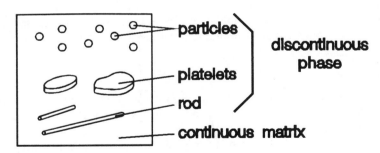

Figure 1.4 Continuous and discontinuous phases in a composite material.

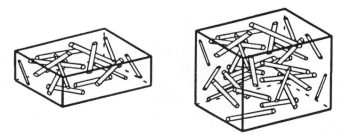

Figure 1.5 Fibres random-in-plane and random in 3-dimensions.

A moment's reflection raises important questions about the likely usefulness of constituent materials in a composite. Particles of sand may be good at resisting compressive loads, but how can you apply a tensile load to a single sand particle? How can you apply a tensile force to a bubble of a gas? However stiff and strong long slender fibres or rods may be in tension, how practicable is it to apply axial compressive loads – surely the rods will bend or buckle?

Reflections of these kinds underline the challenges of, and the reasons for, developing composite materials, which are to exploit the best properties of the constituents and suppress the worst, as far as possible. Thus in rod-based composites the rods are supported by the surrounding matrix, so that the rods can carry much larger loads in axial compression than they could alone.

Particulate composites

Examples of particulate composites include concrete, where sand or larger aggregate particles are held in place by a hydrated cement matrix; furnishing upholstery foam consists of gas bubbles surrounded by a polyurethane matrix; rich fruit cake, where raisins or sultanas are embedded in the 'cake mixture' (which is itself a composite). Steel and aluminium alloys where additional elements alter crystalline structure and may increase strength; butter and chocolate, where the fat has a partially crystalline texture. Some uniform polymers, such as polyethylene or polyamide (nylon), are able to develop small partly-crystalline regions (having a maximum dimension of up to 1 mm) within an amorphous matrix. On a still smaller scale of particle size we have toughened polystyrene, used to make vending cups, which consists of spherical particles of a rubber about 1 μm diameter within a matrix of polystyrene.

Platelet composites

Examples of plate- or platelet-reinforced materials include bone, crystalline fine structure in polyethylene; mica and tortoiseshell; strata of rocks in the earth's crust; flaky pastry.

By broadening our definition of a composite somewhat, so that both phases are continuous but only in the plane, most of the laminates described later in this book are themselves plate-based composites.

Rod composites

Examples of rod-reinforced materials pervade the material world, and we shall concentrate on some of these, in sheet or shell form, in this book.

The types of synthetic composite materials range from boron fibre reinforced aluminium, steel rod reinforced concrete, glass or carbon fibre reinforced thermoplastics or crosslinkable plastics; steel cable or textile fibre reinforced rubber.

Polymer/fibre composites are a growing market. Examples include pipes, pressure vessels and bulk storage tanks; belting for conveying and for power transmission and reinforced hose; aircraft structural components such as sailplane wings, military aircraft fusilages, and helicopter rotor blades; marine applications such as dinghy, boat and ship hulls, and rail transport. Uses in road transport from body panels through pneumatic tyres, air inlet manifolds for internal combustion engines and braided hose for coolant supply, to driveshafts and leafsprings. Other examples include cladding for building panels, and sports goods such as tennis rackets, vaulting poles and golf club shafts.

Most of these applications rely on different combinations of many of the following advantages:

- light weight;
- corrosion resistance;
- non-magnetic;
- high stiffness and strength per unit mass;
- low conductivity and thermal expansion;
- ease of manufacture.

Many polymers consists of long chains of repeating units. Depending on the nature of the basic building blocks, polymer chains can be partially lined up into parallel arrays by applying tensile or shear forces. This partial alignment may then be represented by parallel rods in matrix. Many of the principles of composites apply also to unreinforced polymer structures where molecular orientation occurs. Applications which exploit the molecular orientation include stretch blown PET bottles, tubular blown film, plastics strapping tape,

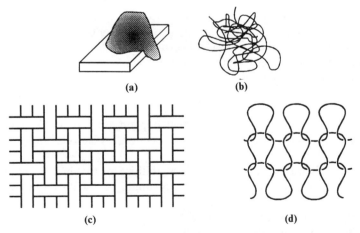

Figure 1.6 Unbonded discontinuous phase (a) heap of sand; (b) tangle of long fibres; (c) plain weave fabric; (d) plain knit fabric.

and cool-stretched reticulated mesh netting structures. On a much smaller scale, the very structure of crystallizable polymers depends on localized close-packing of molecular chains to give regions of dense packing and high stiffness and strength, surrounded by (or interspersed with) potentially more-flexible arrangements of the same polymer.

The natural world is responsible for its own polymer-based stiff rod-like structures such as cotton, wool and silk fibres. Many materials in the animal kingdom or the plant world rely on polymeric fibres made from such materials as collagen (based on proteins) and cellulose (based on saccharide (glucose) polymers). The matrix may consist of hydrated forms of the same sorts of polymers with different flexible structures.

The behaviour of many natural or fabricated materials can be understood, and behaviour designed, using the insights of composites. Examples include wood, and many plant structures; and tendons, muscles and arteries in man and animals, and woven or knitted fabrics used in industry or in garment manufacture.

1.1.4 Review

It may be helpful at this stage to bring these examples of rods-in-matrix composites together with some unifying remarks, before going into yet further detail. If we concentrate on those composites where the discontinuous phase (the rods) has a reinforcing (or toughening) effect, we may summarize the extremes of behaviour as the following:

1. All reinforcement and no matrix can give a very fragile material: a heap of dry sand, or a loose arrangement of fibres (Figure 1.6(a)(b)). One exception is where the reinforcement is woven or knitted: friction between the fibres or yarns ensures stiffness, strength and durability in fabrics for making garments or for other industrial uses (Figure 1.6(c)(d)). Another exception is where the discontinuous single fibres are twisted together; again frictional forces confer structural integrity in (say) wool, but only as long as the structure remains twisted.
2. All matrix and no reinforcement gives a uniform material which is not then a composite.
3. The properties of the composite will depend on the properties of the constituents and the proportion (and direction, where relevant) of each of them.
4. Conceptually there is a spectrum of possibilities for the reinforcement from continuous rods through short rods to roughly cubical or spherical particles.
5. Usually it is beneficial to try and achieve a good bond between the matrix and the reinforcing rods or particles.

Table 1.1 The scale of polymer fibre composites

	Range of representative dimension					
	pm	nm	μm	mm	m	km
Atom	**					
Polymer molecules		***				
Biological high polymers		****				
Crystallite		**				
Spherulite			***			
Fibre diameter			*			
Lamina thickness			***			
Laminate thickness				***		
Laminate length				*****		
Usable structure					*****	

1.1.5 The scale of composites

To put polymer fibre composites in perspective, it is helpful to examine representative dimensions of typical 'building blocks' encountered within polymer engineering, as shown in Table 1.1.

1.2 UNIDIRECTIONAL COMPOSITES: MANUFACTURE AND MICROMECHANICS

1.2.1 The basic rods-in-matrix model

The emphasis in the remainder of this book will be on long fibre reinforced polymers. Before delving into detail, it is well worth outlining a general model of these materials. The most useful concept is a regular parallel array of long stiff strong 'rods' well-bonded to and totally surrounded by a more flexible and weaker matrix (Figure 1.7). For most synthetic materials the 'rods' alone also have a lower coefficient of thermal expansion than the matrix alone. This 'sheet of aligned-rods', or 'unidirectional ply' has a range of properties, many of which we can understand almost intuitively without going into the technicalities. The unidirectional ply forms the basis of most analysis of laminates, and can be used to explain the behaviour of a wide range of combinations of materials.

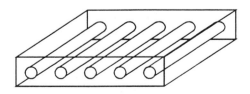

Figure 1.7 The basic rods-in-matrix model.

The purpose of the rods is to transmit and withstand loads. The rods can only carry loads along their length, and not across their diameter. Examples of fibres widely used in plastics composites are glass fibres and, to a lesser extent, carbon and aromatic polyamide (e.g. Kevlar) which are more expensive. The diameters of individual glass or carbon fibres is about 8 to 10 μm, most often used in bundles or yarns. Rods used in rubber composites include steel wire, rayon or polyester tyre cord and polyamide.

The purpose of the matrix is to hold the rods in place, to protect the rods from damage, and to transfer applied loads into the rods. The polymer matrix may be either a crosslinkable rubber, a crosslinkable plastic, or a thermoplastic.

The density of the unidirectional ply depends on the densities of the rods and the matrix, and their proportions, but not on any direction in the composite. To predict how a structure will behave, we shall see that the designer expresses the proportion of fibres in the composite using the concept of the volume fraction of fibres. The practical constructor prefers to work with the weight fraction of fibres in the composite, because he specifies the ingredients by weight rather than by volume. The volume fractions and weight fractions are related by the densities of the material making up the composite, and we shall discuss this in Section 1.2.6.

For stiffness, strength and thermal expansion along the direction of the fibres, it would seem reasonable to suppose that the proportion of fibres in that direction, and their properties, dominates the behaviour of the ply. Except at very low proportions of fibres this is found to be so.

But properties in the plane of the aligned sheet perpendicular to the fibres are governed by the properties of the matrix, except at very high proportions of fibres. The fibres contribute very little to the stiffness in the perpendicular direction, and often reduce the strength considerably. The ratio of the moduli in the along and across directions, the 'anisotropy ratio', can be in the range 3 to 100 for plastics composites, and more than 10 000 for rubber composites.

There are several important applications of unidirectional composites, such as conveyor belting and strapping tape, where the loads are applied along the direction of alignment or orientation, and the stiffness and strength across the alignment is satisfactory.

The electrical analogue is the flat ribbon cable used to connect computers and associated hardware. This offers insulated multiwire electrical connections along the wire, and the required electrical insulation between the conductors in the transverse direction. There is mechanical strength along the connector but it splits fairly readily because of poor transverse strength. The connector bends easily along its length because each electrical conductor is made from twisted wires.

But in most applications of the concept of fibre reinforced composites, loads are applied in several directions, and the dramatic range of directionality of properties of a unidirectional sheet is a major drawback. This leads the designer to line up the rods or fibres to resist best the directions of the applied forces. New structures are made, and we can think of them as based on bonded stacks of aligned plies with rods in different layers lined up in different directions; plywood is a simple and widely-used example. We shall discuss the behaviour of these laminates in Section 1.4, but now shall return to the theme of the single aligned sheet.

1.2.2 Design and manufacture

In many traditional technologies we take a particular material and shape it into the desired product. We may cast or machine a metal, or mould or extrude a thermoplastic.

In contrast, the designer of articles made from composites often designs the material as well as the structure he is interested in, and he must also be clear how the product is to be made. Questions specifically relating to composites uppermost in the designer's mind include:

- Where best to place the fibres?
- What amounts/proportions of fibres and polymer are needed?
- How will the fibres be put into place in the product?
- How will the fibres be held in place?

This then leads to two main approaches to design.

To make the best use of fibres, high proportions of fibres are used, for example in pipes and pressure vessels. To exploit the fibres, the designer has to use complicated design procedures, and exacting manufacturing techniques. Materials made like this are sometimes called 'advanced composites'. Most of the emphasis in this book is on this type of composite.

The alternative is to use random arrangements of fibres which permits only low proportions of fibres. The properties are not very good by the standards of the advanced composites, but the design for stiffness and strength can be based largely on conventional solid body mechanics for uniform materials, which is more readily understood. In fact the random arrangements of fibres (in the plane) look particularly promising for mass production methods, and advances in manufacturing techniques could well lead to a major new market penetration. One example is the car body panel, where much experience is being acquired in areas such as making the panels, assembling them to make doors, and achieving an acceptable surface finish which will also accept conventional automotive hot-painting treatment. Early production runs look most encouraging.

Note that for both advanced composites and random composites the ability to transfer loads across the interface between adjacent plies must always be checked: the interlaminar properties are essential for maintaining structural integrity. Fibres are seldom aligned with the through thickness direction, so delamination is a risk unless the design takes this into account. This idea is developed in Section 1.4.5.

1.2.3 How to make unidirectional composites

There are four main ways to make unidirectional composites: these are to hand lay-up unidirectional fabrics; to calendar rubber sheet; to pultrude; and to wind a large tube. There are many other techniques. Our objective here is to outline the main concepts rather than go into exhaustive detail. All of these techniques can also be used to make laminates with reinforcing fibres in more than one direction in the sheet, and this will be discussed in Section 1.4. Here we are only concerned with making unidirectional sheets.

Hand lay-up technique

This is a manual process, using simple inexpensive items of equipment, and suitable for crosslinkable plastics resins. The normal basis of the method is a 'unidirectional fabric' where most of the yarns or bundles of fibres are in the warp direction, with just a few weft yarns to provide location (Figure 1.8(a)). It is therefore a flimsy fabric which is easy to distort.

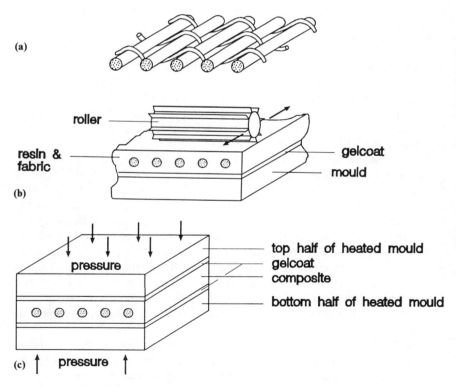

Figure 1.8 (a) Unidirectional fabric; (b) the principle of hand lay-up; (c) basic press (compression) moulding.

The simplest mould surface is a flat polished sheet of glass, metal or wood. The waxed sheet is covered with a thin coat of unreinforced crosslinkable resin (called a gelcoat), then covered with the fabric, taking care to ensure that the fibres are as aligned as possible. Resin (with added catalyst and accelerator) is then poured on, and ribbed rollers are used to wet out the cloth, coating each fibre, removing air pockets (voids), and consolidating the lay-up (Figure 1.8(b)).

In such a basic process it is difficult to ensure uniformity of thickness and volume fraction of fibres, and to ensure that the fibres are parallel. An alternative is to transfer the mould to a press (this increases the capital needed by a factor of at least a thousand), and squeeze out excess air and resin, to obtain a more uniform product with a defined thickness (Figure 1.8(c)).

Calendering

This process uses very robust (and hence expensive) equipment to withstand the high pressures, and is normally restricted to making rubber composites. The basic calender consist of three or four rigid short counter-rotating rollers

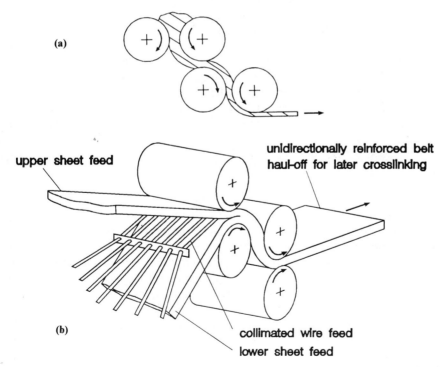

Figure 1.9 (a) The basic 4-roll calender; (b) the principle of making a unidirectional fibre reinforced belt.

('rolls') with horizontal small gaps ('nips') between them (Figure 1.9(a)). A rubber feedstock is dragged through the first gap, adheres to one roll as a sheet, and is taken through successively smaller gaps to control the thickness and surface finish of the sheet. The uncured sheet can then be rolled for subsequent processing.

To make a unidirectional sheet, a parallel array of primed 'rods', e.g. high tensile twisted steel wires, is sandwiched between two sheets of uncured rubber and pulled through the gap between two rolls of a calender to produce a compressed sandwich construction (Figure 1.9(b)). The sandwich is subsequently cured (crosslinked) at about 160 °C. This is the basis for making bracing plies for radial tyres for cars, and heavy duty conveyor belting for coal mines. Long lengths are possible, limited by the available length of the wires. Volume fractions of reinforcement are usually small, about 0.1 to 0.2.

Pultrusion

Pultrusion is used to make long lengths of sheet, rod, tube, I-section, channels, angle sections, or profile usually from crosslinkable plastics. The shape of the product is defined by the die. Fibres or yarns from many large-capacity spools (or creels) are pulled through a catalysed liquid resin bath and the excess resin from the wetted fibres is removed. The wetted fibres are brought together and pulled through the die. The die compresses the fibres, squeezes excess resin out of the entry of the die, and during the passage through the heated die the resin is crosslinked to produce a stiff strong unidirectional composite with a high volume fraction of fibres, typically in the range 0.5 to 0.7.

Pultruded sections are used to make ladder rungs, handrails, beams for footbridges, grid flooring, ducting, and large-scale guttering for industrial roofing.

Prepreg techniques

A prepreg is a convenient thin sheet of fibres traditionally pre-impregnated with a slightly crosslinked resin to hold the fibres in place. If the fibres are lined in one direction the prepreg is rather flimsy in the transverse direction. But overall a prepreg is much easier and cleaner to handle than any wet lay-up process, and the proportion of fibres is already precisely fixed, and usually high, typically in the range 0.5 to 0.7. Prepregs are usually destined for the manufacture of high-performance precision laminates. It is also possible to make prepregs based on one of a small number of speciality thermoplastics.

One basic process is to pull fibres from many large capacity spools through a resin bath, remove excess resin, and then wind the fibres helically on to a large diameter drum at a very small angle to the circumference. In this way a very thin-walled tube is made with the fibres arranged in the circumferential direction. The tube is then cut along its length, removed with great care from the drum, and flattened between the platens of a heated press. The heat causes the resin to just begin to crosslink (the 'B'-stage), so it loses much or all of its stickiness and can then be conveniently handled.

1.2.4 General behaviour of unidirectional sheets

The general response of composites to external loads or changes in temperature can be fascinating and often quite different from the response given by the special case of the uniform material which forms the backbone of the conventional educational experience. The new insights gained refresh existing knowledge too.

For example, under a uniformly-applied load acting in just one direction, a square element made from a composite may show any of the deformational responses given in Figure 1.10, one of which is not encountered in uniform materials. Contrary to casual expectation, it may come as quite a surprise to find that it is quite in order for cracks to grow in the direction of applied loads as well as perpendicular to them, as shown in Figure 1.11.

On heating up a composite, it may change dimensions in the ways shown in Figure 1.12, and circular holes in the composite panel may become ellipses on heating, as shown in Figure 1.13.

> **Problem 1.1:** If the thermal expansion coefficient of the fibres is much less than that of the matrix, and the responses shown in Figures 1.12 and 1.13 are unidirectional sheets, suggest the alignment of the fibres.

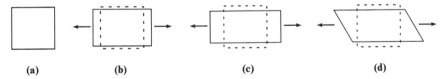

(a)   **(b)**   **(c)**   **(d)**

Figure 1.10 Three types of deformation under uniform uniaxial tensile load.

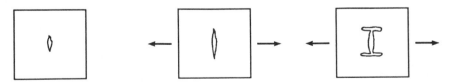

Figure 1.11 Crack growth in a unidirectional sheet (two possible modes).

Figure 1.12 Effect of change of temperature on dimensions of unidirectional sheet.

Figure 1.13 Change of circular hole (exaggerated) on heating a unidirectional ply.

1.2.5 Vocabulary of unidirectional composites

So far we have described materials as uniform or as a fibre composite with properties which vary with direction. It is now appropriate to introduce rather more precise and detailed descriptions.

The concept of a uniform material can be split into two categories which are not synonymous (although many people in casual speech fail to distinguish them):

1. Isotropic: the material has the same properties when measured in any direction from any point (Figure 1.14). Window glass is a good example.
2. Homogeneous: the material has uniform properties throughout, but may vary locally from point to point (Figure 1.15). A rubber or plastic containing a modest proportion of well-dispersed or well distributed particulate filler may have uniform isotropic properties, but the properties of the polymer and of the filler are different, and this is apparent on a very small scale.

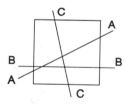

Figure 1.14 Isotropic material has the same properties in each direction AA, BB, CC.

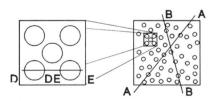

Figure 1.15 Homogeneous material: On the large scale properties are uniform along AA, BB. On the small scale properties in particles DD are different from those in the matrix DE.

The important point is that isotropy does not depend on any dimensional scale (except at the atomic scale outside the scope of this chapter) whereas homogeneity does. Sensible discussions of homogeneity must therefore include a statement of scale below which differences between constituents are acknowledged. A brick and mortar wall for a two storey building looks homogeneous from afar, but not close up. The brick itself appears homogeneous until you look closely at it.

As far as this book is concerned, *inhomogeneous* or *heterogeneous* means the opposite of homogeneous, and denotes distinct phases of material on the scale of interest. A brick wall looks inhomogeneous on the scale 10 to 1000 mm, and has non-uniform properties:

- Anisotropic means not-isotropic: properties differ in all directions from a point. A woven or knitted fabric shows this characteristic.
- An orthotropic sheet has three mutually perpendicular planes of symmetry, and properties are different in three mutually perpendicular directions from any point. A common example is shown in Figure 1.7.
- A ply or lamina is a flat sheet of material. In most of this book the ply is normally taken to be a flat unidirectional array of fibres uniformly spaced in and well-bonded to a matrix. Other forms such as random arrangements of fibres in chopped strand mat, or sheets of foam, can also be described as plies in certain technologies. Plies can also be isotropic, as used in foam-cored sandwich structures, or in packaging foils.
- A laminate is a stack of plies or laminae bonded together (Figure 1.16). Plywood is a good example.

In designing with composites engineers pay most attention to two scales of significance:

1. Micromechanics focuses on the small scale in the order of micrometres, and examines the interaction between the different constituents on a microscopic scale. This will be discussed briefly in Section 1.2.6.
2. Macromechanics focuses on the larger scale (millimetres to metres) in which the composite is presumed homogeneous, and that the properties of the composite are averaged apparent properties which are anisotropic. The macromechanics of a single unidirectional lamina is discussed in Section 1.3, and the macromechanics of laminates follows in Section 1.4.

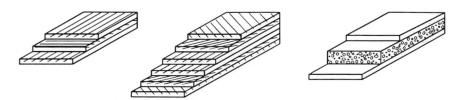

Figure 1.16 Three different laminates.

1.2.6 Micromechanics of unidirectional lamina

The basic building block in composites is a sheet or ply consisting of a regular array of parallel stiff strong 'rods' held in place by a more flexible and often weaker matrix (Figure 1.7). This unidirectional ply is orthotropic, using the vocabulary of Section 1.2.5. The rods can be steel wires, or fibres of glass or carbon, or can be fibres of polymers such as cellulose, cotton, or nylon (polyamide). (In this book the 'rods' or 'fibres' may indeed be single rods or fibres; but we commonly use these terms loosely to mean bundles of rods or bundles of fibres. Sometimes the bundles of rods or fibres are parallel, sometimes loosely or even quite tightly twisted, and each arrangement gives different properties. In this chapter we shall distinguish arrangements only where it is really necessary to do so.) The matrix can be stiff like concrete or aluminium, more flexible such as epoxy or crosslinkable polyester, or very flexible like rubber. There are two extremes of behaviour; all rods gives a structure which falls apart, and all matrix gives a uniform isotropic sheet.

We wish to examine the behaviour of the unidirectional orthotropic lamina. The effectiveness of the composite depends on a good bond between polymer and fibre (the interfacial bond), the length of the fibres, the properties of the fibre and the polymer, and the proportion of fibres and their arrangement.

It is the dream of the materials scientist to predict the properties of a fibre composite using only the properties of the fibres, the matrix and any other ingredients, together with their proportions and arrangement. Some properties can be reasonably estimated in this way, but many cannot be predicted with any useful precision, so it is good practice to measure the properties before making calculations of performance.

The bond between rods and matrix

The achievement of a good bond at the interface (Figure 1.17) between the surface of the fibre and the polymeric matrix is important for full exploitation of the fibres within the composite.

In many composites a good bond is only achieved by coating the fibres with a thin layer of a primer or 'size' which is chemically compatible with both the fibre and the polymer matrix. To give a simple example, a vinylsilane size is

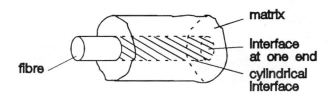

Figure 1.17 Interfaces between fibre and matrix.

applied to the surface of glass fibres used to reinforce unsaturated polyester resin. The vinyl groups promote adhesion to the resin, and the silane groups are compatible with the glass. A vinylsilane size is not suitable for epoxy resin and glass fibres, but an aminosilane will do the job, because the amino groups are compatible with the epoxy resin. Another example concerns the reinforcement of rubber with a cloth woven from nylon fibres. Normally rubber and nylon will not bond at all, but coating the nylon with a toluene di-isocyanate (using a solution in methylene chloride) permits a strong bond.

The area of the interface can be large for even a modest amount of fibre, not least because of the small diameter of the fibres. This calls for care in wetting out the fibres with polymer, especially at the high speeds used in commercial production.

Perhaps we should note that for good energy absorption during the failure of a structure we may wish to make a composite with a controlled interfacial strength, but we shall not explore this here.

Problem 1.2: At what length to diameter ratio does the area of the ends of a fibre become less than 1% of the area of the cylindrical area of a fibre of circular cross-section?

Problem 1.3: What is the surface area of 20 g of continuous glass fibres? The density of the fibres is 2540 kg/m^3, and the fibre diameter may be taken as 10 μm.

Fibre length

We have already mentioned that fibres are only effective when loaded along their length. It is seldom practicable to apply loads to fibres in composites by direct tension. Normally the load is applied to the matrix, and this load is transferred by shear through the interface to the fibre.

As an example of this principle, it is easy to apply a substantial axial load to a long pencil by gripping with both hands in pulling along the axis of the pencil. But with a short pencil there is not much length over which a grip can be achieved, and trying to apply too much axial force via the fingers merely results in the pencil slipping away by shear.

With very long fibres, the load can be effectively transferred by this shear mechanism, and apart from a region near the ends of the fibres, most of the fibre is stressed to take its full share of the external load, which is the main objective. But for short fibre lengths, there may be insufficient length over which the shear can be transferred (Figure 1.18). The minimum effective length depends on the aspect ratio (length/diameter ratio) of the fibre and the strength of the fibres and the interface. For many glass fibre/plastics composites the critical fibre length is typically only a few hundred micrometres.

Short fibre reinforced thermoplastics are sold with fibre lengths of a millimetre or so, sufficiently above the critical length to give useful enhance-

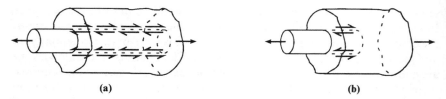

Figure 1.18 The interface transfers load from matrix to fibre: (a) sufficient length; (b) insufficient length.

ment of mechanical properties, and reduced thermal expansion coefficient, in the fibre direction, while retaining the versatility of, and ability to be processed into end-products by, conventional mass-production injection moulding technology. (This is an oversimplified view: the moulding process reduces the length of the fibres somewhat, which reduces property enhancement. Much research work is currently going on aimed at retaining longer lengths in the fibres by reducing breakdown of fibres.)

Sheet and dough moulding compounds based on unsaturated polyester contain fibre lengths in the range 25 to 50 mm, and are moulded by compression techniques; DMC can also be processed by a form of injection moulding. Compression moulding does not cause undue reduction in fibre lengths, but the in-plane movement of the charge as the press closes may introduce uneven movement of the fibres leading to an uneven composition from point to point.

The properties of fibres and polymers

We shall discuss the terminology of properties and the properties of individual materials in more detail in Chapter 2, but it is helpful to give some brief descriptions and definitions here, so that the properties of materials can be described.

For a bar of isotropic material of cross-sectional area A_x loaded uniformly in the x direction with a uniform force F_x perpendicular to the area A_x (Figure 1.19), the direct or tensile stress is $\sigma_x = F_x/A_x$. The longitudinal strain

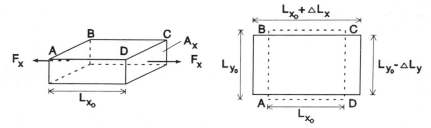

Figure 1.19 Application of tensile force to a bar, and the resulting deformations.

in the bar ε_x is defined as the change in length ΔL_x divided by the original length L_{xo}: $\varepsilon_x = \Delta L_x / L_{xo}$. For our purposes here (but it is not universally true), the material is assumed to have a linear stress–strain relationship up to failure. The tensile modulus of the material E is defined as $E = \sigma_x / \varepsilon_x$. The bar also contracts in width from L_{yo} to $L_{yo} - \Delta L_y$, and so the lateral contraction strain is $\varepsilon_y = -\Delta L_y / L_{yo}$, the negative sign indicating a decrease in relative length. Under the stress σ_x, the lateral contraction ratio, usually called the Poisson's ratio, v, is defined as $v = -\varepsilon_y / \varepsilon_x$ and is a positive quantity. The material fails in tension when it reaches the maximum stress (the strength), σ_{max}, which the material can withstand.

If uniform shear forces F_x act on faces of the same bar perpendicular to the z axis and having area A_z and separated by a distance h (Figure 1.20), then the shear stress τ_{xz} is given by $\tau_{xz} = F_x / A_z$. These shear forces induce a small displacement u shown on the upper surface in Figure 1.19, and hence the relative displacement of the two loaded surfaces is defined by the angle $\gamma = u/h$. (Strictly this should be tan γ, but we assume small angles.) We define the shear modulus G as the ratio of shear stress to shear strain: $G = \tau_{xz} / \gamma_{xz}$, where the

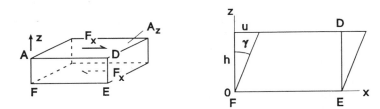

Figure 1.20 Application of shear force to a bar, and the resulting deformation.

Table 1.2 Typical properties of fibres and crosslinkable plastics used to make load-bearing resin-bonded fibre composites

Property	High modulus carbon fibres	High strength carbon fibres	E glass fibres	Epoxy resin	Unsaturated polyester
Density (kg/m³)	1950	1750	2560	1200	1400
Diameter (μm)	8	8	10	–	–
Tensile modulus (GPa)	390	250	75	3–6	2–4.5
Tensile strength (MPa)	2200	2700	1750	35–100	40–90
Break strain (tension) (%)	0.5	1	2–3	1–6	2
Linear expansion coeff (10⁻⁶/K)	−0.5 to −1.2	−0.1 to 0.5	5	60	100–200
Cure shrinkage (%)	–	–	–	1–2	4–6

stress–strain relationship is assumed linear. The shear strength τ_{max} is the maximum shear stress the material can withstand.

Stresses and strains are not properties of a material. The terms modulus, Poisson's ratio, and strength are properties of a material, and for isotropic materials do not depend on the direction in which the load is applied.

For reference the properties of selected representative fibres and polymers are summarized in Table 1.2.

Proportions and arrangements of fibres

There is a variety of possible arrangements of fibres in practical composites. The high-technology composites calling for high volume fractions of fibres are usually based on unidirectional plies. These may be laminated as unidirectional laminates or, as we shall see in Section 1.4, much more commonly as crossply or angleply laminates with the plies precisely laid up in a variety of directions to best resist loading conditions. It is much more convenient (easier to handle) to lay down woven cloth with a variety of weave designs (though more difficult to position with great precision), but these have reduced volume fractions of fibres after impregnation with resin compared with laminates based on unidirectional plies. (Why do you think this should be so?) An intermediate process is available for making products with axial symmetry. Pipes (and spherical shapes) can be wound with a helical criss-cross array of fibres or tapes; tubes can be braided. Knitted fabrics can also be used as preforms for subsequent impregnation with a polymeric matrix. The third distinct class is based on mats where fibres are 'short' (typically 25 to 50 mm) and are oriented in all possible directions in the plane, thus providing only small volume fractions of fibres.

We trespass into laminates in Table 1.3 in indicating typical volume fractions of fibres in different situations.

Problem 1.4: What is the maximum possible volume fraction of cylindrical fibres in a unidirectional array? Compare your answer with the representative volume fractions in Table 1.3.

Problem 1.5: A panel is to be made from a dough moulding compound having the following proportions of constituents by weight: 22% glass fibres (density 2560 kg/m^3), 53% polyester resin (density 1200 kg/m^3), and

Table 1.3 Typical volume fractions of fibres in composites

Unidirectional continuous fibre ply	0.5–0.7
Multidirectional cloth laminate	0.4
Random orientation in-plane	0.2
Short fibres in thermoplastics	0.1–0.3

25% calcium carbonate filler (density 2700 kg/m³). What is the volume fraction of fibres in the panel?

1.3 MACROMECHANICS OF A UNIDIRECTIONAL SHEET

For an array of aligned fibres in a matrix, which is orthotropic and certainly not isotropic, the co-ordinate system defining the principal directions is shown in Figure 1.21; by convention the fibre axis is called the '1' direction (or the longitudinal 'L' direction), and across the lamina in the plane is called the '2' direction (or the transverse 'T' direction). We shall apply loads, and hence stresses in these principal directions. We assume that stiffness behaviour of the fibres, the matrix and the composite lamina are all linear up to failure.

1.3.1 In-plane stiffnesses in the principal directions

The purpose of the fibres is to stiffen and strengthen the composite (at least in the direction of the fibres). It is not too surprising therefore that there are two values of modulus under uniaxial stress, one for along the fibres (E_1) and one across (E_2).

Micromechanics suggests that the longitudinal modulus E_1 of the unidirectional lamina is related to the properties of matrix E_p and fibres E_f, and the volume fraction of fibres, V_f by the rule of mixtures:

$$E_1 = V_f E_f + E_p(1 - V_f) \approx V_f E_f \quad \text{where} \quad E_p \ll E_f$$

Thus the greater the volume fraction of fibres, the higher the longitudinal modulus, E_1. The Poisson's ratio v_{12} also follows the rule of mixtures:

$$v_{12} = v_f V_f + v_p(1 - V_f)$$

The transverse modulus E_2 does not follow the rule of mixtures and correlation between simple theory such as

$$1/E_2 = V_f/E_f + (1 - V_f)/E_p$$

and experimental results is not usually very accurate. Except at high volume fractions of fibres it is perhaps simplest to regard E_2 as having a similar value to that of the matrix. This is not unreasonable as we have already explained

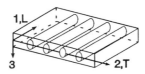

Figure 1.21 Co-ordinates in a unidirectional sheet.

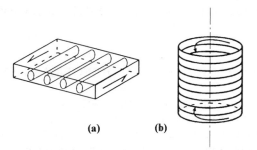

(a)   (b)

Figure 1.22 Applications of: (a) in-plane shear force; and (b) torque to a unidirectional sheet.

that it is difficult to transfer much load into short fibres, and the effective length of the fibre in the transverse direction is less than one diameter. (Try transferring any tensile load between your hands by gripping a pencil radially: the maximum load transfer is via half a circumference, and it is very difficult to get the fingers to grip on such a small surface unless you have very small (and sticky) fingers.)

Values of in-plane shear modulus G_{12} (due to the action of in-plane shear forces or torques, Figure 1.22) are difficult to predict accurately from the values for the fibres and the matrix, and are usually only slightly larger than those of the polymer matrix. Values calculated from

$$1/G_{12} = V_f/G_f + (1 - V_f)/G_p$$

do not correspond too well with measured values.

1.3.2 Bending behaviour in principal directions

When a bending moment is applied to a unidirectional sheet, the longitudinal bending stiffness is much greater than the transverse stiffness. This behaviour can be simply seen by taking a piece of corrugated card faced on one side only. The difference is bending stiffness in the two principal directions is much more apparent than the in-plane stiffnesses. On wide specimens one can see substantial anticlastic curvature, as shown schematically in Figure 1.23.

Figure 1.23 Anticlastic saddle-shaped curvature when a single uniform bending moment is applied to a unidirectional sheet.

1.3.3 Strength of unidirectional lamina in principal directions

Throughout this brief and simplified account we shall assume that the fibres are all of the same length and strength, unless otherwise stated. By 'strength' we mean the maximum stress which the material can withstand; in composites the strength depends on the type of stress (e.g. tensile, compressive or shear) and on the direction of the applied stress to the direction(s) of the fibres.

Longitudinal tensile strength

In the fibre ('1') direction of a void-free unidirectional lamina, theory suggests that the tensile strength of the ply is given approximately by $\sigma_{1\max} = \sigma_{f\max} V_f$ except at low fibre volume fractions, assuming that the fibres are much stronger than the matrix. What happens depends on which constituent stretches less before it breaks, the fibre or the matrix.

If the fibre can stretch more than the matrix, then the matrix will break when the matrix failure strain is reached. If there were no interfacial bond between matrix and fibre, then all the modest load shed by the matrix is readily taken by the fibres, which do not break until the load is increased to the value corresponding to the fibre breaking strain. With a good interfacial bond, although the matrix breaks at its maximum allowable strain, load is still shared between fibre and matrix away from the first matrix crack by shear transfer, so multiple transverse cracking can occur in the matrix in a good lamina.

If the fibre breaks at a lower strain than the matrix, then when the fibres reach their breaking strain, the substantial load from the fibres has to be taken by the weak matrix, which therefore also fractures with no further increase in load (Figure 1.24).

In practice the strengths of fibres are not uniform along their length, and fractures of individual fibres do not always coincide with the largest matrix crack. This leaves fibres bridging the crack plane. To achieve complete separ-

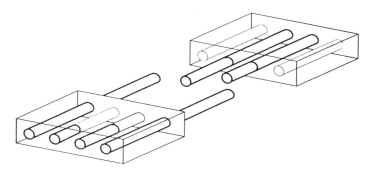

Figure 1.24 Fibre pull-out.

ation, it is necessary to break the interfacial bond and then overcome friction to pull the fibre out of its socket. The process of breaking the interface creates a great deal of new surface compared with just breaking a fibre across its diameter and hence absorbs a great deal of energy.

Poor interfacial strengths can also lead to the absorption of considerable amounts of energy before complete fracture occurs, and this can be beneficial in applications where it is desirable on safety grounds to absorb a lot of energy before failure is completed.

Transverse tensile strength

We have already mentioned (in Section 1.3.1) that it is difficult to transfer loads radially into the fibre, which therefore is ineffective at reinforcing the lamina when transverse loads are applied. The fibres tend to concentrate the load in the surrounding matrix thus locally raising the intensity of stress nearer to the maximum which the material can withstand, as shown in Figure 1.25. We therefore speak in shorthand of the fibres being sources of 'stress concentration'. For many composites the transverse tensile strength is only about one-third of the matrix strength. Three failure mechanisms are bond failure, matrix failure, and fibre failure, as shown schematically in Figure 1.26.

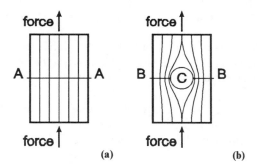

Figure 1.25 (a) The uniformly spaced lines represent a uniformly distributed load over cross-section AA; (b) round a circular hole C the lines of force are concentrated near the edge of the hole along BB.

Figure 1.26 Three mechanisms for transverse failure.

Although a particulate composite rather than a fibre composite, bars of cool chocolate containing whole hazelnuts will normally break at the interface between sound nuts and the chocolate matrix, when bending loads are applied to break the bar. Try it and see.

Tensile strengths in the longitudinal and transverse directions can differ by at least one order of magnitude, often much more. This is known for wood (which shows less extreme differences), for celery and for banana skins, all of which are easy to split in the longitudinal direction, and all of which show substantial pull-out of fibres when any attempt is made to try and break them by applying longitudinal loads (Figures 1.27 and 1.28).

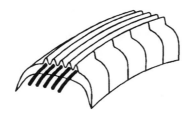

Figure 1.27 Pull-out failure in celery.

Figure 1.28 Longitudinal and transverse failure in a banana skin.

Compressive strength

There is no satisfactory account of the relation between properties of constituents and of the composite: the compressive strengths of composites are sensitive to details of manufacturing conditions. For glass and carbon fibre reinforced plastics the longitudinal compressive strengths are often compar-

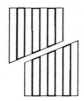

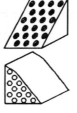

Figure 1.29 Modes of compressive failure.

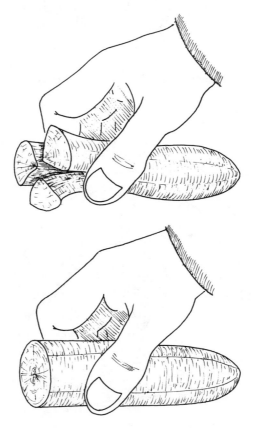

Figure 1.30 Compressive failure in banana fruit.

able with their tensile counterparts, but the longitudinal compressive strength of Kevlar composites is notoriously only about a quarter of the tensile strength, due to easy yielding and subsequent kinking of the fibres.

Transverse compression usually leads to a shear type of failure, as indicated in Figure 1.29, and the failure may involve just matrix failure, or failure by debonding between fibre and matrix. Either way the strengths are low, being comparable with those of unreinforced matrix. The planes of weakness are evident in the radial compression of a skinned banana (Figure 1.30).

Shear strength

The weakest shear direction is parallel to the fibres. It is quite possible for the in-plane shear strength to be less than that of the matrix alone because of the stress concentrating effect of the fibres.

1.3.4 Stress–strain relationships for off-axis in-plane loading

For many practical reasons discussed later in this book, we need to be able to relate loads applied in coordinates (x, y) which are at an angle θ to the principal directions $(1, 2)$ to the corresponding changes in length or angle in the (x, y) directions. The modulus E_x varies with θ as shown in Figure 1.31. When $\theta = 0°$, $E_x = E_1$, and when $\theta = 90°$, $E_x = E_2$. It is apparent that E_x varies widely. As shown in Figure 1.32, the shear modulus G_{xy} reaches a maximum value at $\theta = 45°$.

The direct stress caused by the in-plane off-axis loading also results in a shear strain as well as longitudinal extension and lateral contraction, as shown in Figure 1.33: we shall see later that this distortion is of great significance in the behaviour of laminates. This distortion does not arise in isotropic materials.

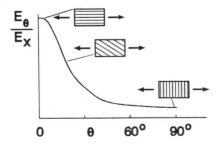

Figure 1.31 Variation of longitudinal modulus of unidirectional ply with direction of loading.

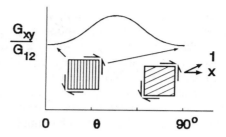

Figure 1.32 Variation of shear modulus of unidirectional ply with directional of loading.

Figure 1.33 Shear of unidirectional ply under off-axis tensile load.

1.3.5 Bending in non-principal directions

Essentially, we transform the bending resistances in the applied (x, y) axes to those in the principal directions $(1, 2)$. We then expect to see differences in curvature in the two principal directions, which is accompanied by anticlastic curvature. Less obvious is the development of twisting curvatures, as indicated in Figure 1.34, though resolution of applied moments into the principal directions explains them quite readily.

1.3.6 Strength under in-plane off-axis loading

We shall examine here just the tensile strength of a unidirectional lamina as a function of angle of test to the fibre direction. At very small angles the failure is dominated by longitudinal strength as shown in Figure 1.35. At angles larger

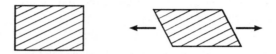

Figure 1.34 Applying a bending moment to two edges of a unidirectional sheet cut at an angle causes twisting about axes AA and BB as well as anticlastic curvature.

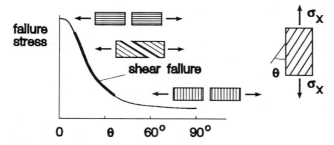

Figure 1.35 Failure under tensile load in-plane but at an angle θ to the fibres in a unidirectional sheet.

than about 40° failure is due to the resolved component of the applied load in the transverse direction. At angles in the range 5° to 40° a shear mode is often responsible for failure. It is apparent from the slope of the failure curve at small angles that the measured strength is very sensitive to the alignment of fibres: a clumsily aligned specimen could give an unjustifiably low value of longitudinal tensile strength.

1.3.7 Main conclusions for unidirectional lamina behaviour

Unidirectional laminae can offer high values of longitudinal modulus and tensile strength. Where a polymeric matrix is used, there is the added advantage of low density and light weight compared with metals. These points help to sell composites to both the designer and the accountant.

Applications of a unidirectional lamina or laminate rely on loads being almost entirely along the length of the reinforcement. Examples include untwisted polypropylene string and strapping tape, most synthetic fibres, steel wire reinforced rubber conveyor belting, glass fibre reinforced rubber power transmission belting, glass fibre reinforced crosslinkable plastics leaf springs for car and lorry suspension systems, and much steel reinforced concrete. The most traditional example is wood, and established design and craftsmanship ensures that the material is stressed mainly along the grain, and that stresses across the grain are minimized. Other examples of natural materials include wood, human hair, tendons, dandelion stems, banana skins and celery.

The major limitations is that values of transverse stiffness and strength, and shear modulus and strength, are very poor, only comparable with that of the unreinforced polymer. The same comment applies to the values of thermal conductivity and coefficient of thermal expansion. Where a structure carries in-plane loads in widely different directions, it is necessary to construct laminates to make the best use of composites and to offset the poor transverse properties. The following section outlines the main features of laminates.

1.4 MACROMECHANICS OF LAMINATES

Laminates consist of a bonded stack of individual thin sheets of material: they comprise a composite in their own right irrespective of the nature of each of the sheets. The modern kitchen worktop is often a laminate of a melamine (-formaldehyde) surface bonded to and supported by a composite thick sheet made from wood chips held in place by a crosslinked resin.

A stack of sheets is not in itself a laminate. The pages of a book are bound at one edge; the curved petals of a rosebud or the layers in the head of a cabbage; the clothes we wear together, such as a vest, shirt, pullover and jacket; none of these assemblies is a laminate because the arrangement of layers can be taken apart.

The bond between layers is normally good in synthetic composites for technical reasons, and we shall assume this in the following discussion. In nature a sufficiently strong bond obtains in laminated systems such as the concentric rings in tree trunks, or the layers of slate or mica. On the grander scale the earth's crust consists of various layers of geological structure which are bonded together: we can see evidence of this laminate structure in some cliff faces.

But a strong bond is not strictly essential: weakly bonded shells are seen in the bulb of an onion, in the concentric cylinders forming the stem of grasses or young onions. These are undoubtedly laminates. The food scientist and the cook use laminates to achieve appealing textures. Flaky pastry, and swiss rolls, are weakly bonded laminates, though not normally thought of in quite these terms. Sandwiches made from slices of bread, butter or margarine, and a suitable filling (particulate or sliced) are a convenient temporary laminate.

1.4.1 The vocabulary of laminates

A wide variety of patterns of deformation in laminates in possible. What actually happens under a given set of loading conditions depends on the properties of each ply (as described above), the numbers of each kind of ply, and the stacking sequence and angle between the principal direction in the ply and the reference direction in the laminate.

There are several terms which recur in describing the plies and their arrangement within a laminate. A brief description of them follows:

- Regular: the laminate consists of plies of the same thickness.
- Symmetric: the nature, material and orientation of any ply at any co-ordinate $-z$ above the midplane is exactly matched at co-ordinate $+z$.
- Anti-symmetric: the orientation of plies at $-z$ is matched by orientation of opposite sign at $+z$.
- Balanced: a laminate is balanced when there is the same number of plies in each orientation.

1.4.2 The advantage of symmetric laminates

Insisting on symmetry prevents many nasty problems arising. Symmetry occurs in the design of a laminate when any ply above the midplane is matched by an identical ply (including its orientation) in the mirror image position below the midplane, as shown in Figure 1.36. The particular advantage of symmetry is that flat panels stay flat after hot curing, or under in-plane loads, or both.

1.4.3 Stiffness and manufacture of simple symmetric laminates

Unidirectional laminate

Stiffness
The simplest laminate consists of a bonded stack of identical plies all oriented in the same direction (Figure 1.37). This is really a thicker version of the unidirectional lamina, with the same problems such as poor transverse and shear modulus, and poor transverse and shear strengths. It is no surprise that the unidirectional laminate is not widely used, except in special circumstances where its advantages can be exploited and disadvantages safely ignored.

Manufacture
To make unidirectional laminates with precision and with high volume fractions of fibres, unidirectional prepregs are laid in a flat mould and pressed

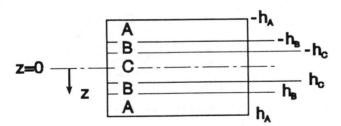

Figure 1.36 A symmetrical laminate.

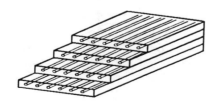

Figure 1.37 A unidirectional laminate.

together with enough heat to crosslink the resin (or cause the thermoplastics matrix to fuse); alternatively (but this is not really a laminate) the die gap controlling the thickness of the pultrudate is opened up and additional rovings are fed in. Where low volume fractions are acceptable, or very large items are to be made, the hand lay-up technique (Section 1.2.3) may be used: after the first layer of 'unidirectional' fabric has been wetted out, carefully add further layers, wetting out after each layer.

Crossply laminate

Stiffness

The next laminate is the crossply laminate, with half the plies at 0° and half at 90° to a reference direction (Figure 1.38). This increases the transverse modulus and strength considerably, although only half the fibres act in each direction. Many composites are made from woven glass rovings (Figure 1.39). The variation of modulus with direction is much less than for a single ply, as shown in Figure 1.40. The maximum value of tensile modulus in a balanced woven cloth laminate can never be greater than half the longitudinal modulus of a unidirectional cloth reinforced laminate because only half the fibres (often less) are resisting the applied load. The maximum extensibility (lowest modulus) occurs at 45° to the reference direction.

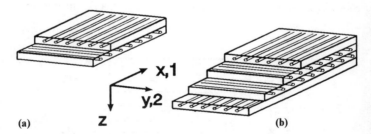

Figure 1.38 Crossply laminates: (a) non-symmetric; (b) symmetric.

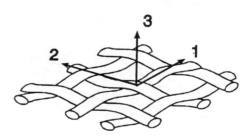

Figure 1.39 Balanced plain-weave cloth.

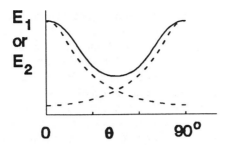

Figure 1.40 Variation of longitudinal modulus with angle of loading to weave direction.

The severe directionality of woven (or knitted) cloth without any resin matrix is well known. Sail manufacturers rightly insist that the free edges of sails are cut along one of the principal directions. Umbrella manufacturers do the same when they use woven cloth. On the other hand, to achieve maximum drape, designers of some ladies' dresses and skirts ensure that loads are applied in the 45° direction, i.e. along the bias direction of the fabric. Crossply laminates show these patterns of behaviour, though to a much lesser extent. The directions of warp and weft in rubber-coated fabrics for inflatable boats matches the directions of stresses after inflation, to confer longitudinal stability in the tubes making up the inflated structure.

Most plywood consists of an uneven number of unidirectional plies bonded together in a crossply arrangement, and is therefore an unbalanced laminate. In practice this is not too disadvantageous because plywood is used under circumstances where only modest loads (or changes in temperature) are applied, and edges are usually fixed to prevent warping becoming too apparent.

Manufacture
The simplest procedure is to stack unidirectional sheets of prepreg with fibre alignment alternating at 0° and 90° to a chosen reference direction, *x*, and bond them together in a heated press. Some examples of crossply construction are shown in Figure 1.38.

The transverse strength of pultruded sections can be improved substantially by feeding in some woven fabric with the continuous unidirectional rovings.

Where less precision and more modest volume fractions of fibre can be accepted, it is more convenient to use woven fabrics as the basic reinforcement. Where the fabric has equal proportions of fibre in the warp and weft directions, each layer of fabric may be modelled as providing both a 0° and a 90° alignment, with half the fibres in each direction. A single ply made in this way by hand layup is orthotropic (see Section 1.2.5), and is, in essence, a one-layer crossply laminate. Additional stiffness and strength may be achieved by adding additional layers of fabric. The type of weave influences the ease of processing, the achievable fibre volume fraction and the behaviour of the laminate, and we shall discuss aspects of these features in Chapter 5.

The calendering process (Section 1.2.3) may also be used to make a kind of crossply laminate: woven cloth is fed between rubber sheets and pressed together. In practice this involves two stages: one rubber sheet and fabric are fed between rolls, and to ensure penetration of the weave, the roll carrying the rubber rotates faster than the roll carrying the fabric; the fabric with the single coating is then passed with the other sheet of rubber through a second set of rolls. The rubber is then crosslinked. Applications include light-duty conveyor belting, waterproof and abrasion-resistant fabrics for inflatable dinghies, and inflatable buildings.

It is also possible to make a sort of crossply from a knitted fabric (Figure 1.6(b)). This is becoming of increasing interest for complicated shapes, where the ease of drape in placing the fabric is attractive. The efficiency of the fibres in the knitted form must be very low because of the low fibre volume fraction and the pronounced curvature of all the fibres arising from the stitch form.

Problem 1.6: Describe the following crossply laminates using as may as possible of the terms in Section 1.4.1. All laminates have a total thickness h, and are described from the top layer downwards. (a) $h/2 @ 0°$, $h/2 @ 90°$; (b) $h/3 @ 0°$, $h/3 @ 90°$, $h/3 @ 0°$; (c) $2h/3 @ 0°$, $h/3 @ 90°$, (d) $h/3 @ 0°$, $2h/3 @ 0°$; (e) $h/3 @ 0°$, $2h/3 @ 90°$.

Problem 1.7: Why must a crossply laminate based on woven cloth have a lower volume fraction of fibres and be weaker than one based on prepregs, assuming the same volume of fibres is used in each construction?

Problem 1.8: What is the maximum possible volume fraction of cylindrical fibres in the following arrangements: (a) crossply of unidirectional plies at $0°$ and $90°$, (b) plain weave, (c) satin weave with eight over eight construction. Assume that the fibres cannot be stretched or compressed in the radial direction. Compare your answers with the representative volume fractions in Table 1.3.

Problem 1.9: What is the likely maximum fibre volume fraction of a plain-knitted fabric? Assume the knitted fabric is made from a single fibre which retains its circular shape when deflected. Comment on your result.

Angleply laminate

Stiffness

The simplest symmetric angleply laminate consists of half the plies set at angle α to the reference direction, and half at $-\alpha$ (Figure 1.41). This system is widely used in the arrangement of bracing plies for radial pneumatic tyres, filament wound pressure pipes, and helicopter rotor blades. Although wood has much

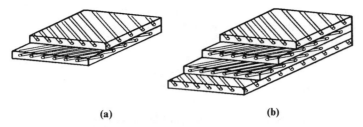

(a)   **(b)**

Figure 1.41 Angleply laminates: (a) nonsymmetric; (b) symmetric.

of the characteristics of a unidirectional ply on the macro-scale, the small scale plumbing tubes in wood act as if they were in part filament wound from natural fibres. Some muscles acting over large areas in animals are laid out and behave much as if they were angle-ply laminates. Veins and arteries behave as if the reinforcing fibres were filament wound and provide an efficient way of containing blood under pressure with minimal amounts of material. Muscle structures in large mammalian amphibians wind helically round the outside of the flesh, and are contracted or relaxed to influence the direction of motion.

In the symmetric angle ply laminate, the modulus in the reference direction is $E_{ref} \approx E_1 \cos^4 \alpha$.

In angle ply laminates it is sometimes possible to achieve an unusual effect: the design of a filament wound tube which does not change its diameter under internal pressure, such as may be needed to contain a solid fuel propellent for a rocket. Indeed it is entirely possible to make tubes which *decrease* in diameter under internal pressure; and releasing them from a solid mandrel after construction is then a major problem.

In the sailplane ('glider') wing, the designer must ensure that the wing does not bend too much, and therefore uses plies with fibres along the wing. But there are twisting forces also acting on the wing, and the greatest resistance to torsion is achieved with layers set at $+45°$ and $-45°$ to the length of the wing as shown in Figure 1.42. This use of plies in the most appropriate directions is commonplace in the design of effective fibre composite structures. In power

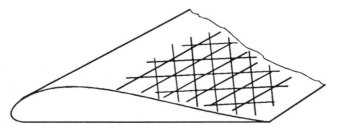

Figure 1.42 The main fibre directions in a sailplane wing.

driven planes there will also be additional fibres laid up in the direction from leading edge to trailing edge of the wing to transfer loads into the wing from flaps and other control surfaces.

Manufacture
Angle ply laminates can be precisely made by carefully lining up unidirectional prepreg sheets in the chosen directions of $+\alpha$ and $-\alpha$ to the reference direction, x, and pressing between heated plates to crosslink the resin (or fuse the thermoplastic matrix). The basis of pressing is shown in Figure 1.8(c). Two arrangements are shown in Figure 1.41.

An alternative procedure for structures having axial symmetry involves filament winding – rather like winding coils of electric wire, or bobbins for sewing machines. In filament winding (Figure 1.43), fibres or rovings are drawn through a catalysed liquid resin bath to wet the fibres, excess resin is removed, and the wet mass is guided onto a rotating mandrel (former) to describe a helix which makes an angle $+\alpha$ to the circumferential direction. When the fibres reach the end of the mandrel, the mandrel continues to turn and the axial line of advance is reversed, so that a reverse helix at $-\alpha$ is achieved. Successive passes completely cover the mandrel with overlapping layers which have much of the character of a weave with rovings interlaced at $+\alpha$ and $-\alpha$ to the reference direction. The resin in the completed winding is then allowed to crosslink, after which the mandrel is removed and the winding trimmed. The angle, or angles, can be readily chosen within wide limits to meet the loading conditions on the product.

Filament wound products include rocket motor casings, fire extinguisher casings, helicopter rotor blades, and pipes for pressurized oil or gas supply. Precise placement of fibres is necessary in order to obtain the best performance.

An alternative method of construction is braiding, adapted from the process used to make shoe laces. Braiding is widely used to reinforce hose in diameters from about 10 to 100 mm diameter. The core on which the yarns (primed or sized as necessary) are wound are an integral part of the final product; the braid is held in place by an additional layer of unreinforced polymer. The polymer forming the core and the sheath is usually either a rubber or a thermoplastic. The core is extruded, the braiding applied, and the two are then passed through a second extruder to achieve the outer sheath or covering.

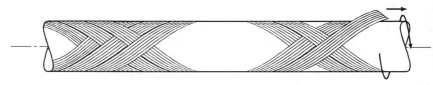

Figure 1.43 Helical filament winding.

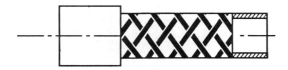

Figure 1.44 Interlacing in a braided hose – note the incomplete coverage.

The interlacing effect in braiding is much the same in general character as that achieved in filament winding, but sometimes no attempt is made in braiding to cover the core completely with yarns. Rather, a net-like structure is made (Figure 1.44) and covered with a thick layer of matrix, so the volume fraction of fibres is low, and compensated by the use of strong reinforcing fibres or wires.

Typical applications of flexible continuous braided hose include petrol hose, brake fluid hose and engine coolant hose for cars; fire hose for mains water; hose for washing machines, and compressed air supply lines for pneumatic drills.

It is also possible to hand lay-up angle ply laminates using unidirectional cloth; the special case of $\alpha = 45°$ is a crossply laminate, and can also be made from balanced woven fabric.

It is quite possible to make angleply laminates where proportions of fibres are at $\pm \alpha$, and at $\pm \beta$ to the reference direction, to suit loading conditions. Indeed, for some applications it is convenient to set $\beta = 0$ and lay up a laminate having suitable proportions at $+ \alpha$, 0, and $- \alpha$. Circumferential windings can be readily achieved, but not axial windings. There are similar limitations on braiding.

> **Problem 1.10:** Describe the following angleply laminates using the vocabulary of Section 1.4.1. Plies are unidirectional unless otherwise stated, angles are those which a principal direction makes to the chosen reference direction in the plane. Each laminate has a total thickness h. (a) $h/2 @ 45°$, $h/2 @ -45°$; (b) $h/3 @ 30°$, $h/3 @ 0°$, $h/3 @ -30°$; (c) $h/3 @ +45°$, $h/3 @ -45°$, $h/3 @ +45°$; (d) $h/3 @ +45°$, $h/3 @ 0°$, $h/3 @ -45°$, (e) $h/3 @ \pm 45°$ (woven), $h/3 @ 0°$ (unidirectional), $h/3 @ \pm 45°$ (woven); (f) $h/4 @ 0°$, $h/2 @ 90°$, $h/4 @ 0°$, (g) $h/4 @ 0°$, $h/4 @ 90°$, $h/4 @ 90°$, $h/4 @ 0°$; (h) $h/4 @ 45°$, $h/4 @ -45°$, $h/4 @ -30°$, $h/4 @ +30°$; (i) $h/4 @ +45°$, $h/4 @ -45°$, $h/4 @ +45°$, $h/4 @ -45°$.

Random laminate

Stiffness
The random laminate may be modelled as a symmetric stack of plies aligned at all possible angles to the reference direction (Figure 1.45), and has a uniform

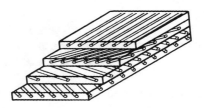

Figure 1.45 Representation of a notional random laminate.

modulus in the plane given by $E_{ref} \approx (3/8)V_f E_f$. This value of modulus is bound to be low because it is impossible to achieve any close packing, and volume fractions of fibres are usually only about 0.1 to 0.2. In practice random mats are made from chopped glass fibres some 50 mm long which are then impregnated with a crosslinkable resin, and sufficient mats are used to give the required stiffness and strength.

This random laminate is widely used in light-duty contructions such as boxes and covers for services, and in home-made canoes, and in boat and ship hulls. The material is isotropic in the plane, and this facilitates design. But the material is not reinforced through the thickness, and it is therefore essential to check that the proposed structure will resist any interlaminar forces.

There is a temptation to regard structures made from SMC, DMC, and the short fibre reinforced thermoplastics as if fibres are orientated randomly in the plane, but this is by no means always true. There is usually some directionality of fibres caused by displacement during the processing operation. This is extremely difficult to predict quantitatively and is the subject of active research. There may also be significant variation of volume fraction of fibres through the moulded product, especially in SMC and DMC, because of preferential flow of resin between the fibres.

Examples of materials outside the scope of this chapter which can be represented by the random mat model include many types of paper, and felt for hat manufacture and craft work.

Manufacture of random-in-plane laminates
There are several ways of making random laminates where fibres, or bundle of fibres, are aligned randomly in the plane of a flat sheet.

Contact moulding is the cheapest technique and uses a smooth waxed mould to which a resin gel coat is applied. There are then two simple variants for incorporating the fibres.

The first uses a chopped strand mat consisting of a random array of fibres some 25 to 50 mm long held together with a suitable binder (which must stick to both the fibres and later to the resin). The mat is cut to shape, laid on the gel coat, and covered with resin which is rolled by hand to give good impregnation of the mat, wetting out of the fibres, expulsion of voids and bubbles, and a compact sheet. The principle is the same as that shown for unidirectional sheet in Figure 1.8(b). The moulds are of relatively lightweight construction, because they do not need to withstand high pressures. Items range from chair shells through cladding panels for buildings, to dinghy hulls, and even hulls for minesweepers (where the non-magnetic qualities of glass fibre reinforced plastics are essential).

The second variant of contact moulding consists of spraying a mixture of chopped fibre rovings and catalysed resin on to the mould surface. This is faster than hand lay-up, and local areas can be thickened at will. The mass of composite still has to be rolled out and compacted, and tolerances on thickness are kept to acceptable (but wide) tolerances only by the use of skilled labour.

Higher pressure moulding techniques are used particularly for more complicated shapes to be made quickly in large quantities, where tolerances on wall thickness and other dimensions are critical, where good surface finish is needed on both surfaces, and where bosses, ribs and holes are required. The higher pressures call for more robust and more expensive equipment.

One feedstock is a resin-impregnated random fibre sheet, called a sheet moulding compound (SMC). This rather sticky sheet is cut to shape, and one layer (or more overlapping layers) laid in an open heated split metal mould. On closing the mould under modest pressure, the charge flows to fill the cavity, and the matrix crosslinks to give a stable shape. The product can be removed hot, and excess material is trimmed away. This process is called compression moulding (Figure 1.46). the volume fraction of fibres in SMC is usually only about 0.1, and the mixture usually contains a substantial proportion of a particulate filler such as magnesium hydroxide or aluminium trihydrate, which provides in effect a light permanent crosslinking action, thus increasing the viscosity of the compound.

An alternative feedstock is a dough moulding compound (DMC) which has a similar composition to SMC, but in which the fibres are aligned randomly in

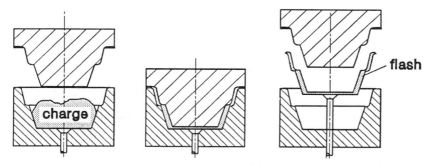

Figure 1.46 Compression moulding of SMC.

three dimensions. It too can be compression moulded. But it can also be injected under high pressure into a closed heated split mould. The flow of material into the mould does introduce some lining up of fibres, so the product is not really made of fibres orientated randomly in the plane. Typical DMC injection moulded products include car headlight reflectors, gas and electricity meter housings, and space-filling panels and boxes for equipment.

The SMC and DMC compositions are extremely viscous and only flow under high pressure. Resin injection is an alternative approach. Cut dry random (or woven) fibre mat to shape and place it carefully in a split mould, and then inject a very mobile resin which under modest pressure will displace the air and wet-out the fibres. When full impregnation and wet-out is achieved, the resin is allowed to crosslink to lock the fibres in position and fix the shape of the product. The fibre volume fraction is still modest.

Where the effect of randomly aligned fibres in the plane really calls for high fibre volume fractions in the range 0.5 to 0.7, the only technique is to use a stack of unidirectional plies notionally aligned in all possible directions in the plane. In practice we shall see (in Chapter 5) that only a few directions are needed to achieve the random effect. $(+60°/0°/-60°)$ is the minimum sequence (Figure 1.47(a)), laid up in a mirror image to preserve symmetry, and $(0°/90°/+45°/-45°)$ is another useful building block (Figure 1.47(b)). These arrangements give quasi-isotropic laminates which are isotropic in the plane but not isotropic under bending loads.

The reader will be quick to realize that if woven cloth is used, then a medium fibre volume fraction composite having the character of $(0°/90°/+45°/-45°)$ can be readily made. Triangular weave fabric (Figure 1.48) is also commercially available (at some expense) for making quasi-isotropic laminates where its qualities can be justified.

General stiffness behaviour of laminates

We can now extend the discussion from Section 1.2.4 from how a ply behaves to how a laminate might behave. For laminates it is entirely possible (and may

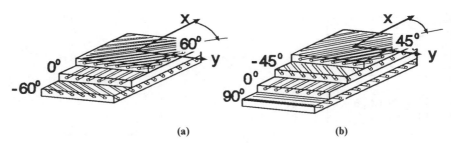

(a)   (b)

Figure 1.47 Basic building blocks for a quasi-isotropic laminate: (a) $(60°/0°/-60°)$; (b) $(45°/-45°/0°/90°)$.

Figure 1.48 Triangular weave fabric.

well even be desirable) to obtain any of the responses in Figure 1.49 when a load is applied in one direction in the plane of the laminate.

Changing the temperature can induce changes in linear dimension in the plane (Figure 1.50), and can also lead to out-of-plane bending and twisting (warping).

Judged from the viewpoint of isotropic materials, most of the responses in Figures 1.49 and 1.50 seem to break the laws of physics. But when the direc-

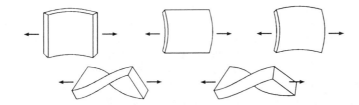

Figure 1.49 Out-of-plane responses to a tensile load.

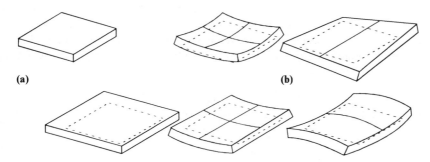

(a)　　　　　　　　　　　　　　　**(b)**

Figure 1.50 Some responses to a change of uniform laminate temperature: (a) original; (b) heated sheet with broken lines indicating the original.

tionality of properties is taken into account, the relevant laws are obeyed, and they readily explain the observed phenomena.

Problem 1.11: What is the maximum volume fraction of cylindrical fibres in a triangular weave fabric? Assume that the fibres cannot be compressed or extended in the radial direction.

1.4.4 Strength of wide symmetric laminates

It may seem strange to specify 'wide' laminates at this late stage in the chapter. The reason for so doing is that in narrow laminates there is an emphasis on edge effects (within a width approximately equal to the laminate thickness) which can sometimes lead to splitting and delaminations. The detailed explanations for these effects lie outside the scope of this chapter; they are summarized in Chapter 6 and are clearly described in some detail by Hull and by Datoo.

Failure of symmetric crossply laminates

We shall assume that the laminate is made from a stack of identical laminae, that the stress–strain curve for each principal direction in the lamina is linear to failure, and that the strain to break in the transverse direction of the lamina is less than that to break in the longitudinal direction.

Consider first a stack of plies which are not bonded together, but which are uniformly extended when stressed in the x direction (Figure 1.51(a)). When the applied strain reaches the failure value for the transverse ply, it breaks. This ply can no longer sustain load, and the stiffness of the laminate decreases, giving a single knee in the stress–strain curve, as shown in Figure 1.51(b). The lateral contractions in the two sets of plies are different. The transverse ply does not wish to contract along its length very much because of the minute minor Poisson's ratio. The longitudinally loaded plies have a much larger major Poisson's ratio and contract much more.

Now consider a similar stack of plies which are well-bonded together. When the applied strain reaches the failure value for the transverse ply, there will be a crack at a suitable point in the transverse ply. The stiffness will decrease, but it will still be possible to transmit some load via the transverse ply because of the

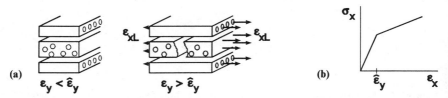

Figure 1.51 Failure of wide crossply laminates with unbonded plies: (a) development of transverse cracks; (b) stress strain curve.

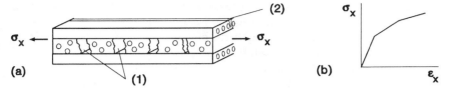

Figure 1.52 Failure of wide crossply laminate with good bond between plies: (a) 1. multiple cracking of transverse plies, 2. longitudinal cracking; (b) stress–strain curve.

interlaminar bond. At some higher overall load, more cracks develop, and hence the stress–strain curve shows a series of steps as shown in Figure 1.52(a). Being bonded together, the transverse plies restrain the desired contraction of the longitudinal plies. If the longitudinal strain becomes excessive, the longitudinal plies may crack in the direction of the applied load, as shown in Figure 1.52(b).

1.4.5 Interlaminar failure in bending

In isotropic materials, shear stress is developed most in the midplane when a bending moment is applied; this shear stress is usually modest compared with the bending stress and can be ignored unless the beam or plate has a very short span. Few laminates are made which have reinforcing fibres passing between adjacent plies. The bond between plies is therefore not reinforced and, in particular, is weak in shear. In the plane the reinforced composite is usually strong, and well able to resist bending stress, but the interlaminar bond is weak and even thin beams or plates of large span are prone to interlaminar failure unless this is checked during design. The difference between simple bending and the shear developed when the interface breaks is shown in Figure 1.53.

1.4.6 Behaviour of non-symmetric laminates

Lack of symmetry is responsible for a flat laminate warping out of plane under the action of in-plane loads, or as the result of a change of temperature such as might occur when cooling down after manufacture.

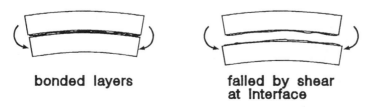

bonded layers

failed by shear at interface

Figure 1.53 Failure in bending by interlaminar shear.

A laminate made from two isotropic layers each having different properties will bend out of plane under an in-plane load or under a change of ambient temperature – the latter is exploited in the bimetallic strip. A crossply laminate made from two identical plies will behave in a similar manner.

An unbonded pair of unidirectional plies with fibres cut at $+\alpha$ and $-\alpha$ will both shear when under load in the reference direction. But if they are bonded, the opposite shears will cancel and cause a twist out of plane. A similar effect is caused in the bonded pair by a change of temperature.

A deliberate application of the characteristics of a non-symmetric laminate is in the design of a wind turbine blade so that it twists under the action of applied loads. The objective is to provide power control whilst avoiding the use of mechanisms for rotating the blade or blade tip. There are three approaches. The first is to choose fibre orientation to couple bending loads with blade twist. The second is to use a different lay-up to give coupling between an axial applied load and twist. The third achieves blade twist by inducing internal pressurization of a load carrying spar.

FURTHER READING

Holloway, L. (ed) (1990) *Polymers and Polymer Composites in Construction*, Telford, London.

Hull, D. (1981) *An Introduction to Composite Materials*, Cambridge University Press. A most readable and well written book.

Phillips, L.N. (ed) (1989) *Design with Advanced Composite Materials*, Design Council, London and Springer Verlag.

. Strong, A.B. (1989) *Fundamentals of Composites Manufacturing: Methods, Materials and Applications*, Society of Manufacturing Engineers, Michigan 48121, USA.

Most of the literature dwells on mathematical analysis and you may find the following heavy going, at this introductory stage. They are relevant to later chapters.

Agarwal, B.D. and Broutman, L.J. (1990) *Analysis and Performance of Fibre Composites*, Wiley.

Ashbee, K.H.G. (1989) *Fundamental Principles of Fiber Reinforced Composites*, Technomic.

Carlsson, L.A. and Pipes, R.B. (1987) *Experimental Characterisation of Advanced Composite Materials*, Prentice-Hall.

Datoo, M.H. (1991) *Mechanics of Fibrous Composites*, Elsevier.

Jones, R.M. (1975) *Mechanics of Composite Materials*, McGraw-Hill.

Vinson, J.R. and Sierakowski, R.L. (1986) *The Behaviour of Structures Composed of Composite Materials*, Kluwer.

Walter, J.D. (1981) Cord reinforced rubber, Chapter 3 in S.K. Clark (ed), *Mechanics of Pneumatic Tires*, US Government Printing Office, Washington DC 20402.

Introduction to solid body mechanics

2

OVERVIEW OF ELEMENTARY SOLID BODY MECHANICS

The main preoccupations of solid body mechanics are seeking answers (for a particular product) to two types of questions, typified by: (a) will it deform within acceptable limits under the applied loads?; and (b) under what values of loads or deformations will it break? Most of this chapter, and indeed the book as a whole, is concerned with (a) not least because it is more straightforward and therefore easier to discuss at the introductory level of the book. Strength is a complicated, and often rather controversial, topic, and we shall leave the main ideas to Chapter 6.

Solid body mechanics is therefore concerned overall with the deformation response to applied forces or moments; or with the forces or moments induced by applied deformations. Our interest focuses on the response to static loads. Although impact and cyclic or fatigue loads are important, they are more complicated to handle, and lie outside the scope of this chapter (and indeed this book).

Applied forces may be tensile, compressive or shear; the loads or moments may be concentrated at a point, applied along a line or over a defined area. They may be uniform along the line or over the area, or they may vary from point to point.

The applied forces set up within a given body *internal forces* or moments which are not necessarily in the directions of the applied loads. To take account of the dimensions of the body, it is necessary to transform the forces into stresses which are independent of the material from which the body is made.

Experiment shows that there are well-defined *relationships between stress and strain*, a particular case explored in this book being linear elasticity. Linear elastic behaviour can apply to many isotropic materials (this chapter) and to composite materials (later chapters); the relationships between stress and strain are simplest for isotropic materials, and more complicated for composites. Each material has its own unique stress–strain relationship which represent materials properties which vary from material to material, and in deformation problems are the only quantities which depend on the materials used. Stresses and strains themselves are each independent of materials. From the stresses at any point it is now possible, using these relationships, to calculate the strains in the body at the same point.

From the *strains*, which are the dimensionless deformations, it is possible to use the dimensions of the body to calculate the actual *deformations* or *displacements*, which may be tensile, compressive, bending or twisting (or a combination). Strains and curvatures are related, so it is also possible to relate loads or moments to *deflections* in beams and plates.

The simple mechanical responses outlined above can be analysed using three major principles together with the boundary conditions relevant to the problem. This still leaves plenty of algebraic manipulation which facilitates the mechanics but is only the servant of the three major principles. These principles are:

1. Conservation of equilibrium. This is a statement of Newton's first law for static applied loads. For a body in equilibrium under applied forces and moments, the action of the applied forces and the reaction of the internal forces and moments are equal and opposite. If action and reaction are not equal, either the body changes shape until they are equal, or some linear or angular accelerations develop, and continue to develop for as long as internal and external forces are out of balance.
2. Compatibility. All the material in the body under consideration must hold together under the applied loads and deformations. This is usually expressed in terms of desired or necessary displacements, deflections, edge conditions, strains, curvatures and other ways of describing change of dimensions or shape.
3. Relationships between stress and strain. These have already been mentioned above.

Where it is necessary to ascertain whether the article will break under the applied loads (or how near to breaking it is), it is necessary to use a failure criterion. This failure criterion can be expressed in terms of a maximum stress (or strain), or a maximum value of a combination of stresses (or strains), or in terms of maximum deflections. The failure criterion is therefore not usually a part of the relationship between stress and strain. Stress-based failure criteria usually involve either ductile yielding or brittle fracture.

Analysis of all but the simplest situations involves making assumptions and approximations: these simplify the real product into structural idealizations which facilitate analysis of the behaviour.

This chapter is written for those who have not covered a course in solid body mechanics in the undergraduate curriculum. It can only introduce the main ideas and vocabulary, and will not delve into fine detail. It is hoped that the reader will be able to understand simple examples, be aware of the common assumptions and approximations used in engineering, and will understand some of the limitations of the approach; and then proceed to read one of the recommended texts for further information, if so wished.

In this chapter there is emphasis on a range of simple structural elements such as bars and beams because these illustrate the simple principles involved. In later chapters the emphasis is an aspects of the behaviour of plates and plate-like elements (including the special cases of thin-walled tubes and shells).

Those readers who already have a grasp of solid body mechanics for isotropic materials may find useful the opportunity for revision of familiarity with nomenclature and sign conventions. In particular there is an emphasis in this chapter (and also in the early parts of Chapter 7) on certain matters which are important in the later analysis of composites but which are not usually given great prominence in elementary texts on solid body mechanics or strength of materials. There is particular emphasis on the description of twisting curvature.

2.1 BASIC STRUCTURAL SHAPES FOR ENGINEERING

2.1.1 Introduction

There has long been the tradition of building from simple elements a complicated structure. Examples include architecture (cathedrals, houses), steam locomotives, ships, motor cars, aircraft, furniture, clothing, and most household appliances.

Most of us have something of the engineer or technologist in us, though some may prefer to call themselves craftsmen or artists. Most readers probably indulge in some do-it-yourself – many enjoy doing so, some on a selective basis, others enforced. Many activities involve a constructional element of building up a complicated structure from simple elements. Examples of activities involving assembly include:

1. assembling a sailboard from substructures which fit onto a car roof;
2. knitting a cardigan or pair of gloves or socks;
3. tailoring a jacket or dress;
4. making a model to represent something;
5. building a garden wall;
6. assembling a cupboard unit or greenhouse from a kit of parts;
7. playing with interlocking assembly puzzles;
8. mixing ingredients and cooking bread, cake or pastry.

Whether made from natural or synthetic, homogeneous and isotropic, or composite materials, many artefacts and manufactured products are complex – we often cannot understand them and the way they work completely in one go. To achieve this understanding, we break them down notionally into component parts if they have more than one part. We may conjure up in our minds exploded view diagrams. We may then further break down each component into structural elements, idealizing them into the basic forms such as:

1. 1-dimensional roads, planks, fibres (where the long dimension is much greater than cross-sectional dimensions);
2. 2-dimensional flat plates (where the thickness is assumed negligible compared with in-plane dimensions), on which emphasis is placed in later chapters of this book;
3. 2.5-dimensional thin curved shells and tubes, also discussed in later chapters;
4. 3-dimensional substantially solid blocks, cylinders, prisms, cuboids (where dimensions in all three perpendicular dimensions are comparable).

Each technology or craft has its own specialist terms and jargon for these elements, which adds a deliberately exclusive colour to the discussion at the expense of clarity to the outsider.

Sometimes we have to make fairly drastic approximation to the real structure to make any progress in discerning basic forms – then modify the original idealization.

In this chapter we shall be looking at the functions of structural elements, which normally include one or more of the following:

1. transmit loads and moments;
2. maintain shape or deform to desired extent;
3. store energy;
4. break under defined conditions (what does this tell the user about causes of fracture?);
5. often seek to avoid failure or fracture in engineering.

We shall try to understand how the various parts work in isolation. Function apart, one needs to specify and choose the correct elements made from suitable materials. It may be necessary to make from scratch the basic elements or may be possible to use standard off-the-shelf parts. One then needs to decide how they fit together (which may modify behaviour). Joints can be rigid such as welds, adhesives, bolts and rivets, or flexible such as hinges and folds.

We should always provide the context and the scale of interest when identifying components or substructures.

On one scale we describe a house as a hollow box, or series of hollow boxes. On a slightly more detailed scale we look at walls as slabs; then look at walls as bricks and mortar, or windows as frame elements, wood, glass, catchment stays, etc., glue and nails. Roof as plate, subframes, rafters and purlins, slates, nails.

In an aircraft we would find substructures such as the nose, centre fusilage and rear fusilage, together with the mainplane or centre section of the wing, the outer section, the wing tips, and the flaps and ailerons; not to mention the horizontal tails and the vertical tail, with tips, rudder and elevator.

2.1.2 Examples of rods

- Axles, screwdrivers, tree trunks, silk fibre, pencils, spaghetti, fishing rods, venetian blind slats;
- cats' whiskers to detect external influences – sensitive to tension, compression, bending, buckling.

2.1.3 Examples of tubes

- Bicycle frames and tyre inner tubes, pipes for gas or water, vacuum cleaner hose, tubular frames for scaffolding and bridge assemblies, side walls of inflatable boats;
- arteries in mammals, sea anemones (hydrostatic skeleton), stems for dandelions, fine structure in wood through which sap flows; tree rings (which are a series of concentric tubes). Skin on arms and legs; socks, stockings and tights on the leg and foot.

2.1.4 Examples of flat plate-like structures

- Roofs, glazing, doors, tree leaves, gramophone records, paper.

2.1.5 Examples of curved shells

- Inflated sails for boats, yachts, dinghies, sailboards; fabric panels of open umbrella; car body panels and windows; finger nails, drinking cups; surface of dressmaker's model dummy; egg shells, beer barrels.

2.1.6 Examples of 3-dimensional solids

- Bricks, nuts, seeds such as dried peas or rice, plastics granules

    Problem 2.1: Identify which category or categories (rod, tube, flat or curved panel, or three-dimensional solid) you would assign to examples from your own experience of everyday and other items. Examples may be taken from the following lists:

    (a) woodwind and string musical instruments, compact discs;
    (b) sleeves on jacket or shirt; general form of skirt or kilt;

(c) paperclips, cardboard boxes;
(d) cellular microstructure of bread or furnishing foam, slice of bread, cans and bottles for carbonated or non-carbonated liquids, potato crisp, soup cubes;
(e) drinking straws, bubbles, balloons, stemmed wineglass, bottle corks;
(f) shafts of arrows, vaulting poles;
(g) tree branches, orange skins, daffodil bulb or flowers, (pelletized) sheep droppings, hairs on plants or mammalian skin surfaces;
(h) wool, mammal tendons, octopus or squid tentacles, worms, bacteria, veins in humans, oesophagus and digestive tract in animals and humans, snake skins, mollusc shells;
(i) floorboards, stones, carpets, nails, curtain rails, fluorescent lamps, ceilings, glazing, kitchen work surfaces, bookshelves, wallpaper, stretch fabric covers on upholstered furniture;
(j) railway track, spokes on traditional cycle wheels, helicopter rotor blades, ball bearings, inner tubes for pneumatic tyres, exhaust pipes for cars, wing of aeroplane.

2.2 FORCES IN BODIES

2.2.1 Idealized structural elements

Many bodies can be represented approximately by one or a group of simple idealized elements which have a one- or two-dimensional form. We shall concentrate on some of these in this book. Much of the skill in doing simple structural analysis lies in the wise choice of which element to use, and how it interacts with its neighbours.

One-dimensional elements

One dimensional elements may be represented by a slender rod or bar of constant cross-section. What the element is called depends on how it is used. Under longitudinal axial load, a 'tie bar'; under axial compression, a 'strut'; under torsion a 'shaft'; and if very flexible a 'cable'. Under transverse load a 'beam', and if only supported at one end, a 'cantilever (beam)'.

The cross-sectional shape depends on the type (or types) of load being carried. Typical shapes include rectangular, I-section, T-section, channel, angle, circular and tube.

In this chapter the emphasis will be on the use of Cartesian coordinates, (x, y, z) usually aligned with the important axes of the element. In some problems it is more convenient to use cylindrical polar coordinates.

2.2.2 Forces and moments

Forces and moments are vector quantities, and may be resolved into components of different magnitudes and directions. Unless a force passes through a point in the body being considered, it can always be replaced by the same force through the point together with a moment about that point. The moment is the product of the force and the perpendicular distance from the line of action of the force to the point.

2.2.3 Statics and equilibrium

Equilibrium

The cornerstone of statics is that a body is in equilibrium if the resultant forces F and moments M acting on it are zero: the body is at rest and does not accelerate.

In x, y, z co-ordinates equilibrium obtains when:

$$\Sigma F_x = \Sigma F_y = \Sigma F_z = 0;\ \text{and}\ \Sigma M_x = \Sigma M_y = \Sigma M_z = 0$$

If all the forces act on one plane, say $z = 0$, then equilibrium satisfies the two-dimensional equation:

$$\Sigma F_x = \Sigma F_y = \Sigma M_z = 0$$

Remember that you may need to replace a system of forces applied to a body by a resultant force through and a moment (or couple) about the chosen point.

In the diagrams a single arrow represents a force or moment resultant. If the arrow represents a point force, the context will make it clear. A series of parallel arrows denotes a distributed force, stress or pressure.

Joints and supports

The idealized structural elements often have to be joined together, and the way they are joined strongly influences the 'hidden' forces and moments which

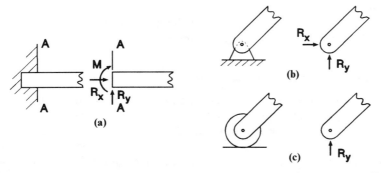

Figure 2.1 End conditions at joints: (a) built-in; (b) simply supported; (c) roller.

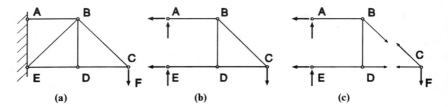

Figure 2.2 Reactions, and two free body diagrams, for a pin-jointed frame.

need to be transmitted through the joints. Those shown in Figure 2.1 are commonly encountered, and are paired with the equivalent reaction system which must be determined during problem solving.

Built-in ends (encastred, fixed) prevent rotation (Figure 2.1 (a)), and therefore include a moment (to oppose the applied moment).

Pin connections (simply-supported) permit rotation and are assumed to be friction-free (Figure 2.1 (b)), so do not include an internal reacting moment.

A roller support permits linear motion (Figure 2.1 (c)).

Free body diagrams

In a pin-jointed frame one needs to calculate reactions and forces in each member. Reactions may be identified from a sketch of the complete structure; internal forces may be identified from a free body diagram which isolates part of a structure and then assigns appropriate reactions to the cut surfaces, as shown in Figure 2.2.

Figure 2.2 (a) shows the complete body, but the reactions at A and E are implied by the wall to which the frame is fixed.

Figure 2.2 (b) shows a free body diagram which isolates the supports to the wall, and draws the reacting forces as positive in the direction shown. It does not matter if the sign of the reactions is incorrect, when initially drawing the free body diagram, because detailed calculations of their magnitude will also determine their direction.

Figure 2.2 (c) shows the free body diagram for joint C, which permits calculation of forces in members CB and CD.

Statical determinacy and indeterminacy

If a system is statically determinate, then the number of equations available from statements of equilibrium is the same as the number of unknown forces and reactions: all the forces can be calculated from equilibrium alone.

Suppose a pin-jointed plane frame has j joints, m members, and r reactions. Under plane loading, there are two equilibrium equations at each joint, so the frame is statically determinate if $m + r = 2j$.

If a system is statically indeterminate, then the number of unknown forces or internal reactions is greater than the number of equilibrium equations available, and to solve the problem one needs to use additional equations by considering either the displacement or deformation of parts of the body, or the properties of the material. Solving even simple statically determinate problems will be discussed much later in this book.

2.2.4 Statics of bending

Bending, and the associated curvature(s), is an important mode of deformation, particularly in products made from plastics or composites. Assessment of the resistance of a structure to bending is a major task for the designer. The transverse forces which cause bending are therefore important.

Consider a bar subject to transverse loads: to prevent rotation and linear movement it is necessary to apply a couple and force resultant at each end, as shown in Figure 2.3 (a). The end forces have been resolved into one perpendicular to the axis of the bar and one parallel to it. The internal forces are found by cutting the bar and drawing in the necessary forces and moments to achieve equilibrium. The couple M is the bending moment. The transverse force Q is called the Shear force, and P is the longitudinal force.

In isotropic materials the most important stresses and deflections are caused by the bending moment, M, (which is directly related to any applied transverse loads) and only to a lesser extent by P and Q. Texts in the Further reading section show that M and Q are closely related, but this will not concern us at this stage.

Sign convention

It is important to be consistent in the analysis of internal forces in bending. There are many different sign conventions, all of them arbitrary. Figure 2.4 shows the positive sense of quantities used in this book to describe bending.

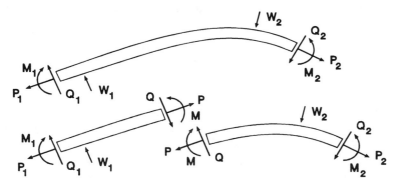

Figure 2.3 External forces and internal forces in a bar.

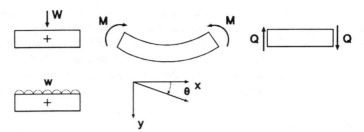

Figure 2.4 Sign conventions for bending, showing directions of positive quantities.

Shear force and bending moment diagrams

Both stresses and deflections are directly related to shear force (SF) and bending moment (BM). It is therefore desirable to know the distribution along a bar under transverse forces and hence where the maximum values occur.

Common modes of support

A simply-supported beam rests on rollers or knife edges (Figure 2.5 (a)). Under transverse load the support load is only a transverse reaction and there is no restraining couple. There is therefore no deflection at the support, but the beam slopes at the support when under transverse load. The applied transverse load may be a point load, or a distributed load along part of or the complete length of the beam. More than one transverse load may be applied.

A beam with built-in ends (fixed or encastered ends) requires both transverse reactions and a couple, and both the slope and the deflection at the ends are zero. One example is a cantilever as shown in Figure 2.5 (b), which is a beam built in at one end and free at the other. There must be a reactive force and a moment at the built-in end.

Cantilever with concentrated transverse load at free end

(The system is shown in Figure 2.6) Note that the length of the beam is only that part which projects from the wall. The first step is to write in the reactions and moment at the built-in end: $R_0 = W$ and $M_0 = - WL$. The second step is

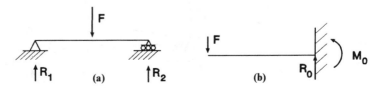

Figure 2.5 (a) Simply supported beam; (b) cantilever.

to cut the beam at any section XX at x measured from the free end, and draw a free body diagram for the left hand end, adding the necessary internal forces and moments in the positive sense according to the sign convention in Figure 2.4. (Using common sense rather than the convention will cause endless trouble, it is best to start with a good habit.)

The procedure is to consider vertical equilibrium, i.e. using $\Sigma F_y = 0$, with forces taken as positive in the direction of positive y:

$$W + Q_x = 0 \quad \text{or} \quad Q_x = -W \text{ for any value of } x$$

Hence the shear force diagram in Figure 2.6. Taking moments positive anti-clockwise about the centroid of the cut section B:

$$W \cdot x + M_x = 0 \quad \text{or} \quad M_x = -W \cdot x$$

and the BM diagram confirms that the maximum bending moment occurs at the support.

Problem 2.1: A weightless horizontal cantilever carries a uniformly distri-buted transverse load over (i) all, (ii) half its length measured from the free end. Sketch graphs of shear force and bending moment over the length of the beam for (i) and (ii).

Problem 2.2: For a horizontal cantilever under a load which varies along its length, show that the shear force Q is related to the bending moment by $Q = dM/dy$, where y is the longitudinal axis of the beam.

Problem 2.3: What is the shear force at any point along the length of a beam which is subject only to pure bending?

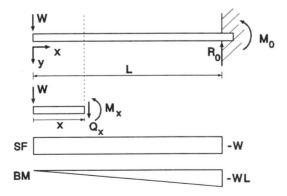

Figure 2.6 Shear force and bending moment diagrams for an end-loaded cantilever.

2.3 STRESS AND STRAIN

2.3.1 Direct stress

Having determined the forces in a simple body, we now need to study the influence of cross-sectional dimension and shape.

For the axially-loaded bar shown in Figure 2.7 the applied force resultant at each end is transmitted through the bar as an internal force. On a cut section in a free body diagram the internal force dF_i on any representative element dA gives an internal force per unit area dF_i/dA. The normal or direct stress at a given point is $\sigma = dF_i/dA$ as $dA \rightarrow 0$. The average stress, σ, assumed uniform over this bar is $\sigma_x = F_x/A_x$. The normal or direct stress acts perpendicular to a plane, and when acting outwards is tensile with a positive sign. Direct stress acting inwards is compressive, and has a negative sign.

Stress occurs inside a body, and cannot be measured directly.

For a given force, the smaller the area the higher the stress. Compare two different designs of shoe heel. The large-area heel in Figure 2.8 (a) gives a large area of contact so the stress is small and there is little risk of damage to the floor; but for the same force F, the heel in Figure 2.8 (b) has a small area of contact, so there is a large stress on the floor, with a risk of damage to the floor and the heel. Indeed when ladies' stiletto heels were in fashion, they did much damage to dance floors. With a drawing pin (Figure 2.8 (c)), the point offers a minute area and the high contact stress enables the point to penetrate quite readily depending on the hardness of the surface material, even though only modest loads are applied.

A house 10 m high exerts about 250 kPa on its foundations. If a two storey detached house has a plan size of about 10 × 10 m; perimeter 40 m, wall thick-

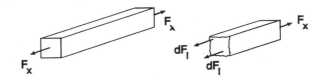

Figure 2.7 External and internal forces in a tie bar under uniaxial load.

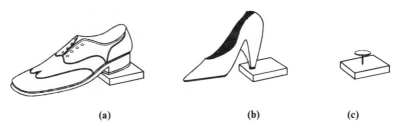

(a)   (b)   (c)

Figure 2.8 Three different contact areas.

ness 0.25 m for double brick cavity wall. Projected area of wall on foundations about 10 m², so the force on the foundations is about 2500 kN, i.e. about 250 tonnes force. (More if you include all the (thinner) internal walls.)

Problem 2.4: Calculate the pressure exerted by a 20 m high brick building on its foundations. Assume the bricks have a density $= 2500$ kg/m³.

Problem 2.5: Consider rectangular ducting 1 m square and 30 m long, containing air at an internal pressure of 1 kPa above atmospheric (about 0.01 atmosphere pressure, or 7.6 mm Hg). What is the force acting on one long side of the ducting?

2.3.2 Tie bar under axial load

The tie bar is the simplest example of a statically determinate stress system: the external forces are balanced by the internal force (Figure 2.9), so that for a normal (perpendicular) cross sectional area A_x,

$$\sigma_x = F_x/A_x$$

It is also possible to cut the bar by a plane which makes an angle θ to the perpendicular to the axis, thus giving components of stress normal to and parallel to the cutting plane. This will be discussed in Chapter 3.

Definitions of compressive and tensile stress are identical, and the behaviour of bars of chunky section under either stress is similar in character. But when sections become slender (long and thin), e.g. the aspect ratio of the bar becomes large, say greater than 10 : 1, structures under compressive loads become prone to buckling collapse and bend or crumple up. We shall discuss buckling in Section 2.8, but as a general rule the reader would be well advised to be always wary of compression and always check against or seek advice on buckling.

2.3.3 Thin shells under internal pressure

Stresses may well act in directions in a body different from those of the externally applied forces or pressures.

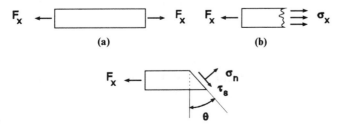

Figure 2.9 Stresses on cut surface of tie bar.

For example, in a thin-walled cylinder under internal pressure, the pressure acts radially outwards, whereas the most significant stresses are in the circumferential (hoop) direction and in the longitudinal (axial) direction. The term 'internal pressure' usually means the difference in pressure across the wall of the vessel or pipe, and not the absolute pressure in the vessel unless so specified.

Thin-shelled bodies under internal pressure occur widely in engineering applications and in everyday experience, as in the following examples:

1. Brake hose in cars, pneumatic tyres, jet engines.
2. The cylindrical walls of rubber toy balloons, and plastics bottles for carbonated drinks.
3. Socks, tights and stretch-fit garments: the body exerts pressure on these garments, causing a change in their size when you put them on.
4. Clarinets, flutes, saxophones, french horns, trombones, tubas.
5. Veins and arteries carrying blood, lungs, swallowing in the alimentary canal, body skin. Inflated structure of hydrostatic skeletons.
6. Osmotic pressure responsible for flow of transpiration fluids in capillaries in trees and cells in dandelion stems.
7. Cylindrical walls of chemical plant pressure vessels; pressure cookers in the kitchen. Potatoes and sausages for cooking whole by frying or baking (unless pricked to relieve internal pressure).

If the shell thickness is less than about one-tenth the radius of curvature we may neglect variations in the tangential and radial stress. We then assume here that the shell acts as a membrane so that it offers no bending resistance and only carries direct stress. Such a body is regarded as 'thin-walled'.

In engineering the term 'thin-walled' refers to the wall thickness being less than about 10% of the radius; the reader may encounter a plastics pipe for pressurized cold water supply, with a wall thickness of 40 mm and a diameter of say 800 mm. There is no doubt that the wall is 'thick' in absolute terms, but the pipe is still analysed as 'thin-walled'.

It must be emphasized that the following is a simplified treatment, and in practice joints and other restraints can cause major bending stresses which must not be ignored: their analysis is outside the scope of this book.

Thin-walled cylinder

The cylinder is blanked off at each end (Figure 2.10(a)) and contains an internal pressure p. If the internal radius is R_i and the wall thickness is h, then in the free body diagram Figure 2.10(b) the force acting on the end wall is

$$F_A = p\pi R_i^2$$

By equilibrium, this is balanced by the axial (or longitudinal) force acting

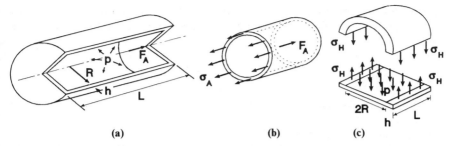

Figure 2.10 (a) Pipe under internal pressure, and free body diagrams for calculating; (b) axial stress; (c) hoop stress.

within the wall of the cylinder, i.e.

$$F_A = \sigma_A \cdot 2\pi R_m h \approx \sigma_A \cdot 2\pi R_i h \quad \text{if we assume } R_m = R_i + h/2 \approx R_i$$

Hence the axial stress in the wall is

$$\sigma_A = pR_m/2h$$

The easiest way to determine the hoop stress is to regard an element of cylinder also blanked off on a diameter. The outward force on the diametral blank caused by internal pressure is balanced by the force caused by hoop stress in the walls:

$$p \cdot 2R_i L = \sigma_H \cdot 2hL, \quad \text{hence } \sigma_H = pR_m/h = 2\sigma_A$$

So if we design the pipe to safely withstand the largest stress (i.e. the hoop stress), the wall thickness will naturally be the same in both the hoop and the axial directions. This means that the pipe is overdesigned in the axial direction because the wall thickness in the axial direction is twice what it needs to be, in order to safely withstand the hoop stress.

A spherical rubber balloon initially 0.5 mm thick at the neck has an initial diameter 10 mm. The maximum pressure the lungs can exert is about 0.1 atmosphere $= 10$ kPa. The hoop stress in the neck of the balloon is $\sigma_H = pD/2h = 10 \times 10/1 = 100$ kPa.

The analysis presented above also applies *in principle* when the tube is under external pressure, but the reader *must* check that there is no risk of buckling under radial compression – this is likely to be a major design problem in thin-walled shells under external pressure, because buckling collapse usually occurs under much lower external pressures than those likely to induce excessive hoop or axial deformations. Examples of tubes under external pressure include tubular fluorescent lighting tubes (internal vacuum), drinking straws in use, and some ducting for ventilation shafts in buildings. One example where buckling often occurs is in plastics containers for volatile organic liquids such as creosote: the creosote gradually escapes by permeating the walls thus creating a partial vacuum inside the container, leading to distortion of the wall. Some comments on buckling are made in Section 2.8 below.

Thin-walled spheres

A similar approach may be followed to determine the circumferential stresses in a thin-walled sphere. The reader will expect that the stress in the wall acts in all possible circumferential directions: the important ones are a pair of mutually perpendicular stresses in the circumference. The directions of this pair of stresses are chosen for convenience. Circumferential stresses in all other directions can be resolved into the two principal directions, and so can be ignored in this analysis. The radial stress in the thickness direction can be neglected if the sphere is thin-walled.

Examples of thin-walled spheres include spherical pressure vessels and toy balloons, and to an approximation the cellular structure of such materials as furnishing foams, bread, soap bubbles and beer froth.

> **Problem 2.6:** A hollow sphere has a radius of 100 mm, and a uniform wall thickness of 10 mm, and contains a fluid at a pressure of 1 MPa above atmospheric pressure. Calculate an approximate value of the stress in the wall of the sphere.

2.3.4 Shear stress

Consider a rectangular block (Figure 2.11). We may apply to faces ABCD and EFGH a uniformly distributed shear force F_y which acts in the y direction. The area of ABCD is A_z, the suffix z denoting the direction of the normal to the area. In this book we use the double suffix notation: the first suffix denotes the direction in which the force acts, the second the normal to the area on which the force acts. Thus we define the shear stress τ_{yz} as the ratio:

$$\tau_{yz} = F_y/A_z$$

This shear stress τ_{yz} acts uniformly on any plane parallel to and between ABCD and EFGH. Shear stress τ_{yz} is defined as positive when acting in the

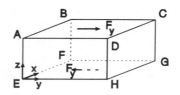

Figure 2.11 Applied shear forces F_y.

positive or negative directions of both suffixes, and negative when acting in the positive sense of one suffix and negative in the other.

Shear stress occurs in the sole of the runner's shoe when the runner accelerates. Shear stress also occurs in modern bridge bearings which consist of a block of rubber interleaved with steel plates. The shear force arises because of horizontal movement in the deck of the bridge arising from thermal expansion. Failure to absorb this horizontal movement would put a large bending moment on to the vertical supports which could cause major structural damage.

Consider now taking moments about an axis JJ in the x-direction which passes through the centroid of the block (Figure 2.12). Static equilibrium only occurs if we notionally apply complementary shear forces F_z in the anticlockwise sense to balance the applied shear forces F_y. The complementary shear forces F_z, induce complementary shear stresses τ_{zy} equal and opposite to τ_{yz}. The shear forces on the surfaces of the block are $F_y = \tau_{yz}\,dy\,dx$ and $F_z = \tau_{zy}\,dx\,dy$. Equating moments of these forces about JJ we have $2F_y(dz/2) = 2F_z(dy/2)$ and hence numerically $\tau_{yz} = \tau_{zy}$. This state of stress, and its constant value through the thickness of the block, is depicted conventionally in Figure 2.13, and the resulting distortion will be discussed in Section 2.3.9. In particular, note that the arrows denoting shear stresses at the corners of the element either point towards each other (as at D and E), or away from each other (as at A and H).

When considering free body diagrams, a cut in any shear plane will reveal two surfaces on which shear forces or stresses will have equal values but opposite signs (Figure 2.14).

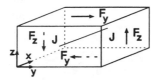

Figure 2.12 Complementary shear forces F_z.

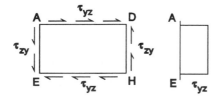

Figure 2.13 Representation of shear stress on planes normal to ADHE.

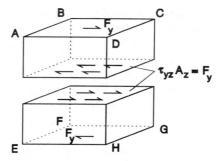

Figure 2.14 Shear stresses on the cut surfaces have equal values but opposite signs.

Problem 2.7: Two rubber blocks have the same plan dimensions, 100 mm square. Block 1 is 10 mm thick, block 2 is 20 mm thick. Shear forces of 100 N are applied to opposite faces. Calculate the shear stress in each block.

Problem 2.8: Does the complementary shear force F_z in Figure 2.12 have the same value as the applied shear force F_y?

2.3.5 Shear stresses in long rods and tubes

Shear stresses occur in long slender rods and tubes under torsion. Drive shafts in cars and lorries are commonplace engineering examples. A golf club is stressed in torsion when the head hits the ball: the reaction induces a twist in the shaft (as well as bending).

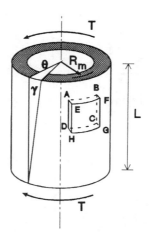

Figure 2.15 Torsion of thin-walled tube.

Long hollow thin-walled tube under torsion

Consider a thin-walled hollow cylinder of mean radius R_m, thickness h and length L (Figure 2.15). 'Thin-walled' has the same meaning here as used in Section 2.3.3, and hence the variation in shear stress across the thickness of the tube wall is ignored. The axial torque T acts on a cross sectional area $2\pi R_m h$, inducing an average shear stress τ:

$$\tau = T/(2\pi R_m^2 h)$$

Problem 2.9: For the element of tube ABCDEFGH in Figure 2.15, identify on an enlarged copy the directions of shear stresses induced by the applied torque.

Long rod under torsion

A torsion T is applied to each end of a rod of length L and radius R (Figure 2.16). We shall now develop a relationship between the applied torque and the shear stress.

Consider an elemental annulus of radius r and thickness dr, on which the shear stress is τ. The force per unit length on the element is $\tau \cdot dr$. The elemental torque about the axis of the rod is $\tau \cdot dr \cdot r$. The resisting torque over the complete elemental annulus is $\tau \cdot dr \cdot r \cdot 2\pi r$. The total internal torque on the complete circular cross-section is $\int \tau \cdot 2\pi r^2 dr$ between limits of $r = 0$ and $r = R$. To achieve equilibrium, this internal torque must balance the externally applied torque T. Hence:

$$T = \int_{r=0}^{R} \tau \cdot 2\pi r^2 \, dr$$

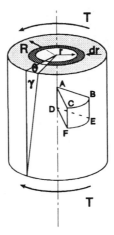

Figure 2.16 Torsion of solid cylinder.

Problem 2.10: For the wedge-shaped element ABCDEF in Figure 2.16 indicate the directions and magnitudes of shear stresses induced by the applied torque.

2.3.6 Strain

In order to predict dimensional changes it is necessary to translate forces into internal stress and dimensional changes into dimensionless strains. Once a relation between stress and strain is determined, changes in the dimensions and shape of the body as a whole can be found from suitable analysis.

Strains are observable as changes in dimensions or shape which can be measured by ruler, dial gauge, micrometer, microscope, strain gauge or other means. The simplest description of strain is as follows, and we shall discuss further details of shear strain in Sections 2.3.9 to 2.3.11, and bending strains in Section 2.5.3.

Consider a bar of constant cross-sectional area and length L_x (Figure 2.17(a)). Under axial force F_x the bar elongates by ΔL_x. The direct tensile strain is $\varepsilon_x = \Delta L_x / L_x$. Direct strain is positive and tensile for an increase in dimension; and negative and compressive for a decrease.

Under shear stress a body changes its shape, and the tangent of the angle through which two initially perpendicular axes rotate is termed shear strain. Thus if in Figure 2.17(b) the top corner of the y axis is displaced in the x direction by an amount u, the shear strain is $\tan \gamma = u/L_y = \gamma$ when γ is small. Shear strain is positive when the angle γ between two perpendicular axes decreases.

Examples of strains

1. Direct strains in metals in use usually small $< 0.1\%$.
2. Direct strains in plastics in use usually less than 1%.
3. Direct strains in rubber up to about 10% in use.
4. Shear strains in rubber block springs usually less than 50%.

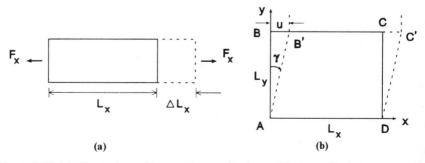

Figure 2.17 (a) Elongation of bar under tensile force; (b) shear displacement under shear force.

5. Direct strains in skin on human elbow or knees up to 50%.
6. Direct strains in fashion tights on the leg about 100 to 400%.
7. Direct strains in inflated toy balloons 500 to 800%.
8. Compressive strains in soft upholstery foam up to 90%.

Strictly speaking, our definitions of strain are only usable where strains are small, of the order of a few per cent. Bending strains are almost always small even though deflections may be large. For large strains, outside the scope of this book, it is necessary to use other definitions of strain, which are discussed in standard texts such as Williams (1973).

Several words which involve aspects of dimensional change need to be distinguished. The following working descriptions will be helpful:

Deformation   Change of dimension or angle
Displacement translation of point A to point B
Deflection   bending displacement (transverse to the long axis)
Strain   dimensionless linear, angular or volumetric deformation
Distortion   usually change of shape, i.e. change of angle

It is often necessary to consider strains acting in more than one orthogonal direction. In tubes and spheres under pressure a state of biaxial strain exists. For example, if the bulb of a toy balloon were originally a sphere of diameter 50 mm, and it increased to 300 mm under internal pressure, the hoop strain is $(300 - 50)/50 = 5$, i.e. 500%. You can see that the strain is biaxial by drawing a small circle on the uninflated balloon and comparing dimensions after inflation.

2.3.7 Lateral strain

If a bar carries longitudinal stress, it extends in the direction of stress and contracts in the transverse or lateral directions (Figure 2.18). This is most

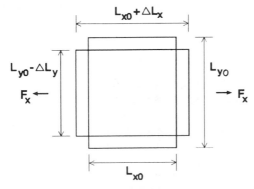

Figure 2.18 Axial extension and lateral contraction under axial load.

Table 2.1 Typical values of Poisson's ratio v

Cheddar cheese	0.5
Flour dough	0.5
Gelatine (80% H_2O)	0.5
Rubber	0.5
Potato tissue	0.49
Apple tissue	0.37
Partially crystalline plastics	0.35 to 0.4
Aluminium	0.33
Amorphous plastics	0.3
Copper	0.30
Steel	0.28
Glass	0.24
Sandstone	0.10
Cork*	0
Flexible foam	0

Note: *Cork is highly anisotropic; values of many properties vary with direction.

readily seen by stretching a rubber band, when it becomes longer and thinner. The deformed dimensions become $L_x + \Delta L_x$, $L_y - \Delta L_y$, $L_z - \Delta L_z$, and the strains become $\varepsilon_x = \Delta L_x/L_x$, $\varepsilon_y = -\Delta L_y/L_y$ and $\varepsilon_z = -\Delta L_z/L_z$. Experiments at small strains confirm that the lateral strain is directly proportional to the longitudinal strain caused by stress. The lateral contraction ratio, usually called Poisson's ratio, is given by $v = -\varepsilon_y/\varepsilon_x = -\varepsilon_z/\varepsilon_x$.

Poisson's ratio is a material property with a value between 0 to 0.5 for isotropic materials. Typical values are given in Table 2.1.

For small elastic deformations in linear elastic material under uniaxial load P_x, the volumetric strain $\varepsilon_v = \varepsilon_x(1 - 2v)$, so

1. $v = 0.5$ means constant volume deformation;
2. $v < 0.5$ increase in volume on deformation; and
3. $v = 0$ increase in volume on deformation.

So Poisson's ratio measures the relative ability of a material to resist change of volume and change of shape.

2.3.8 Thermal strain

The dependence of dimension on temperature change is measured by the coefficient of linear thermal expansion, α, expressed as the direct strain per unit temperature change. If a rod of length L_{x0} at T_0 is heated to a higher temperature T, the change in length is $\Delta L_x = \alpha L_{x0}(T - T_0)$ and hence the thermal strain is $\varepsilon_{xt} = \alpha(T - T_0)$. Increasing the temperature causes expansion and hence a positive strain.

Typical values for linear expansion coefficient are steel 10^{-5}/K, polyethylene or rubber 10^{-4}/K. Further data are given in Chapter 7.

Examples of thermal strain

1. Non-welded metal rails – the gap between the ends is bigger in winter than summer.
2. PVC rainwater guttering on buildings expands when the sun shines, and cools and contracts when clouds pass in front of the sun.
3. Bridge decks get longer in hot weather, so designers must allow for expansion – there are intermeshing combs on the deriving surface, and the deck is mounted on rubber springs.
4. The bimetallic strip consists of two metals having different thermal expansion coefficients: they are firmly bonded together along their length, and a change of temperature causes bending, which is used to actuate a mechanism. This principle is discussed in some detail in Chapter 7.
5. Liquid capillary thermometers indicate a change of temperature based on the difference in volumetric change between the fluid and the volume enclosed by the glass in which it is contained.

2.3.9 Uniform shear strain

We shall now expand on the brief comments already made about shear strain in Section 2.3.6.

We defined positive shear strain as the *reduction* in angle between the co-ordinate axes. This is consistent with our sign convention in Section 2.2.4. More simply, γ_{yz} is the reduction in the right angle between the z and y axes, as represented stylistically in Figure 2.19(a), and the double suffix notation for strain has the same meaning as for shear stress (Section 2.3.4). We could model this distortion by uniformly shearing a pack of playing cards: each card is displaced and in the limit where there were an infinite number of cards of infinitesimal thickness, we have a suitable model for a continuum.

Equally well we could achieve the same shear strain in an isotropic body by considering the shear caused by the complementary force F_z (Figure 2.19(b)). Formally we should take the results of the two sets of shear forces F_y and F_z acting together (Figure 2.19(c)), in which we then define the shear strain as

$$\gamma_{zy} = \gamma_{yz} = \alpha + \beta$$

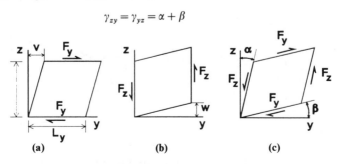

Figure 2.19 Three representations of block under applied shear force.

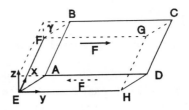

Figure 2.20 Shear deformation in block under shear load.

Figure 2.19(c) makes it clear that shear strain only involves a change of angle and not a change in the length of the side of an element. By convention we usually depict shear using Figure 2.19(a).

We can now examine more carefully the deformations in a uniform block under shear (Figure 2.20), and make some important observations. The loaded faces ABCD and EFGH do not distort or change dimensions, and neither does any parallel plane between them. The faces carrying the complementary shear forces or stresses do not distort. Unloaded planes in the direction of applied loads (e.g. ADHE, BCGF or any included planes) do distort by rotation but sides do not change in length.

It is also apparent therefore that on the distorted face ADHE the diagonal DE is extended and is in a state of uniform tensile strain, whereas AH has contracted and is in compressive strain. The compressive strain is important because if it becomes too large, wrinkling and buckling collapse can occur. Thin-walled structures in torsion are especially prone to this mode of failure.

2.3.10 Shear strain in thin-walled tube under torsion

In a long thin-walled tube of mean radius R_m under opposing torques acting on the ends (Figure 2.21), we assume that the shear stress is uniform across the wall on faces AB and CD. AD and BC are not deformed, but ABCD distorts.

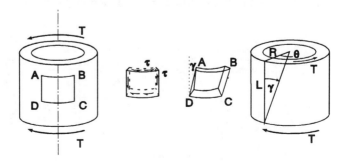

Figure 2.21 Shear in thin-walled tube in torsion.

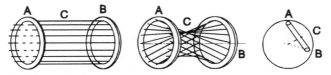

Figure 2.22 Tube modelled as a series of parallel equispaced rods.

Adding the distortion in all possible elements along the length, L, of the tube permits calculation of the amount of rotation of the top surface about the longitudinal axis, and this can be related to the angle of twist θ of the top face relative to the bottom face. By simple geometry, assuming $R_m \approx R_0$:

$$R_0 \theta = \gamma L$$

Strictly speaking, this analysis only applies for small shear strains γ because we use the approximation $\gamma \sim \tan \gamma$.

The adaptation to large strains in this thin-walled tube is quite straightforward, but even the small strain approach shows one reason why twisting a wet dishcloth expels water. In the element ABCD in Figure 2.21, the diagonal AC is in compression. Another reason is also a geometrical one which is readily discerned. Let us model the tube as a series of long stiff rods equally spaced round the circumference and held at each end by a rigid annular manifold as shown in Figure 2.22. With one end fixed, a large rotation of the other end causes the rods to shear and the tube contracts radially into the form or a hyperboloid: the waist is at midlength, C. It is obvious that for inextensible rods the tube must shorten with increase in angle of twist, and that the plane

Figure 2.23 Shear buckling in loose sweater.

end faces which were parallel before twist remain parallel after twist. Engineers use the shorthand phrase 'plane faces remain plane after twist'.

The thin-walled tube (Figure 2.21) compresses in the diagonal AC, so again beware of buckling. A common example is seen in a loose sweater, jacket or shirt; when the shoulders are twisted relative to the hips, the garment on the back develops wrinkles aligned in the direction of tension ('a Wagner tension field') and buckles in the compression direction which is perpendicular to the tension (Figure 2.23).

Problem 2.11: For the element of tube ABCDEFGH in Figure 2.15, indicate how it will strain (distort) under the action of the applied torque.

2.3.11 Shear strain in long slender rod under torsion

For a long slender rod of constant circular cross-section, the approach is similar to that for a thin-walled circular tube. For the same angle of twist θ, (Figure 2.24), geometry requires that $L\gamma_R = R\theta$ and $L\gamma_r = r\theta$, i.e.:

$$\gamma_r/\gamma_R = r/R$$

The parallel plane end faces of the rod are assumed to remain plane after twist. This is a useful approximation for the small angles of twist encountered in most engineering applications.

But construction of a solid rod from a well-spaced parallel array of 'rigid' drinking straws held in place by circular cardboard manifold ends confirms that at large strains the initially-plane circular end faces become spherical caps. The individual axial rod does not change its length whereas those at any distance r from the axis shorten the axial separation of the end plates in proportion to r. There is a limit to the amount of twist before adjacent straws touch, and this is yet further evidence of the efficiency of twisting as a method of removing excess loose water from a wet cloth.

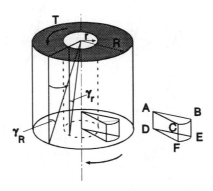

Figure 2.24 Shear in solid circular rod under torsion.

Wool fibres are typically some 25 to 50 mm long. To make a continuous length useful for garment manufacture, these staple fibres are twisted together. The twisted yarns therefore have the structure of a coiled helical spring. The twist may be partially set by thermal treatment, and the twist is responsible for increased friction and structural integrity. Axial tension applied at each end of the yarn causes the helical staple fibres to straighten, and the component of axial force in the radial (inwards) direction helps to lock the staple fibres together. Unrepresentative of normal use, on holding the upper end of a vertical yarn and applying an axial load such that the yarn is free to untwist, the friction evidently decreases with untwisting, and eventually the staple fibres fall apart.

Problem 2.12: For the wedge-shaped element ABCDEF in Figure 2.16, indicate how it will distort under the action of the applied torque.

Problem 2.13: A swan has a neck some 400 mm long with an average surface skin diameter of 20 mm, assumed constant for this problem. If the swan turns its head through 90° to left and right, calculate the maximum shear strain in the neck and state any assumptions you make in arriving at your answer.

2.3.12 Torsion of bars of non-circular section

The analysis of shear stress and strain induced by torsion of a rod of constant rectangular cross-section is beyond the scope of this introduction. We shall merely describe the effects observed and comment on them.

If we consider a parallel array of rigid rods uniformly spaced round the edge of a square tube, held in place by cardboard square manifold ends, then on twisting on end with respect to the other, the corner rods are further from the axis than those midway along the sides and hence they will twist more. The extra twist at the corners distorts the end faces, so they are no longer plane. Engineers say that the corners have warped out of plane. The twist is an important deformation in its own right, and we shall comment on this type of deformation in some detail in Sections 2.5.9 and 2.5.10 in preparation for later discussions on the behaviour of composite plates.

If the reader takes a block of flexible foam and draws a uniform grid of squares on the sides, then on twisting the block, the sides also shear and warp out of plane.

2.4 LINEAR ELASTIC BEHAVIOUR

Depending on loading, a structural element may get longer or shorter, shear or twist, bend or buckle at too low a load. Each of these responses may be desirable or undesirable. Either stiffness or flexibility may be desired depending on the quality you are looking for in the product.

In this section we shall be concerned with the stiffness or deformability (flexibility) under direct stress or shear stress. In Section 2.5 we shall discuss some aspects of stiffness and flexibility in bending.

Examples where stiffness is required

- Tensile stiffness: Space-frame tie bars (e.g. for lifting cranes or for bridge assemblies), suspension bridge cables; cylindrical casings for space rockets or defence missiles.
- Compressive stiffness: Human teeth must be stiff to maintain shape when used to reduce the size of food particles. Foundations for buildings.
- Shear stiffness: Screwdriver shafts must not twist very much when we try to release a stubborn screw. Aircraft wings must not twist too much in flight (neither must they bend too much in flight or on the ground).

Examples where flexibility is required

- Tensile flexibility: Lungs must inflate to permit respiration. Skin must be flexible to permit body movement. Elastic bands to hold things together. Stretch-fit clothes for a good and comfortable fit. Socks to go over the heel while a snug fit at and above the ankle without sag. Muscles flexible over a wide range of movement of adjacent bone assemblies. Strings of musical instruments at a suitable tension to vibrate at desired frequencies.
- Compressive flexibility: Antivibration mountings to absorb unwanted extents of vibration from machinery such as car engines. 'O' ring seals which permit dismantling and reassembly.
- Shear flexibility: Bridge bearings with different flexibilities in different directions. Sealing mastic flexible enough to provide seal between mating surfaces. Suspension threads for galvanometers. Twisting cloth to wring out water after washing. Twisting the body, e.g. shoulders relative to the hips, with feet on the ground.

2.4.1 Hooke's law

For many materials, experiment shows that up to a certain limit the deformation is directly proportional to the applied load: this is a statement of Hooke's law. If on removing the load all the deformation is recovered, the body is called elastic. The slope of the tensile stress–strain curve (Figure 2.25) within the limit of proportionality is defined as the Elastic (Young's) modulus, $E = \sigma/\varepsilon$ and is a fundamental property of the material. Typical values of tensile elastic modulus are shown in Table 2.2. At small strains the compressive modulus has the same value as the tensile modulus.

Values of elastic shear modulus $G = \tau/\gamma$, sometimes called the 'modulus of rigidity', are typically about one-third those for the tensile modulus, provided

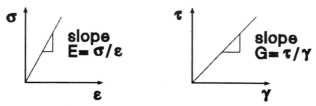

Figure 2.25 Linear stress–strain curves in tension and in shear.

Table 2.2 Typical values of Young's modulus at 20 °C (Pa)

Soft foam rubber	1×10^2
Rubber (crosslinked)	1×10^6
Tendon (wet collagen)*	1×10^9
Most plastics	1 to 3×10^9
Wool*	3×10^9
Dry spaghetti	3×10^9
Lead	1×10^{10}
Silk*	1×10^{10}
Human haversian bone*	1 to 2×10^{10}
Concrete	1.7×10^{10}
Glass	7×10^{10}
Aluminium	70×10^9
Crystalline cellulose*	2 to 3×10^{11}
Steel	2.08×10^{11}

Note: *These materials are anisotropic, properties vary with direction of loading.

Table 2.3 Typical values of shear modulus at 20 °C (Pa)

Gelatine (80% H_2O)	2×10^5
Rubber	3×10^5
Silk	1×10^9
Lead	3×10^9
Concrete	7×10^9
Glass	2×10^{10}
Steel	8×10^{10}

the material is isotropic (Table 2.3). For isotropic linear elastic materials at small strains it can be shown that $G = E/[2(1 + v)]$.

Colloquially the terms stiffness and modulus are used (incorrectly) as if interchangeable. Strictly speaking stiffness is the product of modulus and some function of the thickness over which the stress is applied. For in-plane stresses the stiffness is proportional to thickness, for bending to the cube of the thickness.

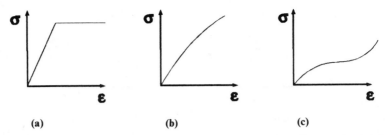

(a)   (b)   (c)

Figure 2.26 Non-linear tensile stress–strain curves: (a) linear–elastic/plastic; (b) non-linear; (c) S-shaped.

In the design of components such as rubber springs, the term stiffness is sometimes used to describe force per unit displacement. For a rectangular block of rubber under shear force, F_s, the shear stiffness would then be $K_s = F_s/u$.

Beyond the scope of this text, there are other relationships between stress and strain. For example, stress–strain curves can be elastic–plastic, non-linear, and J- or S-shaped (Figure 2.26). Thus, natural rubber has an S-shaped stress–strain curve: it is certainly elastic, but not linear elastic.

2.4.2 General stress–strain relationship

The consequence of Poisson's ratio is that applying a single direct stress to an isotropic material will induce not only a strain in the same direction, but also two other strains in directions mutually perpendicular to it. Under a single stress σ_x we have

$$\begin{pmatrix} \varepsilon_x \\ \varepsilon_y \\ \varepsilon_z \end{pmatrix} = \begin{pmatrix} 1/E \\ -v/E \\ -v/E \end{pmatrix} \begin{pmatrix} \sigma_x \\ \sigma_x \\ \sigma_x \end{pmatrix}$$

For a rectangular element with faces aligned in x, y, z, a direct stress σ_x causes strains in all three directions. So under a triaxial stress system the strain response becomes

$$\begin{pmatrix} \varepsilon_x \\ \varepsilon_y \\ \varepsilon_z \end{pmatrix} = \begin{pmatrix} 1/E & -v/E & -v/E \\ -v/E & 1/E & -v/E \\ -v/E & -v/E & 1/E \end{pmatrix} \begin{pmatrix} \sigma_x \\ \sigma_y \\ \sigma_z \end{pmatrix}$$

There is no lateral strain associated with shear strain in an isotropic material.

Each of the terms in the 3×3 matrix may be termed a 'compliance coefficient'. We shall explore the full significance of the compliance matrix in Chapters 3, 4 and 5.

Under a single stress system some texts introduce the term compliance to denote reciprocal modulus. Other texts use compliance to denote displacement per unit load.

2.4.3 Plane stress

In many situations – especially thin-walled structures – the stress in the thickness direction z is zero or assumed negligible. This condition is termed plane stress. It does *not* mean that strains in the thickness direction are zero.

In most of this book we shall look at the state of *plane stress* in thin plates and shells. Direct and shear stresses are applied only in the plane (σ_x, σ_y and τ_{xy} in the x, y plane). There are no stresses through the thickness ($\sigma_z = 0 = \tau_{xz} = \tau_{yz}$). There are strains in the thickness direction ($\varepsilon_z, \gamma_{xz}, \gamma_{yz}$) but these are usually ignored because the dimensional changes through the thickness of a 'thin' plate are usually negligible.

At this stage we shall assume that the stresses and strains are constant across the thickness of the sheet. Later we shall include variation caused by bending.

The strain response to plane stress is given by the following important equation:

$$\begin{pmatrix} \varepsilon_x \\ \varepsilon_y \\ \gamma_{xy} \end{pmatrix} = \begin{pmatrix} 1/E & -v/E & 0 \\ -v/E & 1/E & 0 \\ 0 & 0 & 1/G \end{pmatrix} \begin{pmatrix} \sigma_x \\ \sigma_y \\ \tau_{xy} \end{pmatrix} \qquad (2.1)$$

The zero coefficients in this equation are entirely expected, because in an isotropic material there is no relationship between shear stress and any direct strain, or between shear strain and any direct stress.

One of the clear implications of this equation – which indicates an important consequence of Poisson's ratio – is that under biaxial stress, with $\sigma_x = \sigma_y$, the strain in either loading direction is less by a factor v than it would have been if only a single direct stress of the same magnitude has been applied. In other words Poisson's ratio has a restraining effect under biaxial loading. The amount of restraint depends on the ratio of the two direct stresses.

When the strains in a plane stress situation are specified the stresses in the sheet are found by inverting Equation (2.1).

$$\begin{pmatrix} \sigma_x \\ \sigma_y \\ \tau_{xy} \end{pmatrix} = \begin{pmatrix} E/J & vE/J & 0 \\ vE/J & E/J & 0 \\ 0 & 0 & G \end{pmatrix} \begin{pmatrix} \varepsilon_x \\ \varepsilon_y \\ \gamma_{xy} \end{pmatrix} = [Q] \begin{pmatrix} \varepsilon_x \\ \varepsilon_y \\ \gamma_{xy} \end{pmatrix}$$

where $J = (1 - v^2)$, and $[Q]$ is the reduced stiffness matrix. Because values of Poisson's ratio for most isotropic engineering materials are less than 0.5, the maximum enhancement of modulus corresponds to a factor of $1/0.75$, i.e. an increase in modulus of not more than 1.33, and often much less.

We may ponder what would be a good choice of material for the closure of a wine bottle. Of course we seek a material which has very low permeability to oxygen. But from a mechanical point of view the problem is that when we apply an axial compression to the oversize closure, to insert it into the rigid neck of a bottle, it becomes even more oversize, as it bulges because of the Poisson's ratio effect. The lateral bulging is minimized by choosing a material having $v = 0$. Cork has a small value of v (but only in one direction), as well as other useful attributes, and this is a good reason why it is used to seal wine bottles. The bottle seals because the diameter of the cork is properly oversize.

One might extend this argument to say that the problem of inserting the closure would be even easier if a material having a negative Poisson's ratio were chosen – but the problem of removing the closure would then be even more difficult, because on applying tension the closure would try to expand. So cork, with $v = 0$ in the appropriate direction, is a good compromise.

Problem 2.14: A sheet of polymethylmethacrylate is 8 mm thick, 1m square, and has elastic constants $E = 3$ GPa, $G = 1.11$ GPa and $v = 0.35$. Calculate the in-plane dimensions under the following loading conditions (assumed uniformly distributed): (a) $F_x = 50$ kN; (b) $F_{xy} = 30$ kN; (c) $F_x = F_y = 50$ kN; (d) $F_x = F_y = 50$ kN, $F_{xy} = 30$ kN.

Problem 2.15: A rectangular vulcanized rubber block has edge dimensions 10 mm × 100 mm × 150 mm. The shear modulus is 3 MN/m². The large faces are bonded to steel plates 3 mm thick. A shear load of 6 kN is applied to the metal plates: (a) parallel to the 100 mm face; and (b) parallel to the 150 mm face. Calculate the shear stress in the midplane of the rubber block. By how much are the metal plates displaced (relatively) in the plane?

Problem 2.16: A thin-walled cylinder is subject to an internal pressure p. Which of the following statements is true?

(a) Doubling the pressure doubles the tube diameter.
(b) Doubling the pressure doubles the hoop stress.
(c) Doubling the pressure doubles the hoop strain.
(d) Doubling the pressure doubles the enclosed volume.
(e) The ratio hoop strain/axial strain = hoop stress/axial stress.

Problem 2.17: A pipe made from high-density polyethylene is being tested and installed for the supply of cold drinking water. The pipe is 800 mm outside diameter with a wall thickness of 25 mm, and is 1 km long. The short-term tensile modulus is 0.7 GPa, and the Poisson's ratio is 0.4. (a) What is the increase in length of the pipe under an axial force of 40 tonnes? (b) What is the change in mean diameter and length under an internal pressure of 0.32 MPa?

Problem 2.18: A tube with blanked-off ends made from cast acrylic is 4000 mm long, 5 mm thick and 125 mm diameter. The ends are thick and flat, and it may be assumed that they do not deflect under pressure. The Young's modulus may be taken as 3 GPa at $20\,°C$, and the lateral contraction ratio 0.35. Estimate the fractional change in enclosed volume of the vessel when an internal pressure of 0.5 MPa is applied, and state any assumptions not mentioned in the problem.

Problem 2.19: What is the restraining effect of Poisson's ratio on the strain in (a) a thin-walled pipe, and (b) a thin-walled sphere, when each contains fluid under internal pressure?

Problem 2.20: An initially square thin plate of an isotropic substance, with edges parallel to (x, y), is subjected to one or more of the following in-plane stresses: $\sigma_x, -\sigma_x, \sigma_y, -\sigma_y, \tau_{xy},$ and $-\tau_{xy}$. Sketches of the resulting deformations (suitably exaggerated and not entirely to any scale) have been prepared (Figure 2.27) but unfortunately the loading conditions have been mislaid. The original square plate is shown by broken lines. Which stress or stresses would cause the patterns of deformation shown? It may be assumed that no deformation patterns involve positive and negative values of the same stress.

Are any deformation patterns missing? If so, sketch them, and label them with the applied stresses.

2.4.4 Elastic strain energy

Under a single direct stress σ_x, a linear elastic bar of length L_x will elongate by ΔL. Work is done by the applied load F_x and the body stores the elastic strain energy.

The work done when an incremental force f causes an incremental extension dx is $f \cdot dx$. (Figure 2.28). The total work, or the stored strain energy, will be the area under the load-extension line, $W\Delta L/2$. The volume of the bar is AL and therefore the strain energy per unit volume is $U = W\Delta L/2AL = \sigma^2/2E$.

2.4.5 Rods and tubes under torsion

We shall consider here how torque and twist are related in a long thin rod or a long thin-walled tube of a linear elastic material.

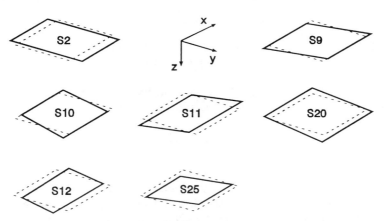

Figure 2.27 Deformations of isotropic sheet under several in-plane loading conditions.

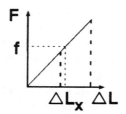

Figure 2.28 Force-extension curve for strain energy.

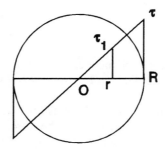

Figure 2.29 Shear stress in rod in torsion.

Long rod under torsion

A torsion T is applied to each end of a rod of length L and radius R (Figure 2.16). We already know that

$$\theta/L = \gamma/R$$

and because in shear in the elastic region Hooke's law is $G = \tau/\gamma$, we arrive at

$$G\theta/L = \tau/R$$

We see that there is a linear relationship between angle of twist, θ, and shear stress τ, and a linear relationship between shear stress and radius for a given twist (Figure 2.29).

Following the derivation of the relationship between torque and shear stress, we can therefore relate torque and angle of twist:

$$T = \int_0^R \tau_{xy} \cdot 2\pi r^2 dr = (G\theta/L) 2\pi \int_0^R r^3 dr = (G\theta/L)(\pi R^4/2)$$

Problem 2.21: On an oil rig, a steel drill shaft 4 km long has an outside diameter of 0.125 m. The shear modulus of the steel is 80 GPa. If for the purposes of this calculation the drill shaft is assumed solid, what is the angle of twist when a torque of 45 kNm is applied?

Long hollow thin-walled tube under torsion

This argument is a logical sequel to that of the rod in the previous sub-section. Consider a thin-walled hollow cylinder of mean radius R_m thickness h and length L (Figure 2.15). We know that

$$\tau = T/(2\pi R_m^2 h)$$

We have already seen for the solid rod under torsion that the angle of twist is

$$\theta = \gamma L/R$$

Recalling that radial shear is accompanied by complementary longitudinal shear, the stress–strain relationship $\gamma = \tau/G$ then gives

$$\tau/r = G\theta/L = T/(2\pi R_m^3 h)$$

Problem 2.22: A circular tube 6 mm thick, 100 mm mean diameter and 1.1 m long is made from an isotropic material having a shear modulus of 1.4 GPa. What torque about the axis of the tube will cause a relative twist of the ends of 4°? What shear strain is present under this loading condition?

2.5 STRESSES, STRAINS AND CURVATURES IN BENDING

2.5.1 Overview

Bending occurs widely in everyday life as well as in engineering structures. We expect structures to be either stiff or flexible.

Examples where bending stiffness is required

Floors in buildings must not sag perceptibly or we lose confidence; the same argument applies to bookshelves; curtain rails must not sag too much or the hemline of the curtain looks unsightly.

Vending machine cups must be sufficiently stiff radially that the contents can be supported by the fingers without the cup slipping through. (The rim plays an important part in achieving the desired stiffness: if you cut off the rim or lip, most vending cups become almost impossible to hold by gripping the sidewalls alone because of the loss of radial bending stiffness.)

Handles on spanners or pincers should not flex too much when trying to undo stubborn fastenings.

Examples where flexibility in bending is important

Bending of fibres in the pile of a carpet, or the blades of grass in a lawn.

Leaves and bigger parts of trees bend to offer less resistance to high winds (and yet stiff enough for horizontal branches not to sag too much under their own mass or under added snow loading).

Plastics or paper bags should be flexible enough to accept articles of suitable size.

Car tyres should be flexible enough to help give a smooth ride over rough ground (as part of the suspension system) while giving local grip on the road to achieve power transfer and steering.

In this section we shall look first at the behaviour of a long slender beam under the action of a bending moment. Our goal is to develop insights into the way strains and stresses vary through the thickness of the beam, and how these relate to the (radii of) curvatures and to the applied moments or bending loads. This will involve a discussion of the second moment of area of the cross-section of the beam, and the concept of bending stiffness in the beam. We shall then adopt a similar approach to the bending and twisting of a thin plate modelled as a wide beam. In Section 2.7 we shall discuss the deflection of beams under transverse load.

2.5.2 Assumptions

Beams are considered to be one-dimensional (Figure 2.30): the length (in the y direction) is assumed to be much greater than the transverse (x and z)

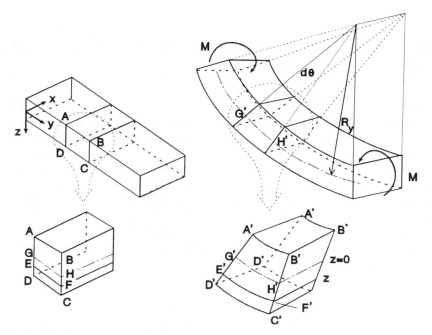

Figure 2.30 Deflection of a long slender beam into a circular arc under pure bending.

dimensions. We make the following assumptions in the following analysis of stresses, strains and deflections in beams:

1. Transverse sections of the beam which are plane before bending remain plane during bending.
2. Transverse sections will be perpendicular to circular arcs having a common centre of curvature.
3. The radius of curvature of the beam is large compared with the transverse dimensions.
4. Longitudinal elements of the beam are subjected only to simple tension or compression giving stress in the y direction, and there is no lateral (width-wise) stress in the x direction.
5. Young's modulus for the beam material has the same value in tension and compression.

Here we shall only discuss the behaviour of beams of symmetric cross-section.

In this text we use the cartesian co-ordinate system x, y, z, with z for the thickness direction of the beam. The corresponding displacements of any elemental volume of material are denoted u, v and w. The symbol M is used in beam theory to denote the conventional sense of bending moment, with units of Nm.

2.5.3 Deformations in beams

We are concerned with deformation along the length of the beam. Under pure bending the beam deforms to an arc (Figure 2.30). The element ABCD (which was originally straight) transforms to A′ B′ C′ D′ and in particular AB compresses and CD stretches. Somewhere in between them is a neutral plane GH = G′ H′, at coordinate $z = 0$ which does not change in length under bending, even though G′ H′ has a radius R_y. The length of G′ H′ is $R_y d\theta$. The faces A′ D′ and B′ C′ include the angle $d\theta$. The length of the straight line element EF at z below the neutral plane is the same as GH and the deformed G′ H′, i.e. $dy = R_y d\theta$. The length of the deformed element E′ F′ is $(R_y + z)d\theta$. Hence the strain in E′ F′ is

$$\varepsilon_y(z) = [(R_y + z)d\theta - Rd\theta]/Rd\theta = z/R = zd\theta/dy$$

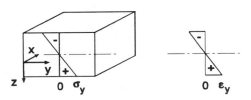

Figure 2.31 Stress and strain profiles in bending.

So the strain varies linearly across the section, with maximum values at the outer surfaces, as shown in Figure 2.31. Strain is entirely independent of the type of material as long as it is linear elastic.

2.5.4 Direct stresses in beams

We have assumed no lateral stress ($\sigma_x = 0$) so for a linear elastic material

$$\varepsilon_y(z) = \sigma_y/E \qquad \text{i.e.} \qquad \sigma_y(z)/z = E/R$$

Hence bending stress varies linearly over the section thickness. Figure 2.31 shows the bending stress profile.

2.5.5 Curvatures in beams

If we apply a pure bending moment M which will make a beam of rectangular cross-section and length L, sag into the arc of a large diameter circle of radius R_y (Figure 2.32), then as we have seen in Sections 2.5.3 and 2.5.4, the strain in a longitudinal element distance z from the neutral axis will be given by $\varepsilon_y(z) = z/R_y = \sigma_y/E$.

The deformed shape of the neutral axis is shown in Figure 2.32. θ is the angle which the tangent at C makes with the x axis. The tangent from D makes the angle $\theta - d\theta$. Normals to the curve at C and D meet at the centre of curvature, 0. The radius of curvature of the element CD is R_y.

Numerically the arc length $ds = R_y d\theta$. Using the sign convention of Section 2.2.4, as ds increases, θ decreases, hence $1/R_y = -d\theta/ds$. The deflection w of the neutral axis is positive downwards. For very small deflections, $ds \sim dy$ and $\theta \sim \tan \theta = dw/dy$, hence

$$-1/R_y = d^2w/dy^2 = -M/EI$$

The radius of curvature R_y is related to the curvature κ_y by

$$\kappa_y = 1/R_y = -d^2w/dy^2$$

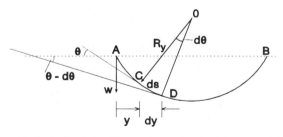

Figure 2.32 Deformation of beam neutral axis under pure bending.

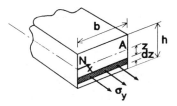

Figure 2.33 Internal stress in element of beam cross-section.

The negative sign of the second derivative of displacement arises because as we move from the left hand end of the beam in the positive y direction along the beam, the slope of the beam decreases, and because the centre of curvature occurs above the beam where z is negative. Our sign convention is that a curvature is considered positive if it is concave upwards. This is consistent with the sign convention for bending moment.

On the shaded elemental strip of the deformed beam (Figure 2.33), which carries a longitudinal stress σ_y of uniform intensity across its width, the moment of the internal force dF_y about the neutral surface is $z\,dF_y = z\sigma_y dA$. The external moment M must be in equilibrium with the sum of the internal moments of all possible strips:

$$M = \int z\sigma_y dA$$

and substituting $\sigma_y = Ez/R_y$ we have

$$M = (E/R_y)\int z^2\,dA = EI/R_y = EI\kappa_y$$

where I is the second moment of area, discussed in Section 2.5.6, which is independent of the material from which the beam is made. EI is termed the bending stiffness of the cross-section.

2.5.6 Properties of areas

Important terms in bending are the centroid, which locates the neutral axis, and the second moment of area.

Consider a body of uniform cross-sectional area A having centroid coordinates (x^*, y^*), Figure 2.34. The first moment of the whole area about an axis must be the same as the sum of the first moments of all the elements of area dA about that axis. Taking moments about the x axis, the distance of the centroid from the x axis is found from

$$Az^* = \int z\,dA \quad \text{or} \quad z^* = (1/A)\int z\,dA$$

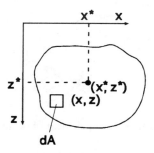

Figure 2.34 Location of centroid (x^*, z^*) in a cross-section.

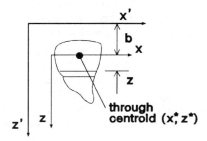

Figure 2.35 Change of axes in relation to centroid.

If the x axis passes through the centre of area, as shown in Figure 2.35, $z^* = 0$ and also $\int z dA = 0$.

The second moment of area of an elemental area about the x axis is $z^2 dA$. The total second moment of area about the centroid is $I_{xx} = \int z^2 dA$, and in older texts is called the moment of inertia.

It is useful to be able to determine the second moment about an axis $x'x'$ parallel to the centroidal axis xx: From Figure 2.35,

$$I_{x'x'} = \int (z + b)^2 dA = \int z^2 dA + 2b \int z dA + \int b^2 dA$$

hence $I_{x'x'} = I_{xx} + Ab^2$ (because $\int z dA = 0$). This is called the parallel axes theorem.

Second moments of area for simple sections

With reference to Figure 2.36 we can derive the following:

(a) Rectangular section
$I_{xx} = bh^3/12 \quad A = bh$
$I_{x'x'} = bh^3/12 + bh^3/4 = bh^3/3$

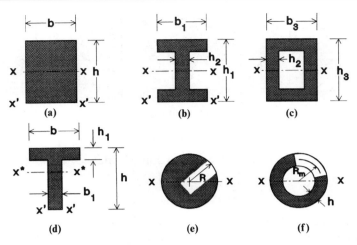

Figure 2.36 Properties of area.

(b) I-section
$$I_{xx} = b_1 h_1^3/12 - (b_1 - h_2)(h_1 - 2h_2)^3/12$$
$$A = b_1 h_1 - (b_1 - h_2)(h_1 - 2h_2)$$

(c) Hollow rectangle
$$I_{xx} = b_3 h_3^3/12 - (b_3 - 2h_2)(h_3 - 2h_2)^3/12$$
$$A = b_3 h_3 - (b_3 - 2h_2)(h_3 - 2h_2)$$

(d) T-section
$$z^* = [bh_1(2h - h_1) + b_1(h - h_1)^2]/\{2(bh_1 + b_1(h - h_1))\}$$
$$I_{xx} = [bh_1^3 + b_1(h - h_1)^3]/12 + bh_1(h - h_1/2 - z^*)^2$$
$$+ b_1(h - h_1)(z^* - (h - h_1)/2)^2$$

(e) Circle
$$I_{xx} = \pi R^4/2$$
$$A = \pi R^2$$

(f) Ring
$$I_{xx} = \pi R_m^3 h$$
$$A = 2\pi R_m h$$

It is sometimes necessary to replace a solid beam of material A of thickness h by one made from two skins of solid material A bonded to a core of another material B of thickness h_c (Figure 2.37). This system is symmetrical about the neutral axis, so for equal bending stiffnesses EI of beams of the same breadth we have

$$(EI)_{\text{solid}} = (EI)_{\text{sandwich}}$$

i.e. $E_A h_A^3 = E_A(h_c + 2h_s)^3 - E_A h_c^3 + E_B h_c^3$

i.e. $h_A^3 = (h_c + 2h_s)^3 - h_c^3 + (E_A/E_B)h_c^3$

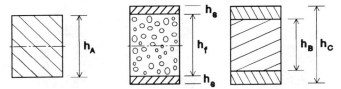

Figure 2.37 Sandwich beams.

Denoting $h_B = \alpha h_c$, where α is the proportion of sandwich thickness which is component B, we find

$$(h_A/h_c)^3 = 1 - \alpha^3(1 - E_B/E_A)$$

Even structures made from isotropic materials but having anisotropic properties are commonplace. For example corrugated sheets made from isotropic or other materials are widely used as roofing or cladding panels. Opaque materials are represented by galvanized steel or asbestos-reinforced cement, translucent materials by polyester reinforced by a low volume fraction of discontinuous glass fibres, and transparent materials by unplasticized PVC, polycarbonate and acrylic plastic. The corrugations confer high bending stiffness per unit weight along their length, which is what is needed to carry applied loads (such as self-weight loads, snow, wind, and people) between supports. But the bending stiffness across the corrugations is simply caused by the thickness of the sheet, which is small compared with the depth of the corrugations.

> **Problem 2.23:** Derive from first principles expressions for the second moment of area of (a) an I section beam, and (b) a T section beam. Sketch profiles of the longitudinal direct bending stress and strain through the thickness (depth) of the beams when subject to pure bending.

> **Problem 2.24:** For an I section beam of uniform wall thickness where the width of the flanges is half the overall depth, what is the contribution of the web to the total second moment of area?

> **Problem 2.25:** A weightless horizontal beam made from an isotropic material of tensile modulus $E = 9$ GPa has a cross-section 12 mm wide $\times$ 5 mm deep. When it is under a uniform positive moment of 0.1 Nm acting in the vertical plane, calculate the curvature of the midplane and the longitudinal strain on the upper surface.

> **Problem 2.26:** Consider two thin rectangular corrugated sheets of the same depth, plan area and wall thickness, made from identical isotropic materials (Figure 2.38). One sheet has a circular corrugation. The other has a square corrugation, i.e. the width of each corrugation is the same as the depth of the sheet, so the period of the corrugation is twice the depth.

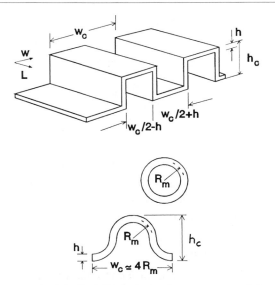

Figure 2.38 Leading dimensions of two corrugated sheets.

Which sheet has the largest ratio of stiffness along to across the sheet? Which sheet has the largest value of stiffness along the corrugation per unit weight? Which sheet is cheaper for a given stiffness requirement?

2.5.7 Equilibrium of forces and moments

We now need a method of finding the neutral plane and neutral axis of a beam. For simple shapes of cross-section this can be ascertained by inspection. Consider a small area dA at y below an arbitrary neutral axis on a beam in pure bending (Figure 2.39). The force on the elemental area dA is $dF_y = \sigma_y dA$. For a positive moment, stresses are compressive for negative z co-ordinates and tensile for positive z. The total longitudinal force $F_y = \int \sigma_y dA = 0$ because in pure bending there is no externally applied axial force present. From $\sigma_y = Ez/R$, we have $F_y = (E/R_y)\int z dA = 0$. As E/R cannot be zero, the integral is zero and hence the centroid corresponds with the neutral axis.

The moment of the axial internal force on the element dA about the neutral surface is $dM = dF_y \cdot z$. The total internal moment M is given by $M = \int z dF_y = \int z\sigma_y dA$. This must balance the external moment M so for equilibrium:

$$M = \int z\sigma_y dA = (E/R_y)\int z^2 dA = EI/R_y = EI\kappa_y$$

This is an extremely important equation. The concept of moment per unit curvature is a measure of how difficult or easy it is to bend the beam. This is

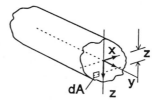

Figure 2.39 Axial stress on element in beam under pure bending.

directly proportional to the bending stiffness, or flexural stiffness, EI of the beam. EI depends on the material from which the beam is made, which controls the value of E, and the cross-sectional shape of the beam, which is totally independent of the material used.

2.5.8 Plate and shell equations

In plate theory we adopt a similar approach to that discussed for beams, but it is customary to use moments per unit width M_x and M_y with units of N. This unit may seem strange until it is expanded to the form Nm/m. The context usually makes it clear whether ordinary moments or moments per unit width are intended, and it is good practice to check on the units of any displayed equations if there is any doubt. These moments per unit width are taken as positive when producing sagging curvatures, as shown in Figure 2.40; the curved plate is concave on the upper surface.

The radii of curvature of the neutral surface are R_x and R_y, corresponding to midplane curvatures $\kappa_x = 1/R_x$ and $\kappa_y = 1/R_y$. On a plane distant z below the neutral surface, the strains are tensile: $\varepsilon_x = z/R_x = z\kappa_x$, and $\varepsilon_y = z/R_y = z\kappa_y$. Under biaxial tension, the relationships between stress and strain for the plate material are

$$\varepsilon_x = \sigma_x/E - \nu\sigma_y/E \quad \text{and} \quad \varepsilon_y = \sigma_y/E - \nu\sigma_x/E$$

Substituting for strains in terms of radius of curvature, and rearranging, gives two sets of equations. The conventional representation in books on solid body

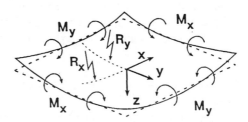

Figure 2.40 Positive bending moments and radii of curvature.

mechanics is

$$\sigma_x = [Ez/(1 - v^2)](1/R_x + v/R_y)$$

and

$$\sigma_y = [Ez/(1 - v^2)](1/R_y + v/R_x)$$

The conventional representation in plate theory and composite mechanics is in terms of curvatures:

$$\sigma_x = [Ez/(1 - v^2)](\kappa_x + v\kappa_y)$$

and

$$\sigma_y = [Ez/(1 - v^2)](\kappa_y + v\kappa_x)$$

Either form shows clearly that stresses are a function of the two curvatures and vary linearly with distance from the neutral surface.

Equilibrium

We can now look at equilibrium between the internal moments caused by the bending stresses σ_x and σ_y on the vertical sides of an element within the plate (Figure 2.41), and the external moments per unit length M_x and M_y.

Proceeding as we did for the beam, we see that the moment M_x on an element of width dy is $\int \sigma_x z\,dy\,dz = M_x dy$ (integral limits from $z = -h/2$ to $z = +h/2$), and for a moment M_y on an element of width dx, $\int \sigma_y\,dx\,dz = M_y dx$.

Substituting for the stresses σ_x and σ_y from Section 2.5.8, and integrating, gives:

$$M_x = [Eh^3/12(1 - v^2)](\kappa_x + v\kappa_y) = D(\kappa_x + v\kappa_y)$$
$$M_y = [Eh^3/12(1 - v^2)](\kappa_y + v\kappa_x) = D(v\kappa_x + \kappa_y)$$

The term $D = [Eh^3/12(1 - v^2)]$ is called the flexural rigidity of the plate, and is closely related to the flexural stiffness EI of a beam corrected by the factor $(1 - v^2)$, recalling that $h^3/12$ is the second moment of area of a rectangular cross-section *per unit width*.

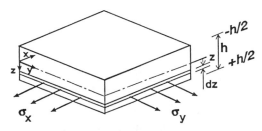

Figure 2.41 Bending stresses within plate under applied moments.

Bending curvature

Rearranging the bending moment equations we obtain expressions for the curvatures:

$$\kappa_x = M_x/[D(1 - v^2)] - vM_y/[D(1 - v^2)]$$
$$\kappa_y = - vM_x/[D(1 - v^2)] + M_y/[D(1 - v^2)]$$

These equations tell us that if we apply only a single positive bending moment per unit width (say M_x), then we obtain a positive curvature κ_x accompanied by a lesser curvature of opposite sign κ_y. The plate then takes a saddle shape as shown in Figure 2.42, sometimes called anticlastic curvature. For the special case of a plate made from a single isotropic linear elastic material, the ratio of curvatures is the lateral contraction ratio v.

This anticlastic curvature is entirely expected from the linear strain distributions given by equations such as $\varepsilon_x = z\kappa_x$ when a single moment M_x is applied. If at some co-ordinate z below the neutral surface the strain in the x direction is ε_x, then the strain in the y direction must be $\varepsilon_y = - v\varepsilon_x$, i.e. of opposite sign to ε_x, and hence the curvatures at z will be of opposite sign in the two perpendicular directions.

Problem 2.27: A sheet of cast acrylic polymer is isotropic and has the elastic constants E = 3 GPa, $v = 0.35$, and G = 1.11 GPa. For a sheet 2 mm thick, calculate the curvatures when pure bending moments per unit width are applied: (a) $M_x = 1$ N; (b) $M_x = M_y = 1$ N. (c) What moment M_y would need to be applied to suppress the anticlastic curvature in problem (a)? What are the stresses and strains on the upper surface of the sheet?

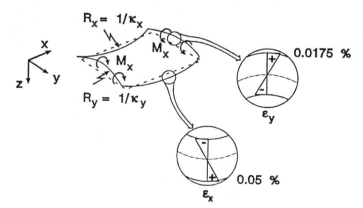

Figure 2.42 Anticlastic curvature and strain profiles in plate under a single bending moment.

2.5.9 Bending and twisting curvatures

We shall now revise the mathematical expression for bending curvature and shall then look at the concept of twisting curvature.

Bending curvature

The relationship between displacement, slope and curvature for a beam undergoing pure bending has already been established as

$$\kappa_y = 1/R_y = -d^2w/dy^2$$

The point A distance y from the left hand end of the undeflected beam has deflected an amount w downwards (Figure 2.43). The slope of the neutral plane of the beam at A is dw/dy, and the tangent to the neutral surface at A subtends an angle θ to the horizontal axis.

At some point B distant $ds \sim dy$ further along the beam the displacement is now $w + dw$, the slope is $dw/dy + (d^2w/dy^2)dy$, and the tangent at B subtends $\theta - d\theta$. The rate of change of slope along the beam between A and B is the curvature $\kappa_y = -d^2w/dy^2$.

This argument also applies for a thin plate where the out-of-plane displacements are very small. Thus under pure bending M_x or M_y we obtain the curvatures $\kappa_x = -d^2w/dx^2$ and $\kappa_y = -d^2w/dy^2$.

Twisting curvatures

We have already discussed axial angular twisting of circular solid and hollow rods in Sections 2.3.10 and 2.3.11. We now consider a flat plate which carries a torsion couple which induces twist. Twisting curvatures are usually glossed over in standard texts, and the following treatment owes much to Jaeger (1964).

Consider a flat plastics comb having a spine of uniform cross-section with a large number of regular closely-spaced teeth, and twist it by applying opposing

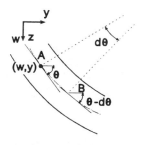

Figure 2.43 Deflection of neutral axis of beam.

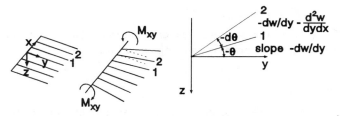

Figure 2.44 Development of twisting curvature under torque.

couples to the ends (Figure 2.44). The teeth in the twisted comb no longer lie in a flat plane, and adjacent teeth are rotated with respect to one another. Looking along the back of the comb AB, and concentrating on two adjacent teeth separated by dx, there is a difference in slope. The slope of the first tooth is $-\theta = -dw/dy$. (The negative sign arises because our sign convention (Figure 2.4) is that positive angles are clockwise, and both z and deflections are positive downwards.) Moving down the comb a distance $+dx$, we realize that the change of slope is only as the result of the change in x. The rate of change of slope is therefore $(d/dx)(-dw/dy) = -d^2w/dxdy$, and the slope of the second tooth is $-dw/dy - (d^2w/dxdy)dx$.

The formula $\kappa_{xy} = -d^2w/dxdy$ is a measure of the twist of the surface with respect to the plane of the teeth making up the comb. The same argument applies to the twist in a continuous thin sheet occupying the (x, y) plane. This twisting curvature may be thought of as having the dimensions of change of

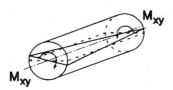

Figure 2.45 Twisting of a diametral plane in a solid rod under torque.

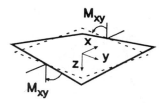

Figure 2.46 A positive twisting moment per unit width induces positive twisting curvatures.

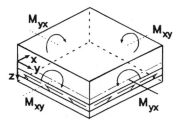

Figure 2.47 Twisting moment, complementary moment and internal shear stress.

slope per unit (perpendicular) length, or of change of angle per unit length. This corresponds to what happens when a rod is twisted along its axis. Look at the longitudinal/diametral plane in Figure 2.45.

The sign convention for positive twisting moment M_{xy} of a plate and positive curvature is shown in Figure 2.46.

Using the concept of complementary shear stress, or simply common sense, it follows that $M_{yx} = -M_{xy}$, as indicated for the shear stress on an element in Figure 2.47.

2.5.10 Significance of twisting curvatures

It is helpful to examine the physical significance of twist in more detail, and the following discussion assists us to sketch the deformations in a flat sheet. Consider a thin flat rectangular plate (Figure 2.48), of side length L_x and L_y, with the origin at the centre of the plate. The twisting angle of the y axis over the length L_x of the x axis is $-\theta$ when a positive bending moment M_{xy} is applied to the sheet.

At any point along the x axis the angle of twist of the plate, α, is given by

$$\alpha = -(\theta/L_x)x$$

so that at $x = L_x/2$, the angle of twist of the y axis is $-\theta/2$.

At any point y the vertical displacement w is

$$w = \alpha y = -\theta/L_x xy$$

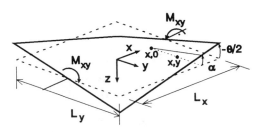

Figure 2.48 Characterizing twisting curvature.

so at the corner $x = L_x/2$, $y = L_y/2$,

$$w = -(\theta/L_x)L_xL_y/4 = -\theta \cdot L_y/4$$

But we define $\theta/L_x = d^2w/dxdy$, hence

$$w = -(d^2w/dxdy)xy = \kappa_{xy} \cdot xy$$

where x and y are measured from the centre of the sheet.

2.5.11 Relation between twisting moment and twisting curvature

In order to derive the relationship between twisting moment M_x and twisting curvature $-d^2w/dxdy$, we first need to relate shear strain to displacements and change of angle, and then take a balance between internal and applied torques.

Consider the element AOB of sides dx and dy (Figure 2.49) which rotates and is displaced to A'O'B' in a state of bi-axial strain in the x, y plane. The rate of change of v in the x direction is dv/dx, so the distance CA' is the rate of change of v in the x direction over the distance dx, i.e. $(dv/dx) \cdot dx$. Hence the angle α is given approximately by

$$\alpha = CA'/O'C \sim dv/dx$$

Similarly DB' is the rate of change of u in the y direction for a length dy, and hence the angle β is approximately

$$\beta \sim DB'/O'D \sim du/dy$$

The shear strain γ_{xy} is defined as the change in angle from AOB to A'O'B', hence

$$\gamma_{xy} = \alpha + \beta = dv/dx + du/dy$$

Let us now relate the shear strain to the deflections of the plate. An element taken through the thickness of the plate will rotate by dw/dx and dw/dy in the xz and yz planes under the action of the twisting moment M_{xy} and the associated moment M_{yx}.

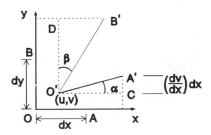

Figure 2.49 Development of (biaxial) shear strain.

Consider Fig 2.50, which shows an element of length dx and thickness h. Because of the rotation, a point E at z below the neutral plane is displaced in the (negative) x direction by an amount $u = -z \cdot dw/dx$, and hence $du/dy = -z \cdot d^2w/dxdy$.

Similarly the displacement v in the y direction is $v = -z \cdot dw/dy$, so $dv/dx = -z \cdot d^2w/dydx$.

We can now write the desired expression for shear strain in terms of the twisting curvature:

$$\gamma_{xy} = -2z \cdot d^2w/dydx = z\kappa_{xy}$$

Consider now the element in Figure 2.51 of width dy and carrying a shear stress τ_{xy}. The total moment of the (internal) shear force about the midplane is countered by the total applied torque $M_{xy}dy$:

$$M_{xy}dy = \int \tau_{xy} \cdot dy \cdot zdz$$

hence

$$M_{xy} = G \int \gamma_{xy} zdz = G\kappa_{xy} \int_{-h/2}^{+h/2} z^2 dz = (Gh^3/12)\kappa_{xy}$$

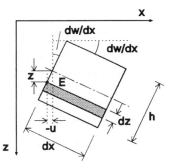

Figure 2.50 Rotational displacement of elemental cross-section.

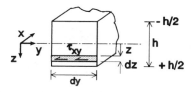

Figure 2.51 Shear stress in element under torsion.

But for an isotropic material $G = E/[2(1 + v)]$, and hence we achieve

$$M_{xy} = (D/2)(1 - v)\kappa_{xy}$$

2.5.12 Review

We are now in the position where we can display in matrix form the general relationship betwen moments and curvatures for thin plates having isotropic properties:

$$\begin{pmatrix} M_x \\ M_y \\ M_{xy} \end{pmatrix} = \begin{pmatrix} D & vD & 0 \\ vD & D & 0 \\ 0 & 0 & (D/2)(1-v) \end{pmatrix} \begin{pmatrix} \kappa_x \\ \kappa_y \\ \kappa_{xy} \end{pmatrix} = [D] \begin{pmatrix} \kappa_x \\ \kappa_y \\ \kappa_{xy} \end{pmatrix}$$

and the inverse form:

$$\begin{pmatrix} \kappa_x \\ \kappa_y \\ \kappa_{xy} \end{pmatrix} = \begin{pmatrix} 1/D(1-v^2) & -v/D(1-v^2) & 0 \\ -v/D(1-v^2) & 1/D(1-v^2) & 0 \\ 0 & 0 & 2/D(1-v) \end{pmatrix} \begin{pmatrix} M_x \\ M_y \\ M_{xy} \end{pmatrix} = [D]^{-1} \begin{pmatrix} M_x \\ M_y \\ M_{xy} \end{pmatrix}$$

where later laminate analysis uses $[D]$ to describe the bending stiffness matrix for the plate.

In this compact format we can now see that a wide variety of forms of deformation of the sheet can result from application of one or more moments which can each have either a positive or negative sign.

2.5.13 Sketching of midplane strains and curvatures

It is quite useful to be able quickly to translate the midplane strains and midplane curvatures of a sheet under various combinations of in-plane and bending loads into a pictorial form. With practice (and there are many opportunities later in this book), this can be done mentally. But to start with, the reader may find the following comments helpful.

1. Sketch a firm grid with broken lines to indicate the original size and shape of the undeformed sheet, and label the origin with the intended coordinate directions (Figure 2.52(a)).
2. Draw lightly the strains ε_x and ε_y, and then adjust for shear γ_{xy}, making the shear parallel to the x axis (Figure 2.52(b)).

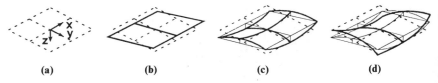

| (a) | (b) | (c) | (d) |

Figure 2.52 (a) Undeformed sheet; sketching; (b) in-plane strains; (c) bending curvatures; and (d) twisting curvatures.

3. Using the lines $y = 0$ and $x = 0$, draw in the curvatures κ_x and κ_y respectively until you reach the edge of the deformed sheet from step 2. It is easiest to ignore any foreshortening of curved lines caused by bending. At the ends of each curved centre line, draw in the edge curvatures parallel to their centre line. The corners should meet if the sketch is to be informative. Figure 2.52(c) shows the result.
4. Where twisting curvatures need to be shown, the curved centre lines remain untouched. What is needed now is a rotation of the edge curved boundaries about their midpoints in accordance with the sign of κ_{xy}. Thus the corners of the sheet will either lift or fall by the same amount. Again we ignore any foreshortening caused by the rotation.
5. The end result can then be drawn in bold lines (Figure 2.52(d)).

Note that in the sequence of sketches, the three strain components are clearly evident in Figure 2.52(b). The curvatures are always clearly seen if a reasonable scale is used. But when curvatures are present, it is often not easy in these rough sketches to discern the in-plane strains. The numerical values on which the sketches are based will make plain any local difficulties.

Problem 2.28: Some modes of deformation under a variety of combinations of bending moments are shown in Figure 2.53, exaggerated for clarity and not entirely to any scale, superposed on the original flat sheet (broken line). Each moment has the same numerical value. Can you assign to each of these figures the moment or combination of moments likely to cause the deformation shown? Are any combinations of moments not depicted? If so,

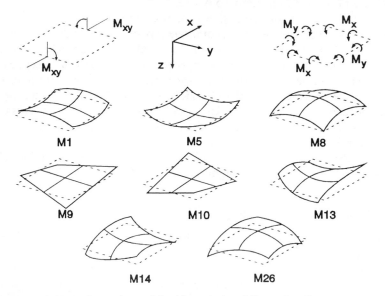

Figure 2.53 Deformations of flat sheet under different bending moments.

identify the moments involved on suitable sketches. It may be assumed that no patterns of deformation involve positive and negative values of the same moment.

Problem 2.29: Strain gauges have been applied to a horizontal thin flat square sheet of linear elastic isotropic cast polyester resin 4 mm thick. Under certain loading conditions the strains on the top and bottom surfaces of the sheet parallel to the sides have been calculated as follows:

Surface	$\varepsilon_x\%$	$\varepsilon_y\%$	$\gamma_{xy}\%$
Top	0.3046	−0.2748	−0.3266
Bottom	0.3121	−0.2335	−0.1239

Sketch the shape of the deformed sheet, and the stress and strain profiles through the thickness. What minimum information about the material would you need in order to determine the ratio of applied in-plane stresses σ_x/σ_y?

2.6 SHEAR STRESSES CAUSED BY BENDING

2.6.1 Introduction to shear in bending

Most beams are long and slender and, when carrying bending moments or transverse loads, the direct bending stresses in and deflections of beams can be calculated using the assumptions and methods of the engineer's theory of bending, as outlined in Section 2.5 above.

In general, when beams bend there are also present shear stresses, and the associated shear strains develop a component of deflection caused by shear. In slender beams made from an isotropic material these shear stresses seldom cause problems unless the beam is short and deep, and their effects can be calculated where necessary. In beams (or plates) made from fibre reinforced polymers, the shear strength of a ply can be low and the shear strength of the bond between plies in a laminate can also be low. It is therefore necessary to check as a matter of routine that shear stresses do not exceed safe values, even if the beam is still quite long and slender.

What we shall now discuss for isotropic materials is the origin of these shear effects, their analysis and the magnitude of the effects in practice.

2.6.2 Origins of shear in bending

There are several ways we can identify by observation the possibility of shear deformation in the bending of beams.

If we support a horizontal stack of normal unbonded playing cards on two pencils parallel to and close to the short edges and apply a vertical load at midspan (Figure 2.54), then the cards will deflect by simple bending, and at the

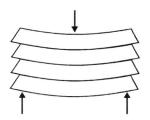

Figure 2.54 Shear develops when unbonded plies carry a bending moment.

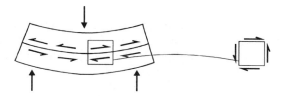

Figure 2.55 Shear stresses needed to prevent shear between unbonded plies.

ends each cards can be seen to have sheared past its neighbour. If the cards have sheared at the ends, they will also have sheared horizontally at any other point along their length. To stop the sliding we would need to apply horizontal opposing shear stresses at the interface between the cards (Figure 2.55). These are actually present in laminated structures as well as in isotropic materials. They are responsible for the bonded block of cards being much stiffer as it now behaves as a block having a second moment of area based on its total depth rather than as a series of cards each having its own depth.

So far we have talked about horizontal shear stresses in a horizontal beam. But recalling the principle of complementary shear stresses introduced in Section 2.3.4, there are also present vertical shear stresses (Figure 2.55). It is less easy to picture these (though easy to describe using the principle of vertical equilibrium, as we shall see below), but if we suppress the shear at the ends of the stack of generally unbonded cards, then in sagging curvature the upper cards will be in direct compression and will buckle as well as try to shear horizontally (Figure 2.56): the buckling shows an obvious vertical shear between the cards.

Figure 2.56 Buckling of unbonded plies when shear at ends is prevented.

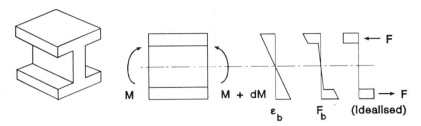

Figure 2.57 Development of shear forces in flanges of I-beam.

It is now worth looking at a more technical example. In discussing the principle of second moment of area in Section 2.5.6, it became clear that for an I-beam, the greater the separation of the flanges, the greater the second moment of area, and the web contributed little (indeed engineers often ignore the web when roughly calculating second moment of area for bending stiffness). The temptation is to separate the flanges as much as possible to maximize bending stiffness per unit mass in an I-section. The limits are that the flange or the web may buckle locally because of the action of the compressive stresses present in bending, or that the web cannot take the shear stresses present. Ignoring the web, there will be a net bending force acting in opposite directions on each flange which will induce shear (Figure 2.57), unless the web prevents it.

Problem 2.30: If we take a paper covered book such as a telephone directory and support it at the free edge and the spine, where does the shear occur when it is loaded in midspan parallel to the supports?

2.6.3 Shear stresses in beam of rectangular section

Let us consider a long slender horizontal beam of constant breadth b and depth h, where $h \gg b$, of which an element of length dy is shown in Figure 2.58. Acting on the vertical end faces of this element are a positive shear force Q at each end, and positive moments M_y and $M_y + dM_y$. Consider a thin horizontal slice of thickness dz parallel to the neutral axis. The bending moment induces a direct bending stress $\sigma_y(z)$ on the top surface, and a larger stress $\sigma_y + d\sigma_y$ at $z + dz$. The shear force will induce a positive vertical shear stress τ_{xy}. Associated with the vertical shear stress is the complementary horizontal shear stress τ_{yz} which is also positive and equal to τ_{zy}. The cause of the presence of shear stress is now obvious. What we seek is how the shear stress varies through the thickness of the beam but we do not yet have quite sufficient information to determine this.

The boundary condition we need can be gained by considering what happens at the lower surface of the beam. There is no external shear force

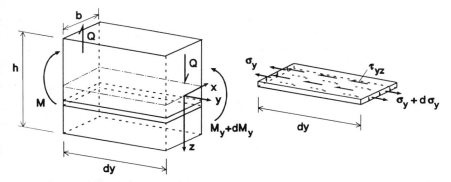

Figure 2.58 Shear stresses in beam under bending moment gradient along length.

along the bottom surface of the beam, and therefore by the principle of complementary shear stress there can be no vertical shear stress at the lower surface of the beam. Similarly there is no shear stress at the upper surface.

We can now proceed to analyse a second elemental strip extending from co-ordinate z_1 to the lower surface of the beam $+ h/2$ (Figure 2.59), so that the boundary condition at $z = + h/2$ can be used. There are two horizontal bending forces acting on some sub-element dA: dF_{yA} at co-ordinate y, and dF_{yB} at co-ordinate $y + dy$.

The horizontal total forces on the element are related to the bending stresses and the areas over which they act;

$$F_{yA} = \int_{z_1}^{h/2} (M_y z/I)dA \quad \text{and} \quad F_{yB} \int_{z_1}^{h/2} ((M_y z + dM_y)/I)dA$$

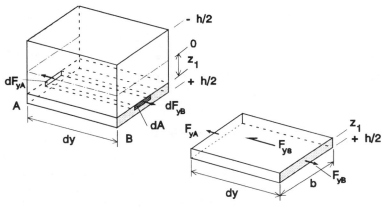

Figure 2.59 Direct and shear forces on element of beam.

where $I = bh^3/12$ is the second moment of area of the beam as a whole. The shear force F_{yS} acts at $z = z_1$ on the strip of length dy and breadth b so that, assuming the shear stress is uniform across the width of the element, the shear stress $\tau_{yz} = F_{yS}/bdy$.

The horizontal forces must be in equilibrium, so we have

$$F_{yS} = F_{yB} - F_{yA}$$

i.e.

$$\tau_{yz} = (dM/dy)(1/bI) \int z dA$$

Substituting the relation between shear force and bending moment discussed in Section 2.2.4, $Q = dM/dy$, we find

$$\tau_{zy} = (Q/bI) \int z dA$$

For a rectangular element, the first moment of area about the centroid, z^* is given by

$$z^* = \int z dA = \int_{z_1}^{h/2} zb dz = (b/2)(h^2/4 - z_1^2)$$

Thus the shear stress profile in a rectangular beam is

$$\tau_{zy} = (Q/2I)(h^2/4 - z_1^2)$$

and gives a parabolic profile which varies with distance z_1 from the neutral axis. The maximum value of shear stress occurs at the neutral axis, with a value

$$\tau_{zymax} = Qh^2/8I = 3Q/2A$$

where $A = bh$ is the area of the section. It is interesting to note that for a rectangular cross-section the maximum value of shear stress is 50% larger than the average shear stress calculated as $\tau_{av} = Q/A$.

Problem 2.31: Compare the stress profiles for bending and for shear at any cross-section in a cantilever beam carrying a point load at the free end.

Problem 2.32: Describe the shear stress profile when a beam carries a pure bending moment.

2.6.4 Shear stress in long slender I-section beam

In bending of a long slender beam of constant I-section, the following analysis shows that almost all the shear stress is taken by the web and hardly any in the flange. This analysis is not complete – for additional details of secondary shear stresses in the flanges the reader should consult standard texts such as Benham and Crawford (1987).

2.7 DEFLECTIONS IN BEAMS UNDER TRANSVERSE LOADS OR MOMENTS

In isotropic materials deflections under transverse loads are almost entirely caused by bending stresses. In polymers, most beams are designed for stiffness rather than strength. In composites bending caused by shear deformation can also be important and a comment on deflection caused by shear is made below.

2.7.1 Bending deflections in beams

We have already studied in Sections 2.5.3 and 2.5.5 the strains, curvatures and the second derivative of displacement in beams under applied moments or transverse forces. In this section we shall calculate the deflection curve of the neutral axis of a simple beam. There are several ways of doing this, and here we shall use the method of double integration and explore the effect of different loading and end conditions. Other methods, and more complicated situations, are discussed in the standard texts, and summarized in handbooks such as Roark and Young (1989).

From Section 2.5.5 the second derivative of the displacement, $w(y)$, has the form

$$d^2w/dy^2 = -M(y)/EI$$

where $M(y)$ is the local moment or the moment of forces acting in the transverse (z) displacement direction.

The slope of the beam at any point along its length is obtained by integration as

$$dw/dy = \int (-M(y)/EI)\,dy + C_1$$

and the displacement (deflection) becomes

$$w(y) = \int\int (-M(y)/EI)\,dy\,dy + C_1 y + C_2$$

C_1 and C_2 are constants of integration which relate to how the beam is supported at one or both ends, and hence are found from known conditions of slope and deflection at suitable points, usually at the ends or midspan.

We shall look at two examples of this approach, and then provide a summary for a range of simple beams and common loading conditions.

Simply supported beam under constant bending moment

Of particular interest in later chapters of this book, which concern plates under constant applied moments, is the example where the moment is constant along the length of the beam (this situation is called 'pure bending'). For an applied

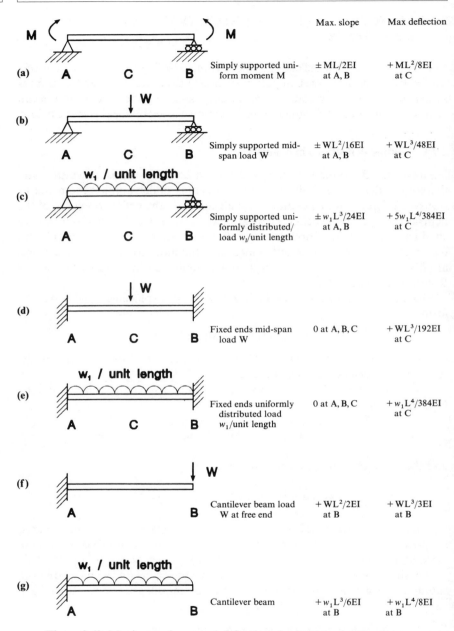

Figure 2.60 Maximum slopes and deflections in simple weightless beams.

constant moment M, integration of the second derivative gives the slope and deflection:

$$dw/dy = -My/EI + C_1, \quad \text{and} \quad w(y) = -My^2/2EI + C_1 y + C_2$$

For a simply supported beam (Figure 2.60), the ends cannot deflect, so the boundary conditions are $(y = 0, w = 0)$, and $(y = L, w = 0)$, so that $C_2 = 0$ and $C_1 = ML/2EI$. The deflection equation for this beam and moment then becomes

$$w(y) = (M/2EI)[y(L - y)]$$

and clearly the maximum deflection occurs at the midspan, $y = L/2$, with the value

$$w_{max} = ML^2/8EI.$$

Simply supported beam with distributed load

The load is w_1 per unit length, as shown in Figure 2.60(c). The bending moment at D is

$$M = w_1 Ly/2 - w_1 y^2/2$$

so

$$EIdw/dx = -w_1 Lx^2/4 + w_1 x^3/6 + C_1$$

and by symmetry $dw/dy = 0$ at $y = L/2$, hence $C_1 = w_1 L^3/24$. Further integration gives

$$w = -(w_1/2EI)[Ly^3/6 - y^4/12 - L^3 y/12] + C_2$$

At $y = 0$, $w = 0$, hence $C_2 = 0$. The maximum deflection at mid-span is

$$w_{max} = +5w_1 L^4/384EI$$

Common beams and the principle of superposition

Figure 2.60 gives expressions for the principal slopes and maximum deflections of weightless beams with common loading conditions. Two factors immediately become apparent, and confirm practical experience. If the same total load is applied, then a distributed load causes less slope and deflection than a concentrated load. Fixing the ends of a beam so they cannot rotate substantially reduces deflections compared with simply-supported ends.

To obtain the slopes or deflections in a given beam under a combination of the loading patterns shown in Figure 2.60, we can simply add the slopes or deflections for the individual loadings. The is known as the principle of superposition, and works well for linear elastic isotropic materials. Thus if a horizontal beam with built-in ends carries a uniformly distributed load w_1

and a concentrated load W at mid-span, the deflection at midspan is
$w_{max} = (L^3/EI)(W/192 + w_1 L/384)$.

Problem 2.33: A horizontal, initially-straight, cantilever beam 150 mm long, 5 mm wide and 10 mm deep is made from a linear elastic material with a modulus $E = 15$ GPa and a density 1600 kg/m³. The beam carries a vertical load of 10 N at the free end. (a) Where is the value of compressive stress a maximum, and what is its value? (b) What is the deflection and slope of the weightless beam at 45 mm from the free end? (c) What is the deflection at the free end if self-weight loading is taken into account?

2.7.2 Deflections in plates

If we wished to explore the deflection, w, of a plate as a result of the applied moment(s), we would need to use the differential form of the curvatures, and integrate the following equations with appropriate boundary conditions:

$$M_x = -D(\partial^2 w/\partial x^2 + v\partial^2 w/\partial y^2)$$
$$M_y = -D(v\partial^2 w/\partial x^2 + \partial^2 w/\partial y^2)$$

Because we are dealing with plates, these are moments per unit width.

Although we shall explore this approach for beams in Section 3.6, the corresponding approach for deflection of plates lies outside the scope of this book. It can be pursued in such books as Jaeger (1964) and Mansfield (1989).

2.7.3 Shear deflections of beams

The shear deflection in a beam is usually small and occurs as the result of the shearing force on transverse sections. Advanced analysis shows that the deflection w_s in a beam of depth d and breadth b has the general form $w_s = kL/bdG$, typical values being:

Beam and load	k
Cantilever, load at free end	$6W/5$
Cantilever with uniformly distributed load	$3w_1 L/5$
Simply supported beam, load at midspan	$3W/10$
Simply supported beam, uniformly distributed load	$3w_1 L/20$

2.8 BUCKLING

Definitions of tensile and compressive stress as force per unit area are very similar: it is just the direction of the applied forces which are different. There is the temptation to think that the behaviour of a structure or material in tension or compression is the same too. This is not generally so. Both the stiffness and the strength in compression can be quite different from the strength in tension.

This section is concerned almost entirely with stiffness. It discusses what engineers call buckling; from other technologies terms such as crinkling, crumpling, wrinkling, folding, draping and gathering describe the same sort of behaviour. Buckling is perhaps the most commonly observed mode of deformation and failure in everyday life. In engineering, buckling describes excessive deformation or distortion in a structure as the result of compressive forces or compressive components of forces, and is usually regarded as a mode of failure to be avoided by design. It is a geometrical effect strongly influenced by slenderness; buckling strain is usually independent of material, though the buckling load does depend on the type of material used.

The subject of buckling is easy to describe pictorially, easy to observe in everyday life, but rather difficult to analyse from first principles without resort to quite complicated and unappealing mathematical analysis. Here we have the space only to present the results of some simple basic analyses, with comments emphasizing their practical significance rather than their derivation. For more detailed treatments the reader is directed to a readable summary by Wainwright, *et al.* (1976), and to other texts listed in the further reading section.

2.8.1 Everyday examples of buckling

A drinking straw is quite stiff in tension; but if you put the stem into axial compression, it bends transversely out of the way under hardly any load. Increasing the load increases the bending until a nasty kink forms, which acts as a hinge and the structure collapses completely. Remove the load before the hinge is formed and the transverse deformation usually disappears.

A sheet of writing paper is quite stiff in tension but under minute compressive loads in its plane it bends out of the way.

Human skin is quite stretchy in tension (think of your lips when you smile broadly), but it wrinkles readily in compression. Look at your wrist when you bend your flat hand towards you, or watch your forehead crease up when you raise your eyebrows in disbelief.

Woven cloth can be quite stiff and strong in tension, but domestic woven fabrics drape and fold when you try to squash them in their plane by applying in-plane forces, e.g. folds in hanging curtains. It is apparent that in the use of fabrics buckling is often highly desirable; most garments made from non-stretch fabrics rely on buckling either to achieve fashion styling features or to permit comfort during wear.

2.8.2 Causes of buckling

There are three main factors which cause buckling in an article. First a compressive strain must be present; second, the article must have a low bending stiffness perpendicular to the compressive strain; and third, the article must be 'slender'.

Direct compressive strain is easy to understand: for example, axial compression of the drinking straw. But compression can occur in two disguised forms. Shear stress and shear strain can be regarded as a combination of tension and compression at right angles to one another and each at 45° to the direction of shear. A thin sheet of uniform material shows this well (Figure 2.17(b)): the diagonal AC extends to AC′ but BD contracts to B′D. A example of this shear buckling is seen in the creases on the legs of trousers when sitting down. Bending provides another example of compression: on bending a beam under an applied moment M (Figure 2.30), it is clear that the upper surface A′B′ is in compression which, if large enough, is responsible for buckling.

Low bending stiffness is the product of the modulus, a property of the material, and the second moment of area of the cross-section to which the compressive stress is applied. A product of the same dimensions made from rubber is more likely to buckle in compression than one made from steel. We have already seen that the second moment of area plays a major role in bending: buckling is essentially a bending phenomenon.

If an article is slender, we mean that it has a small second moment of area in relation to the length (over which the compressive load is applied), like a drinking straw. The corresponding principle applies in a plate which is wide but thin: the thinness contributes to low bending stiffness, as seen in a carpet or a flat sheet of paper under compression in its plane: trying to push a carpet from one edge usually makes it ruck up, unless the carpet is small and rests on a very slippery floor.

2.8.3 Stable and unstable equilibrium

Earlier discussions have assumed equilibrium between internal and external forces. But if an applied force is disturbed by a small displacement of the body, one of two events can occur. Either the internal force will restore the body to its original condition (stable equilibrium), or the internal force will cause further displacement and acceleration (unstable equilibrium).

For a bar which is pin-jointed and has an axial force applied at the free end (Figure 2.61), it is easy to see that if the free end is displaced by a small angle, a tensile force has a component which will ensure stable equilibrium, whereas a compressive force will cause a progressive rotation.

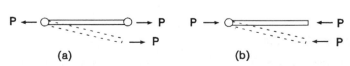

Figure 2.61 Equilibrium in pin-ended strut: (a) stable; (b) unstable.

2.8.4 Euler buckling of slender columns

Consider a long, thin strut of uniform constant cross-section, initially straight, and pin-jointed at each end (Figure 2.62). On increasing the axial compressive load, observation shows that the beam will deform in bending into the arc of a circle of large diameter.

At A, a distance y from the top joint, the displacement u causes a positive bending moment $M_A = Pu$. Hence the bending moment equation $EId^2u/dy^2 = -Pu$, where I is the *least* second moment of area of the cross-section of the strut. Solution of this equation with suitable boundary conditions gives the critical buckling load $P_c = \pi^2 EI/L^2$, where I relates to the minimum second moment of area for the case of axial loading along the centroid. Below P_c the strut is in stable equilibrium, above P_c the strut is in unstable equilibrium and maintaining the load will lead to progressive bulging and collapse. (Strictly, there is a range of solutions with $P_c = n\pi^2 EI/L^2$, with n and integer $1, 2, 3 \ldots$; the higher values of n correspond to larger numbers of wrinkles of shorter wavelength. We assume here that $n = 1$ is the most critical solution. In plates and tubes the higher order wrinkles are more commonplace.)

Other boundary conditions for end restraint in struts can be generalized to give the expression $P_c = K\pi^2 EI/L^2$, where K is defined as follows:

End restraint	K
Pin jointed at both ends	1
One end free, one end fixed	0.25
Both ends fixed	4
One end fixed, one free to rotate	2

Preventing rotation at the ends of the column obviously increases the buckling resistance considerably.

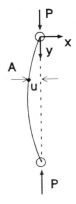

Figure 2.62 Curvature in pin-jointed strut under axial compression.

Figure 2.63 Eccentric loading of a strut.

Real columns obey the Euler criterion quite well under assumed buckling loads only when they are slender, that is with a slenderness ratio (defined below) in the order of 60 or greater.

Perfectly axial loading is rarely achieved in practice, rather loads may be eccentric with the line of action of the load parallel to the centroidal axis of the column (Figure 2.63). Initial curvature or manufacturing defects can have a similar effect. This eccentricity can dramatically increase the bending moment and hence (and of immense importance in design) greatly reduce the buckling load compared with that calculated for perfect axial loading. The effect of combining loads at different locations and of different characters cannot generally be judged by the principal of superposition.

The radius of gyration of a strut is the radius, r, of an imaginary ring at which all the mass of the cross-section is concentrated to have the same second moment of area as the real structure. The radius of gyration is defined as $r = \sqrt{(I/A)}$.

We can therefore express the Euler buckling load for a column under compressive stress in terms of the slenderness ratio (L/r). Substituting for I we have $P_c = \pi^2 EA/(L/r)^2$, or $\sigma_c = \pi^2 E/(L/r)^2$.

Problem 2.34: Show that in a slender column under axial compression the critical buckling strain is independent of the material from which the column is made provided only that it is linearly elastic.

Problem 2.35: By modelling a straight tree trunk as a vertical strut of uniform cross-section under self-weight loading assumed to act at the ends only, derive a relationship between the height of the tree and its average radius, if the tree is not to display Euler buckling collapse.

Problem 2.36: If a column under axial compressive load shows Euler buckling at a slenderness ratio of 60, how long is the column if it is: (a) a thin-walled tube of mean radius R_m and thickness h; (b) a solid rectangular section of breadth $b = 3h$?

Problem 2.37: Compare the buckling loads in a bundle of 1000 straight fibres of modulus $E = 70\,GPa$, diameter $10\,\mu m$, $10\,mm$ long if: (a) the fibres are not bonded together and so free to slip past one another; (b) the fibres are bonded together.

2.8.5 Local buckling

For a given cross-sectional area A, a hollow tube has higher second moment of area than a solid rod, hence the use of tubes in cycle frames and tubular furniture. But to increase structural efficiency, we cannot keep on expanding radius and reducing thickness – the wall would become so thin that there would be a real risk of forming a permanent kink, and the tube would crumple.

We have already described how, under axial compression a drinking straw forms a kink, or hinge, called local buckling. This sets the limit on diameter/thickness ratio which can resist the greatest permissible applied compressive force. The stress analysis is not easy but local buckling in thin-walled tubes under axial compression is expected at about $\sigma_c = kEh/D$, where $k = 1.2$ theoretically, but often only about 0.5 because of localized imperfections.

Problem 2.38: What is: (a) the critical strain; and (b) the relationship between the slenderness ratio and the diameter/thickness ratio for a straight, pin-ended thin-walled tube under axial compression where the transition between Euler buckling and local buckling occurs?

2.8.6 Buckling of flat plates

If loaded edges of a flat plate under uniform in-plane compression are simply supported, and other edges are free to deform, then the plate bends out of plane, or develops wrinkles as Poisson's ratio effects increase, depending on the proportions of the sides of the plate.

If the unloaded edges are supported, then the number of bulges depends again on proportions (Figure 2.64), but above the critical buckling stress the side supports help to restrain collapse, and the failure load then tends to depend rather on the allowable extent of the deflection of the bulges. The message is clear; free edges of thin flat plates under compression are prone to buckling failure at depressingly low values of stress.

Buckling of a thin-walled flat isotropic plate is predicted to occur under uniform in-plane compression at a stress $\sigma_c = KE(h/b)^2$, where b is the loaded width of the plate, and K depends on the aspect ratio of the plate and the edge conditions.

To increase the buckling resistance of a flat plate of a given material, either increase the second moment of area (e.g. by using a thicker panel, a ribbed panel, or a curved or corrugated form), or suppress edge rotation (by using fixed edges rather than simply supported or free).

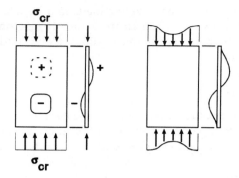

Figure 2.64 Buckling of plate under compression with supported unloaded edges.

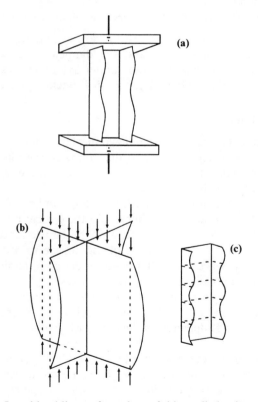

Figure 2.65 Local buckling at free edges of thin-walled columns.

We should, however, remember that adding wide ribs to flat plates creates another free edge. When plate elements of a short length of say I- or channel-section buckle without any overall bending or twisting of the member we usually observe (Figure 2.65): (a) common edges of component plates remain straight; (b) original angle between adjacent plates at common edge is maintained during buckling; (c) wavelength of buckles which occur in all plates simultaneously are the same. Features (a) and (b) are good for load-bearing.

A change in relative dimensions may change mode of buckling from overall to local. Consider an equal angle long strut (Figure 2.66). If h/w is small, buckling of one leg occurs first: sheet buckling, crumpling or wrinkling, generally a local event. If h/w larger, then strut buckles as a complete column. As buckling takes place, the deflections and stresses are *not* proportional to loads even though the material acts elastically.

A compressive bending stress is developed when a beam carries a transverse load, and this has a maximum value at the surface (a 'free edge') most remote from the neutral axis. The T-section beam (Figure 2.67) under the moment M puts the upper surface of the flange in compression: there is a helpful amount of support in the web along the long non-free edge (including some tension), but if the stress is high, or the section is thin or flexible, the free edge A buckles and is

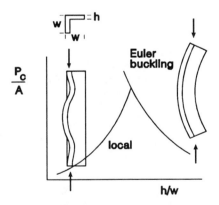

Figure 2.66 Local and Euler buckling in an angle-section strut.

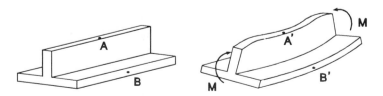

Figure 2.67 Buckling of thin web in beam under moment M.

displaced sideways to A′. The flange B under this moment is in tension and therefore not buckle. (Compare what would happen if the sign of the moment were reversed.)

2.8.7 Buckling of thin-walled cylinders

There are two common important loading situations in thin-walled cylinders where buckling can occur: axial compressive loading, and external pressure.

Axial compression

A perfect thin-walled cylinder in axial compression is in one sense a very wide plate with no free edges, but the curvature (by virtue of its second moment of area) confers good initial resistance to local buckling collapse. Small initial deflections induce either ring buckling (Figure 2.68(a)) or a chessboard pattern of deformation (Figure 2.68(b)) – waves in longitudinal and transverse directions giving rectangular depressions and bulges, which define a (half) wavelength.

We have already seen that buckling collapse of a long thin-walled tube occurs theoretically at $\sigma_c = 0.6Eh/R$. The condition is that the length, L, of such a tube must be several times greater than the half-wave of buckling, $1.72\sqrt{(Rh)}$.

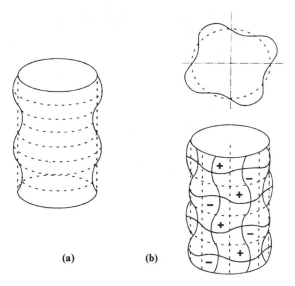

(a)   **(b)**

Figure 2.68 Behaviour of thin-walled cylinders under axial compression: (a) ring buckling; (b) chessboard buckling.

Two of the many examples of buckling readily apparent in clothing, although neither involves uniform loading nor well-defined constraints at the ends or edges.

Consider the back of the leg of a pair of fairly close-fitting new jeans made from non-stretch fabric. When the wearer bends the knee, the cylinder of fabric behind the knee is under bending compression: look at the wrinkles behind the knee.

A lady's court shoe (or the flat equivalent) with flexible sole encloses the toes but not the instep so when this shoe is flexed, the free edges of the side-walls are in compression and buckle locally outwards near the widest part of the foot. Men's shoes are usually more enclosed and the upper over the instep reduces the tendency to bulge sideways, although creases still develop across the width in response to the bending compression: still buckling, but with different edge conditions.

External pressure

Examples of circular thin-walled cylinders under external pressure include tubes or tanks in which a partial vacuum may occur, and pipelines buried in waterlogged soil. Under external pressure, a thin-walled tube develops compressive axial and hoop stresses by the mechanics described in Section 2.3.3. If sufficiently large, these compressive stresses can cause buckling collapse, seen initially as wrinkles and folds in the tube. The length of the tube, as well as the ratio of thickness to diameter, plays a major role in determining the period of the wrinkles and the buckling collapse pressure. We therefore describe failure by the number of lobes e.g. 2, 3, 4 (Figure 2.69). For a given tube, the number of lobes depends on restraint at circumference or edge of the ends of the tube or portion. Long thin-walled tubes develop two lobes, very short tubes lots more.

Estimates (rather than precise calculations) of buckling collapse stresses can be judged from the collapse pressure of an initially straight thin-walled circular tube free from manufacturing defects which has a length L which is at least several times the half wave of buckling $\lambda = 4.9R\sqrt{(R/h)}$. The collapse pressure $p_c = E(h/R)^3/[4(1 - v^2)]$. This formula assumes a linear elastic isotropic material. Ovality reduces the buckling resistance dramatically.

Figure 2.69 Lobe formation in cross-section of initially-circular thin tube under external pressure.

If the tube is shorter than the value of λ calculated above, then Roark and Young (1989) suggest that where the ends are held circular, and for a material of Poisson's ratio 0.3, the buckling collapse pressure may be approximately calculated as $p_c = 0.866(Eh^2/LR)\sqrt{(h/R)}$.

Problem 2.39: A long straight thin-walled tube is 200 mm diameter and 6 mm thick, and is made from a linear-elastic isotropic material with a tensile modulus of 10 GPa. the tube fails under an internal pressure of 5 MPa. Estimate the external pressure for the onset of radial buckling collapse.

2.8.8 Preventing buckling

The designer frequently seeks to use the minimum amount of material in order to reduce the cost of the material used and the cost of manufacture of the product. Where the structure develops compressive strain, there is a risk of buckling which must be assessed. There are several strategies for reducing the likelihood of buckling, which the reader may care to think about.

The chance of buckling can be reduced for a given structure and applied load by reducing slenderness, by increasing bending stiffness, or by avoiding thin free edges to wide stiffeners. For example, bonding together lots of thin struts suppresses relative shear and dramatically improves resistance to axial compressive load. Bundles of bonded fibres are widely used in the manufacture of composites. Putting a restraint round a thin structure helps to reduce the transverse deformation. Much of the modern aircraft may be idealized as a family of thin-walled tubes reinforced with circumferential and other ribs. Encapsulating a thin structure in a lower modulus material can also help to restrain sideways deformations – the bundles of fibres in a resin matrix provide an example from composites technology. If you raise your eyebrows about the usefulness of these proven strategies, the skin on your forehead wrinkles less than it would like to under the compression, because of the support which the underlying skull gives to prevent complete collapse of the skin and tissues attached to it.

Eccentricity is a major problem in the manufacture of thin-walled structures – the thinner the wall, the more difficult it is to make a perfectly straight flat or perfectly circular shape. Carefully introducing local stiffeners can help to improve buckling stability by suppressing local buckling.

Another strategy is to prevent or minimize the compressive stress in a product. This is an attractive principle for which it is easier to find examples outside the realm of engineering then within it. An example taken from architecture concerns the use of saddle-shaped structures: when these are loaded in compression, the load is (perhaps surprisingly) taken mainly by tension, so these are 'smart' structures. The analysis is tricky, and the shapes involved do not always seem applicable in other technologies.

A more obvious way of reducing the compressive strain in a structure during external loading is deliberately to pre-tension the structure. This is widely practised in steel rod reinforced concrete technology. An interesting way of achieving pre-tension in a hollow enclosed structure is to apply internal pressure. An uniflated cylindrical rubber toy balloon is easy to buckle by applying a small bending moment; but when inflated it will withstand an enormous bending moment and eventually fails by local buckling collapse. Stretch-fit garments work on much the same principle.

Tall hollow plant stems show some tendency to Euler buckling under their own mass. Wainwright *et al.* (1976) report that dandelion and wild onion stems can often achieve a slenderness ratio in excess of 150. These plants therefore represent, in engineering terms, slender columns. By the relationship derived in Problem 2.37 they are not therefore expected to fail by local buckling, and it is not uncommon to find that the stems show a gentle curvature consistent with Euler buckling under their own mass. In fact their resistance to Euler buckling is enhanced by their internal structure, which is neither isotropic nor free from internal stress. Vincent and Jeronimidis (in press) describe how their stems consist of close packed cylindrical cells parallel to the stem axis. These cells are under a substantial internal pressure (typically 20 atmospheres), which develops in the cell an axial tension by the mechanics used in Section 2.3.3, and this offsets the compressive stress caused by the mass of the stem.

FURTHER READING

Benham, P.P. and Crawford, R.J. (1987) *Mechanics of Engineering Materials*, Long-man, Harlow, UK. Straightforward but comprehensive undergraduate text ideal for amplifying what is in this book where required.

Brohn, D. (1984) *Understanding Structural Analysis*, Granada. An interesting non-analytical format.

Fenner, R.T. (1989) *Mechanics of Engineering Materials*, Blackwell. A good standard text.

Gere, J.M. and Timoshenko, S.P. (1991) *Mechanics of Materials*, Chapman & Hall. A good standard text.

Hilson, B. (1972) *Basic Structural Behaviour via Models*, Crosby Lockwood. An interesting experimental and non-analytical format.

Jaeger, L.G. (1964) *Elementary Theory of Elastic Plates*, Macmillan. Excellent clearly written introduction.

Mansfield, E.H. (1989) *The Bending and Stretching of Plates*, Cambridge University Press. Chapter 1 provides a well-written account with useful excursions into laminates.

Powell, P.C. (1983) *Engineering with Polymers*, Chapman & Hall. For the mechanical behaviour of polymers.

Roark, R.J. and Young, W.C. (1989) *Formulas for Stress and Strain* (6th edn), McGraw-Hill. A comprehensive handbook.

Vincent, J.F.V. and Jeronimidis, G. (in press) 'The mechanical design of fossil plants', in *Biomechanics in Evolution*, (ed Rayner).

Wainwright, S.A., Biggs, W.D., Curry, J.D. and Gosline, J.M. (1976) *Mechanical Design in Organisms*, Arnold. Despite the title this is a good textbook on mechanics of solids, with fascinating examples.

Williams, J.G. (1973) *Stress Analysis of Polymers*, Longman. Despite the title this is a book solely on solid body mechanics although it does refer to a wide range of situations which are commonly encountered in polymer engineering, and does address large strains briefly. A more advanced style and presentation than Benham and Crawford (1987).

Stiffness behaviour of single ply

3

OVERVIEW

The purpose of this chapter is threefold. First, to introduce the nomenclature and language commonly used in the composites literature. Because many composites are used in the form of flat or curved plates, it is convenient to use concepts such as force resultant per unit width and moment resultant per unit width in addition to the more familiar concept of stress.

Second, to introduce the concepts of stiffnesses in a sheet under in-plane or bending loads per unit width in terms of the properties of the individual ply, together with the associated compliances. This will seem excessively complicated at this stage, but it prepares the way for laminate design in later chapters where the new concepts are essential.

Third, to apply these new concepts to the behaviour of an isotropic sheet of natural rubber, and to a sheet of elastomer reinforced with aligned polyester fibres such that the sheet is stressed or strained in its principal directions, or off-axis.

The chapter provides many examples of the detailed calculations necessary in composites which the reader can follow step-by-step, and problems to try. It is of course recognized that once the reader has grasped these new ideas, the (micro)computer will take over for reasons of speed, reliability and convenience.

The essence of the argument is that to calculate the deformation resulting from a single uniaxial stress, we need two independent elastic constants for an isotropic material (as shown in Chapter 2), four for an orthotropic unidirectional ply stressed in its principal directions, six for an anisotropic ply (e.g. a unidirectional ply stressed in-plane but at an angle to its principal directions, and (as we shall see in Chapter 5) up to 18 for a laminate.

In the early part of the chapter we shall use the co-ordinate system $(1, 2)$ to introduce the nomenclature of stiffness and compliance coefficients. Thereafter (and including all later chapters) we shall use the (x, y) co-ordinate system to portray stresses and strains in global directions or global co-ordinates, i.e. to describe co-ordinates for external loads or strains, and to define a reference 'x' direction for describing fibre directions within the ply. In general the principal directions $(1, 2)$ within a ply make some angle (θ) to the global (x, y) directions. The direction of x is taken as convenient for the problem in hand.

In isotropic materials we do not need to distinguish directions of fibres; in a unidirectional ply loaded in its principal directions, x and 1 directions are the same because $\theta = 0$. In a unidirectional ply loaded off-axis, fibres are aligned at different directions to the reference direction, so the use of global co-ordinates is essential in problem solving. We shall see that in laminates, plies can be oriented in the plane at several angles to the reference direction, and global co-ordinates are always used. In crossply laminates special care is needed.

3.1 STIFFNESS OF ISOTROPIC LINEAR ELASTIC PLY

Our discussion focuses on a single thin sheet of isotropic material (Figure 3.1). We have already seen in Chapter 2 that the stress–strain relationship can be expressed in terms of two independent elastic constants, E and v. Recalling that the shear modulus is related to these two constants by $G = E/[2(1 + v)]$, under applied plane stress we have:

$$\begin{pmatrix} \varepsilon_1 \\ \varepsilon_2 \\ \gamma_{12} \end{pmatrix} = \begin{pmatrix} 1/E & -v/E & 0 \\ -v/E & 1/E & 0 \\ 0 & 0 & 1/G \end{pmatrix} \begin{pmatrix} \sigma_1 \\ \sigma_2 \\ \tau_{12} \end{pmatrix}$$

This equation confirms that under a single uniaxial stress σ_1, the isotropic sheet elongates because of $1/E$, contracts because of $-v/E$, but does not shear. When just a shear stress τ_{12} is applied, the sheet shears because of $1/G$, but does not elongate or contract.

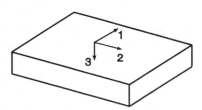

Figure 3.1 Co-ordinate system for a thin isotropic sheet.

3.1.1 Standard and contracted notation

The composites literature uses suffixes some of which can have different meanings from those used in conventional texts on isotropic materials. We therefore need to identify the new contexts and meanings.

For a block of an isotropic linear elastic material at constant temperature, .Hooke's Law in Cartesian co-ordinates $(1, 2, 3)$ under stresses in all three directions takes the form:

$$\begin{pmatrix} \varepsilon_1 \\ \varepsilon_2 \\ \varepsilon_3 \\ \gamma_{23} \\ \gamma_{31} \\ \gamma_{12} \end{pmatrix} = \begin{pmatrix} 1/E & -v/E & -v/E & 0 & 0 & 0 \\ -v/E & 1/E & -v/E & 0 & 0 & 0 \\ -v/E & -v/E & 1/E & 0 & 0 & 0 \\ 0 & 0 & 0 & 1/G & 0 & 0 \\ 0 & 0 & 0 & 0 & 1/G & 0 \\ 0 & 0 & 0 & 0 & 0 & 1/G \end{pmatrix} \begin{pmatrix} \sigma_1 \\ \sigma_2 \\ \sigma_3 \\ \tau_{23} \\ \tau_{31} \\ \tau_{12} \end{pmatrix} \tag{3.1}$$

Terms on the diagonal, e.g. $1/E$ and $1/G$, relate applied stress to a strain of exactly the same type and with the same suffices. Terms on the diagonal, e.g. $-v/E$, relate an applied stress to strains in other directions (or, as we see below, to strains of a different character).

For shear the double suffix notation ij means that the force is applied in the i direction on an area normal to the j direction. Strictly speaking direct stresses and strains should also use the double suffix e.g. σ_{11}, but most texts use the contracted form σ_1.

In composites we need to relate any stress σ_{ij} to any strain ε_{kl}. This would mean that the physical stiffness would need four suffices, e.g. $\tau_{12} = E_{1212}\gamma_{12}$, or more generally:

$$\sigma_{ij} = E_{ijkl}\varepsilon_{kl} \tag{3.2}$$

which is extremely unwieldy and cumbersome. So the convention of contracted suffices has been extended so that shear stresses and strains each have only one suffix, as follows

Standard notation	σ_{11}	σ_{22}	σ_{33}	τ_{31}	τ_{23}	τ_{12}
Contracted notation	σ_1	σ_2	σ_3	σ_4	σ_5	σ_6

$$\tag{3.3}$$

Strains are similarly labelled. Equation 3.1 can now be re-expressed in terms of the compliance coefficients S_{ij} using the new scheme of suffices:

$$\begin{pmatrix} \varepsilon_1 \\ \varepsilon_2 \\ \varepsilon_3 \\ \varepsilon_4 \\ \varepsilon_5 \\ \varepsilon_6 \end{pmatrix} = \begin{pmatrix} S_{11} & S_{12} & S_{13} & 0 & 0 & 0 \\ S_{12} & S_{22} & S_{23} & 0 & 0 & 0 \\ S_{13} & S_{23} & S_{33} & 0 & 0 & 0 \\ 0 & 0 & 0 & S_{44} & 0 & 0 \\ 0 & 0 & 0 & 0 & S_{55} & 0 \\ 0 & 0 & 0 & 0 & 0 & S_{66} \end{pmatrix} \begin{pmatrix} \sigma_1 \\ \sigma_2 \\ \sigma_3 \\ \sigma_4 \\ \sigma_5 \\ \sigma_6 \end{pmatrix}$$

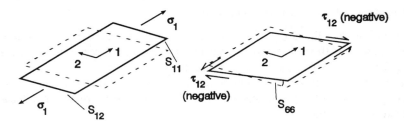

Figure 3.2 Representation of compliance coefficients S_{ij} (undeformed sheet shown by broken lines).

We now write Hooke's law under a single in-plane shear as

$$\gamma_{12} = \varepsilon_6 = S_{66}\sigma_6 = \tau_{12}/G$$

In this book we are concerned only with plane stress $(\sigma_1, \sigma_2, \tau_{12})$ for which the strains are now, for an isotropic material:

$$\begin{pmatrix} \varepsilon_1 \\ \varepsilon_2 \\ \gamma_{12} \end{pmatrix} = \begin{pmatrix} \varepsilon_1 \\ \varepsilon_2 \\ \varepsilon_6 \end{pmatrix} = \begin{pmatrix} S_{11} & S_{12} & 0 \\ S_{12} & S_{22} & 0 \\ 0 & 0 & S_{66} \end{pmatrix} \begin{pmatrix} \sigma_1 \\ \sigma_2 \\ \sigma_6 \end{pmatrix} = \begin{pmatrix} 1/E & -v/E & 0 \\ -v/E & 1/E & 0 \\ 0 & 0 & 1/G \end{pmatrix} \begin{pmatrix} \sigma_1 \\ \sigma_2 \\ \tau_{12} \end{pmatrix} \quad (3.4)$$

The point to note is that S_{12} involves coupling between direct stress σ_1 and direct strain ε_2, or between direct stress σ_2 and direct strain ε_1 (the Poisson lateral contraction), *and not shear*. It is S_{66} which represents shear.

This nomenclature for stiffness or compliance is easy to remember (and the position in the matrix makes it clear). What is perhaps confusing is that the double suffix notation for shear is usually retained. Most texts use τ_{12} as shear stress, and γ_{12} as shear strain, rather than the formal σ_6 or ε_6. Poisson's ratio v_{12} uses the convention which defines the ratio of the strain in the 2 direction to the strain in the 1 direction when a direct stress is applied in the 1 direction.

Figure 3.2 gives a visual representation of elements of the compliance matrix [S].

3.1.2 Reduced stiffness under in-plane stress

We can invert the equation (3.4) to obtain the stresses induced in a sheet under plane stress (Figure 3.2). Ignoring stresses in the thickness directions we obtain

$$\begin{pmatrix} \sigma_1 \\ \sigma_2 \\ \tau_{12} \end{pmatrix} = \begin{pmatrix} Q_{11} & Q_{12} & 0 \\ Q_{12} & Q_{22} & 0 \\ 0 & 0 & Q_{66} \end{pmatrix} \begin{pmatrix} \varepsilon_1 \\ \varepsilon_2 \\ \gamma_{12} \end{pmatrix} = \begin{pmatrix} E/(1-v^2) & vE/(1-v^2) & 0 \\ vE/(1-v^2) & E/(1-v^2) & 0 \\ 0 & 0 & G \end{pmatrix} \begin{pmatrix} \varepsilon_1 \\ \varepsilon_2 \\ \gamma_{12} \end{pmatrix} \quad (3.5)$$

where [Q] is the reduced stiffness matrix such that

$$[Q] = [S]^{-1} \quad \text{or} \quad [S] = [Q]^{-1} \quad (3.6)$$

The term 'reduced stiffness' is used because the in-plane modulus is reduced by the factor $(1 - v^2)$. Q_{12} is positive and S_{12} is negative.

The reader may well ask if there is any difference between applying loads or stresses in-the-plane and applying deformations or strains in a plane stress situation. After all, does it really matter whether you apply a load or a deformation in a test?

It is easy in principle to apply a unidirectional tensile load to a parallel-sided isotropic specimen of uniform thickness, and the response is axial elongation and transverse contraction in width (and we shall ignore here the contraction in thickness, even though it does occur). In such a test we either assume that the grips exert no transverse restraint (e.g. they do not prevent lateral contraction, which is seldom true with bolted or wedge-clamped grips), or we use a long slender specimen and measure strains over a gauge length well away from the grips (which is normal practice). The strains in the gauge length may then be calculated using Equation (3.4).

In contrast, applying an axial extensional deformation alone requires elaborate apparatus. It is clear from Equation (3.5) that just applying ε_1 alone requires a transverse stress $\sigma_2 = Q_{12}\varepsilon_1$ as well as the obvious $\sigma_1 = Q_{11}\varepsilon_1$. We note that $\sigma_2 = + v\sigma_1$, and it is the Poisson's ratio which is responsible for the problem of distinction which we are discussing.

When we talk of elongating a specimen in a normal tensile test, we are not being very precise. In later parts of this book we shall need to spell out the test conditions very carefully. In a conventional tensile test what we actually do is to impose the axial elongation and we allow the specimen to freely contract (apart from the region at the grips). Thus we apply ε_1 and let ε_2 take up a value such that there is no (internal) transverse stress σ_2, because we are neither applying nor developing an external transverse load. From Equation (3.4) or (3.5), the value of the transverse strain is $\varepsilon_2 = (S_{12}/S_{11})\varepsilon_1 = - v\varepsilon_1$.

In the special case (only) of an isotropic ply (or a unidirectional ply loaded in the principal directions), there is no coupling between shear stresses and direct strains, so applying either shear stress or shear strain alone has an identical effect, and Equations (3.4) and (3.5) confirm this.

Problem 3.1: Given the following elastic constants for an isotropic sheet of aluminium, calculate the components of the compliance matrix and the reduced stiffness matrix.

$$E = 70\,\text{GPa}, \quad v = 0.33, \quad G = 26\,\text{GPa}$$

3.1.3 Stiffness matrices for a single isotropic ply

We shall now discuss the basic terms used to describe the stiffness of laminates, focusing first on the behaviour of a single ply. This may seem trivial but it will permit the introduction and application of new concepts to a simple well-known and familiar situation.

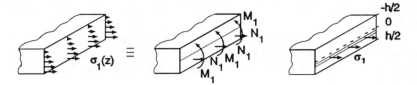

Figure 3.3 Replacement of stress profile by statically equivalent force and moment resultants (per unit width).

It is convenient to replace the stresses in the sheet by a statically equivalent set of force and moment resultants which act on the midplane of the sheet, as represented schematically in Figure 3.3. This equivalent set no longer contains sheet thickness and thickness co-ordinates explicitly. By convention these resultants are described on the basis of unit width of laminate.

For a sheet of thickness h under a stress σ_1, the force resultant per unit width, N_1, is defined as

$$N_1 = \int_{-h/2}^{+h/2} \sigma_1 \, dz \tag{3.7}$$

The sign convention for N_1 is the same as for σ_1. The three force resultants (N_1, N_2, N_{12}) acting on a sheet in plane stress may be written as

$$[N] = \int_{-h/2}^{+h/2} [\sigma] \, dz \tag{3.8}$$

For the special case of uniform in-plane stress(es)

$$[N] = [\sigma]h \tag{3.9}$$

In a pipe of wall thickness 3mm under internal pressure, if the axial stress σ_A is 1 N/mm^2 and the hoop stress σ_H is 2 N/mm^2, then the force resultants per unit width are

$$\begin{pmatrix} N_A \\ N_H \\ N_{AH} \end{pmatrix} = \begin{pmatrix} 1 \\ 2 \\ 0 \end{pmatrix} 3 = \begin{pmatrix} 3 \\ 6 \\ 0 \end{pmatrix} \text{N/mm}$$

and the force resultants are illustrated in Figure 3.4.

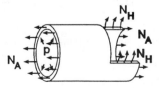

Figure 3.4 Force resultants per unit width in a thin-walled pipe under internal pressure.

The moment resultants per unit width in a sheet are defined by

$$[M] = \int_{-h/2}^{+h/2} [\sigma] z \, dz \tag{3.10}$$

Bending is described by the curvature at the mid plane κ. In-plane strains caused by in-plane loads are taken as uniform across the sheet (which therefore includes the midplane strain $\varepsilon°$). We can show that the strain $\varepsilon(z)$ at any thickness co-ordinate z from a combination of in-plane loads and moments is related to the midplane strain and the midplane curvature by

$$[\varepsilon(z)] = [\varepsilon°] + z[\kappa] \tag{3.11}$$

If the midplane strain and curvature in a given plane are $\varepsilon_x° = 0.01$ and the curvature is $\kappa_x = 2/m$, then on a plate 4mm thick the surface strains will be $\varepsilon_x = 0.01 + 0.002 \times 2 = 0.01 + 0.004 = 0.014$ on the underside and $0.01 - 0.004 = 0.006$ on the upper surface.

The associated stress profile in a material strained in the principal directions is therefore

$$\sigma(z) = [Q][\varepsilon°] + z[Q][\kappa] \tag{3.12}$$

where $[Q]$ is the reduced stiffness matrix e.g. $Q_{11} = E/(1 - v^2)$.

For a single isotropic linear elastic ply the stress and strain profiles have the same form (this is not a generalization for laminates). An example of the strain distribution in a ply carrying a bending moment and direct load is given in Figure 3.5.

Given then the strain distribution in the sheet we can obtain expressions of the force and moment resultants per unit width.

$$[N] = \int_{-h/2}^{+h/2} [\sigma] \, dz = \int_{-h/2}^{+h/2} [Q][\varepsilon°] \, dz + \int_{-h/2}^{+h/2} [Q][\kappa] z \, dz \tag{3.13}$$

For a thin sheet $[\varepsilon°]$ and $[\kappa]$ relate to the midplane only and do not depend on z; and $[Q]$ is assumed to be a set of elastic constants for the given material and

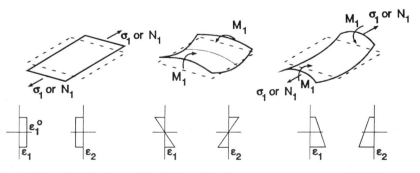

Figure 3.5 Strain profiles induced by direct stress and moment/width.

is also independent of z. We can therefore take these terms outside the integration to give

$$[N] = [Q][\varepsilon^\circ] \int_{-h/2}^{+h/2} \mathrm{d}z + [Q][\kappa] \int_{-h/2}^{+h/2} z\,\mathrm{d}z = [Q]h[\varepsilon^\circ] + (\tfrac{1}{2})[Q](h^2/4 - h^2/4)[\kappa]$$

$$= [A][\varepsilon^\circ] + [B][\kappa] \tag{3.14}$$

[A] is termed the extensional stiffness matrix and, for this single isotropic ply only, is merely the reduced stiffness multiplied by the thickness of the sheet or ply and has the units of N/mm.

$$[A] = \begin{pmatrix} A_{11} & A_{12} & 0 \\ A_{12} & A_{22} & 0 \\ 0 & 0 & A_{66} \end{pmatrix} \tag{3.15}$$

Although [A] is called the extensional stiffness matrix, it also contains terms such as A_{66} describing shear stiffness. A better name for [A] would be 'in-plane stiffness', but this is not in common use.

The coupling matrix [B] is formally defined for a single ply by

$$[B] = [Q] \int_{-h/2}^{+h/2} z\,\mathrm{d}z \tag{3.16}$$

and relates curvatures to in-plane force resultants per unit width. By inspection of this integral [B] = 0, whether the single ply is isotropic or anisotropic. Its main significance occurs in the behaviour of nonsymmetric laminates, and so we shall not pursue it just here.

Moment resultants per unit width can also be expressed in terms of midplane strains and curvatures:

$$[M] = \int [\sigma] z\,\mathrm{d}z = [Q][\varepsilon^\circ] \int_{-h/2}^{+h/2} z\,\mathrm{d}z + [Q][\kappa] \int_{-h/2}^{+h/2} z^2\,\mathrm{d}z$$

$$= [B][\varepsilon^\circ] + (\tfrac{1}{3})[Q][\kappa](h^3/8 - (-h^3/8))$$

$$= [B][\varepsilon^\circ] + [D][\kappa] \tag{3.17}$$

where [D] is the bending stiffness matrix defined for a single ply as

$$[D] = [Q]h^3/12 \tag{3.18}$$

Bearing in mind that for example $Q_{11} = E/(1 - v^2)$ or $Q_{12} = vE/(1 - v^2)$, this neatly corresponds to the relationship previously derived for isotropic panels in Section 2.5.8; the different concepts represent the same physical phenomenon.

Thus for a unidirectional ply loaded in its principal directions both in-plane and in bending, we have

$$\begin{pmatrix} N \\ M \end{pmatrix} = \begin{pmatrix} A & B \\ B & D \end{pmatrix} \begin{pmatrix} \varepsilon^\circ \\ \kappa \end{pmatrix} \tag{3.19}$$

where $[B] = 0$. Hence the application of midplane strains does not induce bending moment resultants and applying curvature does not induce force resultants; in-plane and bending activities are independent.

The fully expanded version of Equation (3.19) applicable to an isotropic single ply is

$$\begin{pmatrix} N_1 \\ N_2 \\ N_{12} \\ M_1 \\ M_2 \\ M_{12} \end{pmatrix} = \begin{pmatrix} A_{11} & A_{12} & 0 & 0 & 0 & 0 \\ A_{12} & A_{22} & 0 & 0 & 0 & 0 \\ 0 & 0 & A_{66} & 0 & 0 & 0 \\ 0 & 0 & 0 & D_{11} & D_{12} & 0 \\ 0 & 0 & 0 & D_{12} & D_{22} & 0 \\ 0 & 0 & 0 & 0 & 0 & D_{66} \end{pmatrix} \begin{pmatrix} \varepsilon_1^\circ \\ \varepsilon_2^\circ \\ \gamma_{12}^\circ \\ \kappa_1 \\ \kappa_2 \\ \kappa_{12} \end{pmatrix} \tag{3.20}$$

In the special case of a single ply we can write

$$[N] = [A][\varepsilon^\circ] \tag{3.21}$$

and by matrix inversion we can obtain the (midplane) strain response to a set of applied uniform force resultants per unit width

$$[\varepsilon^\circ] = [A]^{-1}[N] = [a][N] \tag{3.22}$$

where $[a]$ is the inverse of the matrix $[A]$, with units of mm/N, and may be called the extensional compliance matrix.

We can assign a physical meaning to each of the compliance terms a_{ij} as shown: the change in length or angle of a square sheet under the indicated uniform force resultant is shown in Figure 3.6.

Similarly for the special case of a single ply we can write

$$[M] = [D][\kappa] \tag{3.23}$$

and by inversion

$$\begin{pmatrix} \kappa_1 \\ \kappa_2 \\ \kappa_{12} \end{pmatrix} = [\kappa] = [D]^{-1}[M] = [d][M] = \begin{pmatrix} d_{11} & d_{12} & 0 \\ d_{12} & d_{22} & 0 \\ 0 & 0 & d_{66} \end{pmatrix} \begin{pmatrix} M_1 \\ M_2 \\ M_{12} \end{pmatrix} \tag{3.24}$$

Again sketches of the curvatures resulting from the application of moment

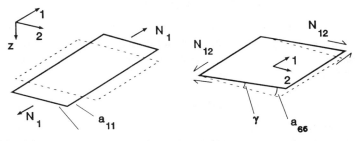

Figure 3.6 Representation of compliance coefficients a_{ij} (undeformed sheet shown by broken lines).

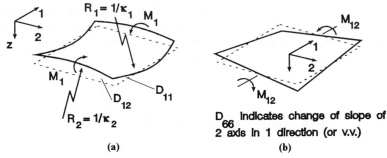

Figure 3.7 Representation of bending compliances d_{ij}.

resultants in the principal directions will give some physical meaning to the elements d_{ij}, as shown for an isotropic ply in Figure 3.7.

Problem 3.2: A weightless horizontal flat plate of uniform thickness 2 mm carries a stress in one direction only, which is constant across the width and which varies linearly through the thickness, being zero at the top surface and 3 MPa at the bottom surface. Calculate the force and moment resultants in the sheet.

Problem 3.3: Construct the likely deformation of a similar square sheet carrying a uniform tensile load N_2 and a negative shear resultant, $-N_{12}$.

3.1.4 Formal summary

The formal summary of lamina stress and strain calculations for an isotropic ply using the language of laminate analysis introduced here is as follows:

1. From elastic constants calculate [Q] (Equation (3.5))
2. Calculate [A] and [D] for the ply
3. Calculate [N] and [M] from applied loads and moments (Equation (3.8))
4. Calculate midplane strains $[\varepsilon^\circ]$ and curvatures $[\kappa]$ (Equations (3.22)–(3.24))
5. Calculate strain profiles (Equation (3.11))
6. Calculate stress profiles (Equation (3.12))

We are now in the position to discuss the behaviour of a uniform isotropic sheet of rubber using the language of laminate analysis. The purpose is to consolidate the vocabulary and concepts while retaining a firm physical insight based on familiar responses.

3.1.5 Lamina of isotropic natural rubber gum vulcanizate

The isotropic lamina is based on the material NR, with properties summarized in Table 3.1, which also shows the reduced stiffness matrix, Q_{ij}. The lamina is cured at room temperature, and Table 3.2 shows the A, B and D matrices, and their inverted forms, for a total thickness of 2mm.

Table 3.1 Properties of lamina NR

$E_{11} = 0.001$ GPa		[Q] in GPa	
$E_{22} = 0.001$ GPa	0.001316	0.0006448	0
$G_{12} = 0.000333$ GPa	0.0006448	0.001316	0
$v_{12} = 0.49$	0	0	0.000333
$\theta = 0°$			

Table 3.2 Stiffnesses and compliances of lamina NR (total thickness 2 mm)

A(kN/mm) B(kN)			B(kN) D(kNmm)		
0.002632	0.00129	0	0	0	0
0.00129	0.002632	0	0	0	0
0	0	0.000666	0	0	0
0	0	0	0.0008773	0.0004299	0
0	0	0	0.0004299	0.0008773	0
0	0	0	0	0	0.000222
a (mm/MN) h (/MN)			b (/MN) d (/MNmm)		
500000	− 245000	0	0	0	0
− 245000	500000	0	0	0	0
0	0	1502000	0	0	0
0	0	0	1500000	− 735000	0
0	0	0	− 735000	1500000	0
0	0	0	0	0	4505000

The [B] matrix is zero, as expected. There is therefore no coupling between in-plane force resultants and bending or twisting curvatures. A_{12} indicates the coupling between longitudinal tensile loading and lateral contraction. A_{16} and A_{26} are both zero, so there is no coupling between direct in-plane loads and in-plane shear strain in an isotropic sheet. D_{16} and D_{26} are also zero, so there is no coupling between bending moments and twisting curvature.

Application of single force resultant $N_1 = 0.1$ N/mm to NR

We must be clear that the test conditions allow free movement to permit contractions to occur fully in the width of the specimen. A simple example of the principle of an experimental set-up is shown in Figure 3.8(a). The normal method of firmly gripping a plate at each end prevents lateral contraction at the grips, as indicated in Figure 3.8(b).

The uniform strain responses (recognized by common sense or by definition of force resultant) are given by

$$\varepsilon_1^\circ = a_{11} N_1 = (500000) \text{ mm/MN} \times 0.1 \text{ N/mm} = 0.05 = 5\%$$
$$\varepsilon_2^\circ = a_{12} N_1 = -2.45\%$$

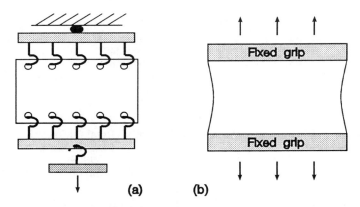

Figure 3.8 End conditions for a 'simple' tensile test.

The sheet becomes longer and narrower as expected: an originally square element deforms schematically as shown in Figure 3.9(a). The states of stress and strain in the 1 and 2 directions are uniform, as indicated in Figure 3.9(b); there is no direct stress σ_2, and there are no shear stresses or strains.

At this stage it seems unnecessary to calculate the stress profiles, but formally

$$\sigma_1 = Q_{11}\varepsilon_1^\circ + Q_{12}\varepsilon_2^\circ$$
$$= (0.001316\,\text{GPa} \times 0.05) + (0.0006448\,\text{GPa} \times -0.0245) = 0.05\,\text{MPa}$$

We can also confirm using Equation (3.9), that

$$\sigma_1 = N_1/h = 0.1\,(\text{N/mm})/2\text{mm} = 0.05\,\text{MPa}$$

Application of single strain $\varepsilon_1 = 5\%$ to NR

It is assumed that lateral contraction is prevented, and hence $\varepsilon_2 = 0$. We therefore have $N_1 = A_{11}\varepsilon_1 = 2.632\,\text{N/mm} \times 0.05 = 0.1316\,\text{N/mm}$ ($\sigma_1 = 0.0658\,\text{MPa}$) and $N_2 = A_{12}\varepsilon_1 = 0.06448\,\text{N/mm}$ ($\sigma_2 = 0.03224\,\text{MPa}$).

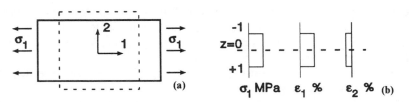

Figure 3.9 Deformation of NR under uniform tension, with stress and strain profiles.

Application of single moment resultant $M_1 = 0.001\ N$ to NR

Natural rubber is very flexible so it is not surprising that the thin sheet bends readily even under a modest moment per unit width.

From Equation (3.24) we find the midplane curvatures

$$\kappa_1 = d_{11} M_1 = (1500000)/(\text{mmMN}) \times 0.001\ N = 0.0015/\text{mm} = 1.50/\text{m}$$

$$\kappa_2 = d_{12} M_1 = (-735000)/(\text{mmMN}) \times 0.001\ N = -0.735/\text{m}$$

The rubber sheet therefore deforms to the shape of Figure 3.10 and clearly shows anticlastic curvature. The reason for anticlastic curvature has been described in Section 2.5.8. We may prefer to imagine the sheet of rubber as notionally split into two subsheets of equal thickness. The top sheet is in (net) compression from the positive applied moment; the lower sheet is in (net) tension (Figure 3.11). In the top sheet the longitudinal net compression $(-\varepsilon_{N1})$ must be accompanied by a net tensile strain (ε_{N2}) induced by the Poisson coupling effect. Hence the saddle shape under a single moment resultant M_1.

The bending stresses and strains vary linearly with distance from the midplane. At the upper surface $(z = -1\ \text{mm})$

$$\sigma_1 = Q_{11} z \kappa_1 + Q_{12} z \kappa_2$$

$$= 0.001316\ \text{GPa} \times -0.001\ \text{m} \times 1.50/\text{m} + 0.0006448\ \text{GPa}$$

$$\times -0.001\ \text{m} \times -0.735/\text{m}$$

$$= -1.5\ \text{kPa}$$

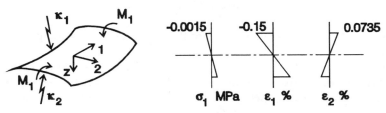

Figure 3.10 Bending response of NR under moment resultant per unit width, M_1.

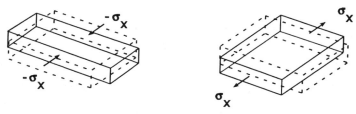

Figure 3.11 Net stresses in top and bottom sheets (taken as uniform for simplicity only) and deformations in the unbonded sheets.

The negative sign confirms that the applied moment is 'sagging' in accordance with the sign convention in Section 2.2.4. The strain profiles follows from Equation (3.11): at the top surface

$$\varepsilon_1 = z\kappa_1 = -0.001\,\text{m} \times 1.50/\text{m} = -0.0015 = -0.15\%$$
$$\varepsilon_2 = z\kappa_2 = -0.001\,\text{m} \times -0.735/\text{m} = +0.0735\%$$

The non-zero in-plane stress and strain profiles are given in Figure 3.10(b), and show zero values at the midplane.

· *Application of a single twisting moment $M_{12} = 0.001N$ to NR*

This results in a positive twisting curvature κ_{12} calculated as

$$\kappa_{12} = d_{66}M_{12} = (4505000/\text{mmMN}) \times 0.01\,\text{N} = 4.505/\text{m}$$

as shown in Figure 3.12(a). The associated stress profile now is

$$\tau_{12} = Q_{66}z\kappa_{12}$$

and at the lower surface we find

$$\tau_{12} = 0.000333\,\text{GPa} \times 0.001\,\text{m} \times 4.505/\text{m} = 1.5\,\text{kPa}$$

The stress and strain profiles are shown in Figure 3.12(b).

Application of other combinations of loading to NR

In a long thin-walled tube under internal pressure, axial tension and applied torque, one might achieve loading represented by $N_1 = N_2 = \pm N_{12} = 0.05\,\text{N/mm}$.

For positive shear we can calculate the midplane strains using equations such as $\varepsilon_1^\circ = a_{11}N_1 + a_{12}N_2$, and $\gamma_{12}^\circ = a_{66}N_{12}$.

We find $\varepsilon_1^\circ = 1.275\%$, $\varepsilon_2^\circ = 1.275\%$ and $\gamma_{12}^\circ = 7.508\%$, which is shown diagramatically in Figure 3.13(a) for an initially square sheet and the response of the tube (ignoring end effects) is shown in Figure 3.13(c). Stress and strain

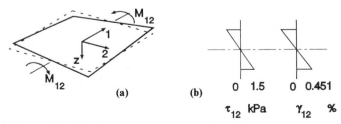

Figure 3.12 Response of NR to twisting moment per unit width, $M_{12} = 0.001$ N.

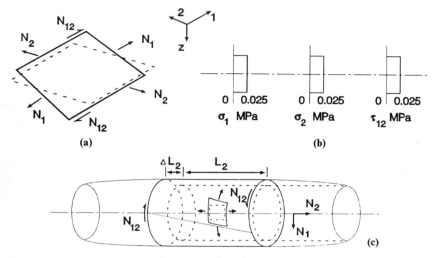

Figure 3.13 (a) Deformation of flat sheet of NR under $N_1 = N_2 = +N_{12} = 0.05$ N/mm; (b) deformation of thin-walled tube under internal pressure.

profiles calculated using $\sigma_1 = Q_{11}\varepsilon_1^\circ + Q_{12}\varepsilon_2^\circ$ are constant across the wall thickness (independent of z), and are indicated in Figure 3.13(b).

The response for negative shear is shown diagramatically in Figure 3.14.

Problem 3.4: Why are values of a_{12} and d_{12} negative when A_{12} and D_{12} are positive?

Problem 3.5: Calculate the midplane strain and curvature responses of ply NR to the application of $N_1 = 0.1$ N/mm and $M_1 = 0.001$ N.

Problem 3.6: Calculate the midplane curvatures in a flat ply of NR under bending moment $M_2 = 0.002$ N and twisting torque $M_{12} = 0.002$ N.

Problem 3.7: A single curvature $\kappa_2 = 1.00/$m is applied to NR. Calculate the moments needed to achieve this.

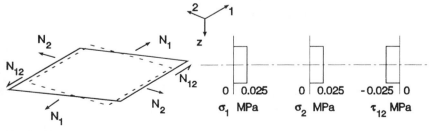

Figure 3.14 Deformation of flat sheet of NR under $N_1 = N_2 = -N_{12} = 0.05$ N/mm.

3.2 STIFFNESS OF A UNIDIRECTIONAL PLY LOADED IN THE PRINCIPAL DIRECTIONS

We have already established the strain response of a sheet of isotropic linear elastic material under in-plane stress, and the curvatures caused by applying bending moments or twisting torques.

We shall now look at the responses of a thin sheet of linear elastic material which consists of a regular array of unidirectional fibres in a matrix (Figure 3.15). The fibre direction is called the longitudinal direction, denoted the '1' or 'L' direction. The in-plane direction perpendicular (or normal) to the fibre direction is called the transverse direction, denoted the '2' or 'T' direction. The 1 and 2 directions are called the principal directions of the lamina. The direction perpendicular to the sheet is called the '3' or 'z' direction.

First we shall look at the effect of in-plane loads applied in the 1 and 2 directions, and the effect of moments applied about axes in the plane of the sheet which are parallel to or perpendicular to the 1 and 2 directions. Then in Section 3.3 we shall look at the 'off-axis' loading problem, where we seek the effect of applying in-plane loads in the x and y directions which are at some angle θ to the principal directions, and the effects of moments. In all of this we shall assume that the sheet, or ply or lamina, is homogeneous from an engineering point of view, but has different stiffnesses in different directions. This approach is called 'macromechanics'.

3.2.1 Examples of unidirectional materials from a range of technologies

Before studying the fine analytical detail, we should recall that the properties of unidirectional structures can vary markedly when measured along the across the fibre directions.

Let us describe this sheet of stuff in terms of a parallel array of stiff strong rods held together by a flexible matrix. Under a stretching force it is stiff along the rods because the rods take almost all the load, but it is flexible across the rods because the load in this direction is taken essentially by the matrix.

The properties of the unidirectional model depend on the properties of the rods, the matrix, and the proportion of rods in the matrix. Stiff strong rods give a stiff strong ply if there are enough of them. The rods can be of steel, glass, carbon, or of a polymer such as cellulose, nylon or rubber. The matrix holds

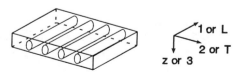

Figure 3.15 Principal directions in the unidirectional ply.

the rods in place, protects them from damage, and often transfers the loads into the rods. The matrix can be stiff like concrete or aluminium, not very stiff like epoxy resin, or flexible like rubber. One extreme is all rods and no matrix, when the array falls apart. All matrix and no rods gives a sheet of material with uniform properties in all directions.

We have already noted (in Chapter 1) that the modulus of the model when stretched along the rod axis direction can be quite high. For a carbon fibre reinforced epoxy ply with 60% volume fraction of fibres this longitudinal modulus is about the same as that of steel, and with a much lower density. This high modulus is achieved because the fibres take almost all the load.

But when stressed across the fibre direction the modulus of a carbon fibre reinforced epoxy ply is only about one-hundredth that of steel, and not much stiffer than the epoxy matrix without the fibres: this is because there is no opportunity to transfer loads from the matrix into the rods. Compare the properties of the rods-in-matrix model with those of a uniform material like glass or acrylic sheet – uniform materials have the same modulus in all directions, which is nice and simple.

Occasionally we can exploit this severe directionality in the unidirectional ply by ensuring that big loads are only allowed in the direction of the rods, and that the loads across the rods are small. This is the basis of the Mathweb lattice structure. The main forces are along the straight bundles of glass fibres in the struts and tie bars of the lattice. The fibres are held together by a crosslinked polyester matrix. This lattice is stiff, strong and light in weight. A radio mast 40 m long and weighing 140 kg can resist an axial compressive load of 4 tonnes. The equivalent steel structure would weigh more than five times as much, and would be more difficult to install, as well as needing painting during its lifetime.

A conveyor belt is a long narrow slab designed to hold material and with sufficient longitudinal strength to withstand the pull of its drive under load. Heavy duty belting uses steel cables which combine strength, low elongation, flexibility, and good adhesion to rubber under cyclic loading. The cables are twisted from thin but very strong steel wires: the twisting gives flexibility in bending as long as there is no corrosion. (Any corrosion causes the individual wires to stick together, which is not wanted.) The belt then consists of a parallel array of cables in a rubber matrix: this is then sandwiched between two rubber skins compounded ('formulated') to give good survival characteristics in service.

The substantial belt used at the coal mine at Selby in Yorkshire consists of about 70 lengths, each some 440 m long, spliced together with 1200 mm overlap. The belt is 28 mm thick, 1300 mm wide, weighs 88 kg/m, with a cord spacing of 22 mm between centres, and the belt can carry an axial load of 10 MN. It delivers about 30 tonnes of coal per minute at a linear speed of 8.4 m/s, and drive motors are rated at 10 MW.

On a much smaller scale, polymer molecules can be represented as chain-like threads containing many mers or repeating units. These molecular chains are long, small and thin. If we represent the molecule by a piece of string 1 mm diameter, the straight piece of string could be 10 or more metres long. We must then scale this down to a real polymer molecule only about 1 micrometre long (if it could actually be straightened out).

Examples of polymers include nylon, polypropylene, PVC and styrene-butadiene rubber, not to mention naturally occurring polymers such as natural rubber, cellulose and silk, and the biopolymers in our own bodies such as collagen (based on polypeptides), chitin (a polysaccharide), and the protein rubbers.

But rather than be straight, polymer molecules have an inborn urge to coil up and tangle together so that the ends of any chain are typically only a few millionths of a millimetre apart. This jumbled nature confers uniform properties on the mass of polymer. But it is possible under suitable conditions to apply a modest tensile force and stretch out bits of these chains. If the new alignment is somehow fixed, we then have an imperfect unidirectional rods-and-matrix model. If there is a high degree of alignment, the matrix all but disappears.

High density polyethylene can be stretched a great deal in a tensile testing machine: it necks down and then the long chain molecules align themselves along the stretch direction (Figure 3.16). This alignment of molecular rods makes the new cross-section much stiffer even though the cross-section in the neck is small. Further extension makes the softer more flexible shoulders of the specimen neck down in spite of the larger area, so the neck grows. The gauge length can increase by a factor of ten. The drawn material is now highly aligned, and if cut out and tested separately would have a modulus some ten times that of the original bar.

On a similar basis, polypropylene is stretched under special conditions to make string. Untwisted polypropylene string is strong enough to tie parcels with because lots of molecules – or bits of molecules – are lined up along the length of the string during manufacture. But the forces between the aligned molecules – the matrix – is weak, and the string can readily be split. Transpar-

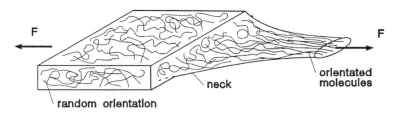

Figure 3.16 Straightening of random polymer molecules during test under increasing tensile load.

ent self-adhesive tape has a pronounced tendency to split along its length, especially when it has been in use for a long time, and manufacturers tend not to draw attention to this unwelcome and frustrating characteristic.

A highly stretched stationer's elastic band consists of highly aligned rubber molecules – the rubber is pale under the strain of up to 5 (500%). The paleness is caused by the formation of many minute crystals which scatter light just as a crystalline solid does. But the band does not fall apart across its width because of the crosslinks in the chemical structure, and it recovers its original dimensions and colour (losing all its crystallinity) on removing the load.

Wet grass leaves can be modelled quite well by a system of high modulus parallel fibres in a low modulus matrix. For example, Vincent (1982) reports that in wet perennial rye grass the modulus of the fibres is about 23 GPa and contributes some 90% to the longitudinal modulus of the leaf, measured as 0.5 GPa, suggesting a fibre volume fraction in the order of 2–4%. The fibres modulus is comparable with that of wet cellulose (hemp about 35 GPa). The measured transverse modulus of this grass is about 14 MPa, giving an anisotropy ratio of about 35. The low transverse modulus may well be a factor in the resistance grass shows to trampling.

Rods-in-matrix models do seem to have the knack of breaking some of the rules of elementary physics for isotropic materials, but they do obey the appropriate laws of physics for anisotropic materials.

Let us recall that if we taken a rubber band and stretch it, it becomes thinner and narrower. The proportion of the thinning down is related to the proportion of the change in length by Poisson's ratio and is the same in both the width and thickness directions. But it is not so simple for a rods-in-matrix model. When stretched across the rods, the matrix does all the flexing, but the rods are very stiff along their length and do not want to contract very much at all. So the Poisson's ratio is really small. But if you pull this model along the rods, it will not stretch much because the rods are stiff – but it will quite happily contract because the matrix is flexible. The difference in Poisson's ratios for stretching in the two directions can be a factor of about a hundred, or even up to ten thousand.

3.2.2 Ply stiffnesses when loaded in principal directions

When a unidirectional ply is loaded in-plane along its principal directions (Figure 3.15), the algebraic form of Hooke's law is identical with that for an isotropic sheet, but values of elements of the compliance matrix [S] or reduced stiffness matrix [Q] have different values in different directions. In particular $S_{11} \neq S_{22}$ and $Q_{11} \neq Q_{22}$. Thus we have

$$\begin{pmatrix} \varepsilon_1 \\ \varepsilon_2 \\ \gamma_{12} \end{pmatrix} = \begin{pmatrix} S_{11} & S_{12} & 0 \\ S_{12} & S_{22} & 0 \\ 0 & 0 & S_{66} \end{pmatrix} \begin{pmatrix} \sigma_1 \\ \sigma_2 \\ \tau_{12} \end{pmatrix} \qquad (3.25)$$

and

$$\begin{pmatrix} \sigma_1 \\ \sigma_2 \\ \tau_{12} \end{pmatrix} = \begin{pmatrix} Q_{11} & Q_{12} & 0 \\ Q_{12} & Q_{22} & 0 \\ 0 & 0 & Q_{66} \end{pmatrix} \begin{pmatrix} \varepsilon_1 \\ \varepsilon_2 \\ \gamma_{12} \end{pmatrix} \tag{3.26}$$

We can relate elements (S_{ij}) of the compliance matrix to elastic constants, and hence assign a physical meaning, by considering the results of simple tests where only one stress is applied (Figure 3.17).

Applying a single uniform longitudinal tensile stress σ_1, to a long strip of a unidirectional linear elastic material produces a direct strain ε_1 and a transverse compressive strain ε_2. Hence

$$\varepsilon_1/\sigma_1 = S_{11} = 1/E_1 \tag{3.27}$$

where E_1 is the longitudinal modulus measured in this test. Similarly we find

$$\varepsilon_2/\sigma_1 = S_{12} = -\nu_{12}/E_1 \tag{3.28}$$

where ν_{12} is the *major* Poisson's ratio defined as the negative ratio of the transverse strain to the longitudinal strain when the specimen carries a uniform longitudinal stress, i.e. $\nu_{12} = -\varepsilon_2/\varepsilon_1$ under σ_1.

Applying a uniform transverse tensile stress σ_2 to a long slender specimen produces a direct transverse strain ε_2 and a longitudinal contraction ε_1. We therefore see that

$$\varepsilon_2/\sigma_2 = S_{22} = 1/E_2 \tag{3.29}$$

where E is the transverse modulus. The *minor* Poisson's ratio ν_{21} is defined as the negative ratio of the longitudinal strain to the transverse strain under a transverse load, i.e. $\nu_{21} = -\varepsilon_1/\varepsilon_2$ under σ_2. In this test we therefore find

$$\varepsilon_1/\sigma_2 = -\nu_{21}/E_2 = S_{12} \tag{3.30}$$

From strain energy arguments it can be shown that $S_{12} = S_{21}$, which is easily remembered in the form of the Maxwell reciprocal relationship

$$\nu_{12}/E_1 = \nu_{21}/E_2 \tag{3.31}$$

More generally we are confirming that [S] is a symmetric matrix.

Applying a uniform shear stress τ_{12} to a suitable specimen (this is not easy

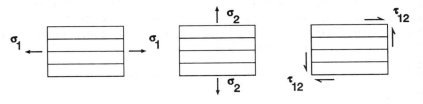

Figure 3.17 Single stress loading of unidirectional ply.

for a flat sheet, but fairly straightforward for a tube with ends under opposing axial torque) produces a shear strain γ_{12}, so we obtain

$$\gamma_{12}/\tau_{12} = 1/G_{12} = S_{66} \tag{3.32}$$

In summary we can now express the compliance matrix for in-plane loading in the principal directions as

$$[S] = \begin{pmatrix} 1/E_1 & -v_{12}/E_1 & 0 \\ -v_{12}/E_1 & 1/E_2 & 0 \\ 0 & 0 & 1/G_{12} \end{pmatrix} \tag{3.33}$$

Equation (3.33) clearly confirms that under plane stress 4 independent elastic constants are needed to relate stresses to strains: E_1, E_2, G_{12} and v_{12}. In particular it is most important to realize that for a unidirectional ply $G_{12} \neq E_1/[2(1 + v_{12})]$.

Equation (3.33) obviously reduces to Equation (3.4) for an isotropic material by putting $E_1 = E_2 = E$, $v_{12} = v$ and $G_{12} = G$.

The reduced stiffness matrix $[Q]$ can be expressed in terms of E_1, E_2, v_{12}, v_{21} and G_{12} by inversion of the compliance matrix $[S]$:

$$[Q] = [S]^{-1} = \begin{pmatrix} Q_{11} & Q_{12} & 0 \\ Q_{12} & Q_{22} & 0 \\ 0 & 0 & Q_{66} \end{pmatrix} = \begin{pmatrix} E_1/J & v_{12}E_2/J & 0 \\ v_{12}E_2/J & E_2/J & 0 \\ 0 & 0 & G \end{pmatrix} \tag{3.34}$$

where $J = (1 - v_{12}v_{21})$. For anisotropic materials v_{21} is usually small, so J has a value which is much closer to unity than occurs for most isotropic materials. Note that $Q_{12} = v_{12}E_2/J = v_{21}E_1/J$.

It is now a simple matter to find the extensional stiffness matrix $[A]$ and the bending stiffness matrix $[D]$, for a unidirectional ply of thickness h loaded in its principal directions, by using the values of $[Q]$ first derived in Equation 3.34 in the expressions for $[A]$ (Equation (3.14)) and $[D]$ (Equation (3.18)).

Thus $A_{11} = E_1h/(1 - v_{12}v_{21})$, and $D_{12} = v_{12}E_1h^3/[12(1 - v_{12}v_{21})]$

Problem 3.8: A single unidirectional ply of carbon fibre reinforced polyetheretherketone (PEEK) is 0.125 mm thick. It has the following values of elastic constants at 20 °C: $E_1 = 134\,GPa$, $E_2 = 8.9\,GPa$, $G_{12} = 5.1\,GPa$, $v_{12} = 0.28$. Calculate leading values of the $[S], [Q], [A], [D], [a]$ and $[d]$ matrices.

Problem 3.9(a): What proportion of the load is taken by the fibres in a unidirectional ply loaded in uniform tension in the longitudinal direction?

3.2.3 Unidirectional lamina NEO loaded in principal directions

This lamina (Figure 3.18) is based on the material NE with properties summarized in Table 3.3. The reduced stiffness matrix Q_{ij} for lamina NE is shown

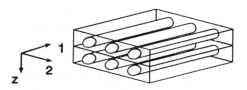

Figure 3.18 The unidirectional ply NE0.

Table 3.3 Properties of lamina NE

$E_{11} = 0.91\,\text{GPa}$	[Q] in GPa		
$E_{22} = 0.00724\,\text{GPa}$	0.9109	0.002609	0
$G_{12} = 0.0018\,\text{GPa}$	0.002609	0.007247	0
$v_{12} = 0.36$	0	0	0.0018
$\theta = 0°$			

Table 3.4 Stiffnesses and compliances of lamina NEO (total thickness 2 mm)

A(kN/mm) B(kN)			B(kN) D(kNmm)		
1.822	0.005218	0	0	0	0
0.005218	0.01449	0	0	0	0
0	0	0.0036	0	0	0
0	0	0	0.6073	0.001739	0
0	0	0	0.001739	0.004832	0
0	0	0	0	0	0.0012

a(mm/MN) h(/MN)			b(/MN) d(/MNmm)		
549.5	− 197.8	0	0	0	0
− 197.8	69060	0	0	0	0
0	0	277800	0	0	0
0	0	0	1648	− 593.4	0
0	0	0	− 593.4	207200	0
0	0	0	0	0	833300

in Table 3.3. The lamina is cured at 20 °C, and Table 3.4 shows the ABD matrices and their inverted form for a total thickness of 2 mm.

Because this unidirectional sheet is stressed in the principal directions, the patterns of deformation and the qualitative internal stress distributions are similar to those of the isotropic sheet NR. The minor Poisson's ratio is much smaller than the major Poisson's ratio. The population of the A, B and D matrices is the same as for NR, but for NEO $A_{11} \neq A_{22}$, and $D_{11} \neq D_{22}$, as expected for a unidirectional orthotropic material.

Application for single force resultant $N_1 = 1$ N/mm to NEO

The strains

$$\varepsilon_1^\circ = a_{11}N_1 = 549.5 \text{ (mm/MN)} \times 1 \text{ N/mm} = 5.495 \times 10^{-4} = 0.05495\%$$

and $\varepsilon_2^\circ = a_{12}N_1 = -0.01978\%$ are shown schematically in Figure 3.19. The major Poisson's ratio $v_{12} = -\varepsilon_2^\circ/\varepsilon_1^\circ = 0.36$ as expected. The midplane strains are much smaller for the applied load then for the unreinforced rubber sheet NR, as expected because of the reinforcing effect of the nylon cords, even though the loading is much higher. The stress σ_1 may be calculated from

$$\sigma_1 = Q_{11}\varepsilon_1^\circ + Q_{12}\varepsilon_2^\circ$$
$$= 0.9108 \times 10^9 \times 5.495 \times 10^{-4} - 2.609 \times 10^6 \times 1.978 \times 10^{-4}$$
$$= 0.5 \text{ MPa}$$

and confirms the more straightforward $\sigma_1 = N_1/h = (1 \text{ N/mm})/2 \text{ mm}$ $= 0.5$ MPa.

Application of single force resultant $N_{12} = 1$ N/mm to NEO

The shear strain $\gamma_{12}^\circ = a_{66}N_{12} = 27.78\%$ is large because the cords have little reinforcing effect on the shear modulus which is matrix dominated. The response is shown schematically in Figure 3.20.

Problem 3.9(b): Calculate the midplane strains and curvatures in an initially flat ply NEO under the moment resultant $M_2 = 1$ MN.

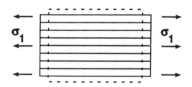

Figure 3.19 Deformation of NE0 under $N_1 = 1$ N/mm.

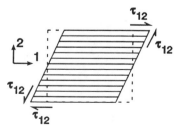

Figure 3.20 Deformation of NE0 under $N_{12} = 1$ N/mm.

Application of single moment resultant $M_2 = 0.01N$

The effect of applying a transverse moment is to induce a curvature $\kappa_2 = 2.072/m$ because the deformation is dominated by the flexible matrix. Under M_2 acting alone we can calculate κ_1 from the minor Poisson's ratio $v_{21} = -\kappa_1/\kappa_2$, i.e. $\kappa_1 = -0.005934/m$, the modest value indicating the stiffening by the cords. More formally we would calculate $\kappa_2 = d_{22}M_2$ and $\kappa_1 = d_{12}M_2$. The curvatures and the strain profiles are shown schematically in Figure 3.21.

Application of single twisting moment $M_{12} = 0.01N$

The curvature $\kappa_{12} = d_{66}M_{12} = 8.333/m$ corresponds to a substantial twist having the character of Figure 3.22(a). The shear stress and shear strain vary linearly through the thickness e.g. $\tau_{12} = zQ_{66}\kappa_{12}$ and $\gamma_{12} = z\kappa_{12}$, as shown in Figure 3.22(b).

Problem 3.10: Calculate the midplane strains in a flat ply NEO under the application of a single force resultant $N_2 = 1N/mm$. What is the value of the minor Poisson's ratio?

Problem 3.11: What in-plane forces are needed in the principal directions to induce biaxial strains $\varepsilon_1 = \varepsilon_2 = 5\%$ in NR?

Problem 3.12: A typical 1650/3 Rayon cord reinforced rubber ply used in the body of an HR78-15 radial tyre has the following properties: $E_1 = 1.74\,GPa$, $E_2 = 14.1\,MPa$, $v_{12} = 0.547$, $G_{12} = 2.5\,MPa$.

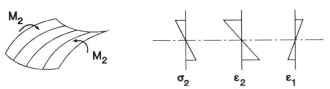

Figure 3.21 Deformation, and stress and strain profiles, in NE0 under $M_2 = 0.01$ N.

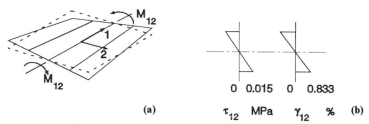

Figure 3.22 Deformation, and stress and strain profiles, for NE0 under $N_{12} = 0.01$ N.

Calculate the value of the minor Poisson's ratio and write down the values of the terms in the compliance matrix [S] when a unidirectional sheet carries in-plane loads in the principal directions.

What ratio of the stresses σ_1/σ_2 would give zero transverse strain?

The cured ply thickness is 1.9 mm. Under test conditions a single ply carries tensile stresses of 20 MPa along the cords, 0.7 MPa across the cords, and a shear stress of 0.1 MPa. What is the strain response?

Problem 3.13: By matrix inversion calculate the elements in the reduced stiffness matrix [Q] for a unidirectional sheet loaded in-plane in the principal directions, in terms of the elastic constants E_1, E_2, v_{12} and G_{12}, where [Q] is given by

$$[Q] = \begin{pmatrix} Q_{11} & Q_{12} & 0 \\ Q_{12} & Q_{22} & 0 \\ 0 & 0 & Q_{66} \end{pmatrix} = [S]^{-1}$$

Problem 3.14: A carbon fibre reinforced epoxy unidirectional ply has an anisotropy ratio of about 40. Highlight (qualitatively) how you would expect such a sheet to deform differently (under in-plane load in the principal directions) from the responses of an isotropic epoxy sheet under the same loadings. Draw sketches of three different characteristic responses to different stress combinations for both the unidirectional ply and the isotropic sheet.

Problem 3.15: Two long thin-walled pipes are to be made from unidirectional plies of fibre reinforced elastomer having a large anisotropy ratio. in one pipe all the fibres are in the axial direction, and in the other pipe all the fibres are in the hoop direction. The pipe and contents are held vertically at the upper end, and may be assumed weightless. For a region well away from the ends, compare and contrast, and explain, the likely deformation responses of each pipe under (a) internal pressure, (b) opposing torques applied to each end of the unpressurised pipe, and (c) axial tensile load.

3.3 STIFFNESS OF UNIDIRECTIONAL PLY UNDER OFF-AXIS LOADING

3.3.1 Overview

If we stretch a rectangular piece of uniform isotropic material, it changes its dimensions by getting longer and thinner – but it stays rectangular in shape. A unidirectional rods-in-matrix sheet loaded along or across the fibre direction behaves in much the same way. But if we load the sheet in the plane at some angle to the axis of the fibres, it not only gets longer and thinner, but it shears as well, as shown in Figure 3.23. We can explain this by resolving the applied load along the fibres and across them. The component of load along the fibres will

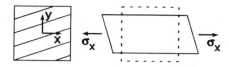

Figure 3.23 Deformation of unidirectional ply loaded off-axis in uniform tension.

give little extension because the fibres are stiff, but the component across the fibres is resisted by the matrix and hence stretches much more. The result of these two effects is the change to a lozenge shape.

When loads are applied to the unidirectional sheet at some angle to the rod direction, the modulus of the sheet takes a value which depends on that angle. If the angle between the rods and the load is small, then the modulus is high. But a small increase in angle gives a dramatic drop in modulus (shown schematically in Figure 3.24), until at 90° the modulus corresponds to that for the across direction. Figure 3.24 is of great importance to those seeking to design in composite materials. Applying the tensile load at some intermediate angle to the rods not only produces extension (and lateral contraction) but also shear: this just does not happen in an isotropic material.

For a single unidirectional lamina stressed in (x, y) at an angle θ to the principal directions $(1, 2)$, as shown in Figure 3.25(d), the approach to understanding the mechanics is broadly similar in character to that used previously.

What we have to do is to re-express the stresses in the applied directions (x, y) to components of stress in the $(1, 2)$ directions. Then from Hooke's law we can find the associated strains in the $(1, 2)$ direction. Finally we transform the strains in $(1, 2)$ into the strains in the off-axis co-ordinates (x, y). The process is shown diagrammatically in Figure 3.25. Each of these operations is conceptually straightforward, but the algebra becomes unwieldy and combersome.

The end result is that we use a transformed reduced stiffness matrix $[Q^*]$, which takes account of the rotations of axes for stresses and for strains, in place of $[Q]$ used for loading in the principal directions. We therefore have the pair

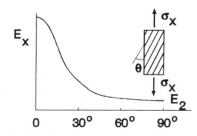

Figure 3.24 Dependence of tensile modulus E_x on angle θ between stress direction and the fibre direction.

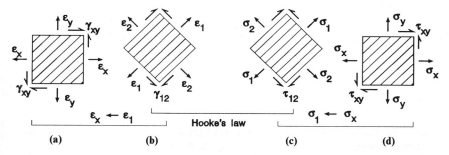

Figure 3.25 The steps needed to determine the off-axis strain responses to applied off-axis in-plane stresses.

of matrix equations:

$$\begin{pmatrix} N \\ M \end{pmatrix} = \begin{pmatrix} A & B \\ B & D \end{pmatrix} \begin{pmatrix} \varepsilon^\circ \\ \kappa \end{pmatrix}$$

(3.35)

where

$$[A] = [Q^*]h \qquad (3.36a)$$

and

$$[D] = [Q^*]h^3/12 \qquad (3.36b)$$

Expect where $\theta = 0°$ or $90°$ we have fully populated matrices for extensional stiffnesses and bending stiffnesses, e.g.

$$\begin{pmatrix} N_x \\ N_y \\ N_{xy} \end{pmatrix} = [A][\varepsilon^\circ] = \begin{pmatrix} A_{11} & A_{12} & A_{16} \\ A_{12} & A_{22} & A_{26} \\ A_{16} & A_{26} & A_{66} \end{pmatrix} \begin{pmatrix} \varepsilon^\circ_x \\ \varepsilon^\circ_y \\ \gamma^\circ_{xy} \end{pmatrix}$$

(3.37)

In addition to the familiar responses from A_{11}, A_{12}, A_{22} and A_{66}, we now find that applying shear strain alone induces direct force resultants per unit width, N_x, N_y because $N_x = A_{16}\gamma^\circ_{xy}$ and $N_y = A_{26}\gamma^\circ_{xy}$. Applying direct strain induces a shear force resultant N_{xy}, e.g. $N_{xy} = A_{16}\varepsilon^\circ_x$ or $N_{xy} = A_{26}\varepsilon^\circ_y$.

Similarly applying a twisting curvature κ_{xy} will induce (direct) moment resultants M_x and M_y. Applying curvatures κ_x and /or κ_y will induce a twisting moment resultant M_{xy}. This can be seen from the fully populated bending stiffness matrix $[D]$ or its inverse $[d]$:

$$\begin{pmatrix} \kappa_x \\ \kappa_y \\ \kappa_{xy} \end{pmatrix} = \begin{pmatrix} d_{11} & d_{12} & d_{16} \\ d_{12} & d_{22} & d_{26} \\ d_{16} & d_{26} & d_{66} \end{pmatrix} \begin{pmatrix} M_x \\ M_y \\ M_{xy} \end{pmatrix}$$

(3.38)

Note that, by convention, we use the suffixes 1, 2 and 6 to identify the stiffness and compliance coefficients even though the global co-ordinates are now (x, y) for off-axis in-plane loading.

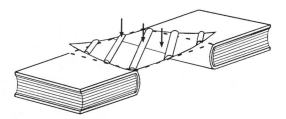

Figure 3.26 Unidirectional stiffened card under transverse load twists as well as bends, when the stiffeners are at an angle to the applied load.

An interesting application of the ideas outlined above is in corrugated cardboard of the kind having one wavy and one plane face. We know that along the direction of the corrugations the sheet is quite stiff–the corrugations act in bending like the rods do in a unidirectional ply in tension. But the corrugated sheet is quite flexible when bent across the ridges. But what happens when you cut out a rectangle of this material with the corrugations at some angle, say 30°, to the long edge, support each end of the beam on a book and apply a vertical load at midspan? (see Figure 3.26). Can you explain what you observe?

3.3.2 Transformation of stresses and strains

Although the programmable microcomputer can certainly relieve the tedium of the matrix manipulation for specific examples, it is well worth making the effort to understand the principles involved, because these principles assume considerable importance when we examine the stiffness and strength behaviour of common forms of laminate. We shall consider the transformation of stresses from (x, y) to $(1, 2)$ and its implications first, then review the transformation of strains.

Transformation of stress

Consider a rectangular element of unit thickness, sides L_x and L_y, aligned in the x and y directions. The axes of the sheet are now rotated by $+\theta$ *anticlockwise* to define the $(1, 2)$ directions. We seek to relate the stresses in the new directions $(1, 2)$ to the applied stresses in (x, y).

Make a cut at θ to the normal to the 1-direction, (Figure 3.27). If the length of the cut is L_2, we see that $L_y = L_2 \cos \theta$ and $L_x = L_2 \sin \theta$. Resolving forces in the 1-direction:

$$F_1 = F_x \cos \theta + F_y \sin \theta + F_{xy} \sin \theta + F_{xy} \cos \theta \qquad (3.39)$$

Expressing the forces in terms of stress acting on edges of the element, and

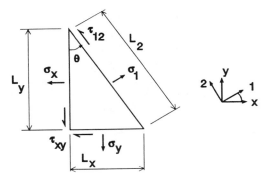

Figure 3.27 Transformation of stresses.

recalling the element has unit thickness, we obtain

$$\sigma_1 L_2 = \sigma_x L_2 \cos^2 \theta + \sigma_y L_2 \sin^2 \theta + 2\tau_{xy} L_2 \sin \theta \cos \theta \qquad (3.40)$$

Resolving in the 2-direction gives an expression for τ_{12}:

$$\tau_{12} = -\sigma_x \sin \theta \cos \theta + \sigma_y \sin \theta \cos \theta + \tau_{xy}(\cos^2 \theta - \sin^2 \theta) \qquad (3.41)$$

To find an expression for σ_2 we may either substitute $(\theta + 90°)$ in the expression for σ_1, or make a cut at θ to the normal to the 2-direction, and resolve forces in the 2-direction. Either approach will give:

$$\sigma_2 = \sigma_x \sin^2 \theta + \sigma_y \cos^2 \theta - 2\tau_{xy} \sin \theta \cos \theta \qquad (3.42)$$

In matrix form we now have

$$\begin{pmatrix} \sigma_1 \\ \sigma_2 \\ \tau_{12} \end{pmatrix} = \begin{pmatrix} \cos^2 \theta & \sin^2 \theta & 2\sin\theta\cos\theta \\ \sin^2 \theta & \cos^2 \theta & -2\sin\theta\cos\theta \\ -\sin\theta\cos\theta & \sin\theta\cos\theta & \cos^2 \theta - \sin^2 \theta \end{pmatrix} \begin{pmatrix} \sigma_x \\ \sigma_y \\ \tau_{xy} \end{pmatrix} = [T] \begin{pmatrix} \sigma_x \\ \sigma_y \\ \tau_{xy} \end{pmatrix} \qquad (3.43)$$

Note that this stress transformation matrix $[T]$ is *not* symmetric.

We are normally interested in values of $[T]$ for angles θ within the range $-90° < \theta < 90°$. For unit values of σ_x, σ_y and τ_{xy}, the corresponding values of σ_1, σ_2 and τ_{12} are shown graphically in Figure 3.28. It is immediately apparent that as the angle θ changes, the contribution which τ_{xy} makes to any stress in 1, 2 varies over double the range contributed by σ_x or σ_y.

We will also be interested in transforming stresses in the (1, 2) co-ordinates into stresses in (x, y). This can be readily achieved by matrix inversion (noting that $|T| = 1$):

$$\begin{pmatrix} \sigma_x \\ \sigma_y \\ \tau_{xy} \end{pmatrix} = \begin{pmatrix} \cos^2 \theta & \sin^2 \theta & -2\sin\theta\cos\theta \\ \sin^2 \theta & \cos^2 \theta & +2\sin\theta\cos\theta \\ \sin\theta\cos\theta & -\sin\theta\cos\theta & \cos^2 \theta - \sin^2 \theta \end{pmatrix} \begin{pmatrix} \sigma_1 \\ \sigma_2 \\ \tau_{12} \end{pmatrix} = [T]^{-1} \begin{pmatrix} \sigma_1 \\ \sigma_2 \\ \tau_{12} \end{pmatrix} \qquad (3.44)$$

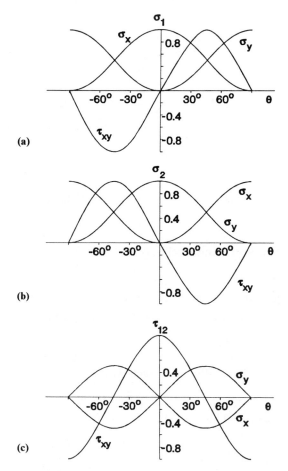

Figure 3.28 Representation of the stress transformation matrix [T].

Note that the convention of defining θ as positive when there is anticlockwise rotation of θ from (x, y) to $(1, 2)$ is widely used in the composites literature. This is consistent with the convention in Chapter 2 if we regard θ as positive when there is *clockwise* rotation of θ from $(1, 2)$ to (x, y).

Problem 3.16: A unidirectional lamina is subject to equal uniform tensile force resultants per unit width, N_x and N_y, where the x axis makes an angle θ to the fibre direction. At what value of θ is the shear stress zero in the principal directions? What is the effect of applying equal compressive force resultants $-N_x$ and $-N_y$?

Problem 3.17: A thin square sheet is 150 mm square and 2 mm thick, with edges aligned with the (x, y) co-ordinates. Uniform force resultants per unit width are applied in the (x, y) directions: $N_x = 5\,\text{N/mm}$, $N_y = 5\,\text{N/mm}$, and $N_{xy} = 1\,\text{N/mm}$. Calculate the stresses in the sheet in the (x, y) directions and in the (a, b) directions at $60°$ to (x, y).

Transformation of axes for strain

We seek to relate the applied strains ε_x, ε_y, γ_{xy} to the coplanar strains ε_1, ε_2, γ_{12} in the $(1, 2)$ directions which make an angle θ to the (x, y) directions.

The approach is to consider an element, $dx \times dy$, under separately applied small strains ε_x, ε_y, γ_{xy}, and find the corresponding strains in the desired directions $(1, 2)$. Adding all the contributions from ε_x, ε_y, γ_{xy} gives the strains ε_1, ε_2, and γ_{12}.

If the element in Figure 3.29(a) is strained by ε_x alone, the diagonal (in the 1 direction at θ to x) increases in length by $\varepsilon_x dx \cdot \cos\theta$. If strained in the y direction alone (Figure 3.29(b)) the diagonal increases by $\varepsilon_x dy \cdot \cos\theta$. If the body is sheared by γ_{xy} alone (Figure 3.29(c)), the diagonal increases by $\gamma_{xy} dy \cdot \cos\theta$. Each of these expressions is an approximation acceptable only for small strains.

If the element is deformed by ε_x, ε_y, and γ_{xy} acting together, the change in length of the diagonal is

$$\varepsilon_1 ds = \varepsilon_x dx \cdot \cos\theta + \varepsilon_y dy \cdot \sin\theta + \gamma_{xy} dy \cdot \cos\theta$$

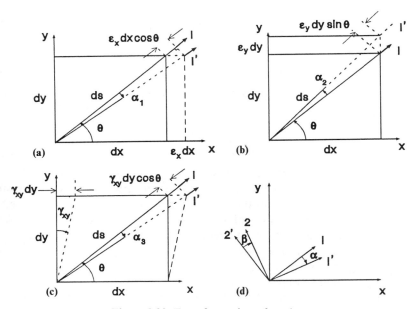

Figure 3.29 Transformation of strains.

Recognizing that $dx/ds = \cos\theta$ and $dy/ds = \sin\theta$, we find

$$\varepsilon_1 = \varepsilon_x \cos^2\theta + \varepsilon_y \sin^2\theta + (\tfrac{1}{2})\gamma_{xy}(2\sin\theta\cos\theta) \tag{3.45}$$

Putting $\theta + 90°$ for θ gives

$$\varepsilon_2 = \varepsilon_x \sin^2\theta + \varepsilon_y \cos^2\theta - (\tfrac{1}{2})\gamma_{xy}(2\sin\theta\cos\theta) \tag{3.46}$$

For shear transformation, consider the net clockwise rotation α of the 1 axis

$$\alpha = \alpha_1 - \alpha_2 + \alpha_3 \tag{3.47}$$

and let the net clockwise rotation of the 2 axis be β (Figure 3.29(d)). We define γ_{12} as the *decrease* in the angle between the 1 and 2 axes (caused by resolution of the applied strains ε_x, ε_y and γ_{xy}).

Hence

$$\gamma_{12} = \beta - \alpha \tag{3.48}$$

Noting that for small angles

$$\alpha_1 = \varepsilon_x \sin\theta \, dx/ds$$
$$\alpha_2 = \varepsilon_y \cos\theta \, dy/ds$$
$$\alpha_3 = \gamma_{xy} \sin\theta \, dy/ds \tag{3.49}$$

Then

$$\alpha = \varepsilon_x \sin\theta\cos\theta - \varepsilon_y \sin\theta\cos\theta + \gamma_{xy}\sin^2\theta \tag{3.50}$$

Putting $\theta + 90°$ for θ gives

$$\beta = -\varepsilon_x \sin\theta\cos\theta + \varepsilon_y \sin\theta\cos\theta + \gamma_{xy}\cos^2\theta \tag{3.51}$$

Thus from Equation (3.48) we obtain

$$(\tfrac{1}{2})\gamma_{12} = -\varepsilon_x \sin\theta\cos\theta + \varepsilon_y \sin\theta\cos\theta + (\tfrac{1}{2})\gamma_{xy}(\cos^2\theta - \sin^2\theta) \tag{3.52}$$

Combining Equations (3.45), (3.46) and (3.52) gives

$$\begin{pmatrix} \varepsilon_1 \\ \varepsilon_2 \\ (\tfrac{1}{2})\gamma_{12} \end{pmatrix} = [T] \begin{pmatrix} \varepsilon_x \\ \varepsilon_y \\ (\tfrac{1}{2})\gamma_{xy} \end{pmatrix} \tag{3.53}$$

where [T] is the transformation matrix already defined in Equation (3.43).

Note carefully the factor of $(\tfrac{1}{2})$ for shear strains, which arises from the derivation of shear strain and the conventional wish to use the same transformation matrix for rotation of axes for both stresses and strains. (We could equally arbitrarily have retained the familiar γ_{xy} and used a different trignometric matrix, which we will show in Section 3.3.3 has the value $[R][T]^{-1}[R]^{-1}$.

Analysis involving transformation of axes needs to relate the two types of strain if the T matrix is used. This is most readily achieved using the Reuter's matrix $[R]$ defined as

$$[R] = \begin{pmatrix} 1 & 0 & 0 \\ 0 & 1 & 0 \\ 0 & 0 & 2 \end{pmatrix} \quad \text{and hence} \quad [R]^{-1} = \begin{pmatrix} 1 & 0 & 0 \\ 0 & 1 & 0 \\ 0 & 0 & \frac{1}{2} \end{pmatrix} \tag{3.54}$$

It follows that the strains are related by

$$\begin{pmatrix} \varepsilon_1 \\ \varepsilon_2 \\ \gamma_{12} \end{pmatrix} = [R] \begin{pmatrix} \varepsilon_1 \\ \varepsilon_2 \\ (\frac{1}{2})\gamma_{12} \end{pmatrix} \quad \text{and} \quad \begin{pmatrix} \varepsilon_1 \\ \varepsilon_2 \\ (\frac{1}{2})\gamma_{12} \end{pmatrix} = [R]^{-1} \begin{pmatrix} \varepsilon_1 \\ \varepsilon_2 \\ \gamma_{12} \end{pmatrix} \tag{3.55}$$

The $[R]$ matrix applies for a cartesian co-ordinate system including the (x, y) set at θ to $(1, 2)$. For example

$$\begin{pmatrix} \varepsilon_x \\ \varepsilon_y \\ \gamma_{xy} \end{pmatrix} = [R] \begin{pmatrix} \varepsilon_x \\ \varepsilon_y \\ (\frac{1}{2})\gamma_{xy} \end{pmatrix} \tag{3.56}$$

Problem 3.18: An element in a sheet of material carries strains $\varepsilon_x = 0.005$, $\varepsilon_y = 0.007$, $\gamma_{xy} = 0.002$ in the (x, y) directions. Calculate the strains in the $(1, 2)$ directions which make $30°$ to the (x, y) axes.

3.3.3 Stress–strain behaviour under in-plane off-axis loading

Stress and strains in (x, y) can be related to those in $(1, 2)$ using the $[T]$ matrix, together with Reuter's matrix for strains, and applying Hooke's law to relate stresses and strains in the principal directions. The sequence of operations relating ε_x to the applied σ_x is as follows:

$$\begin{pmatrix} \varepsilon_x \\ \varepsilon_y \\ \gamma_{xy} \end{pmatrix} = [R] \begin{pmatrix} \varepsilon_x \\ \varepsilon_y \\ \gamma_{xy}/2 \end{pmatrix} = [R][T]^{-1} \begin{pmatrix} \varepsilon_1 \\ \varepsilon_2 \\ \varepsilon_{12}/2 \end{pmatrix} = [R][T]^{-1}[R]^{-1} \begin{pmatrix} \varepsilon_1 \\ \varepsilon_2 \\ \gamma_{12} \end{pmatrix}$$

$$= [R][T]^{-1}[R]^{-1}[S] \begin{pmatrix} \sigma_1 \\ \sigma_2 \\ \tau_{12} \end{pmatrix} = [R][T]^{-1}[R]^{-1}[S][T] \begin{pmatrix} \sigma_x \\ \sigma_y \\ \tau_{xy} \end{pmatrix}$$

$$= [S^*] \begin{pmatrix} \sigma_x \\ \sigma_y \\ \tau_{xy} \end{pmatrix} \tag{3.57}$$

$[S^*]$ is the transformed compliance matrix which relates the strains in (x, y) to the applied stresses in x, y in a unidirectional ply under in-plane off-axis loading. It is a statement of the generalized Hooke's Law for off-axis loading. $[S^*]$ is a fully populated matrix which can be evaluated simply (but tediously) using

$$[S^*] = \begin{pmatrix} S_{11}^* & S_{12}^* & S_{16}^* \\ S_{12}^* & S_{22}^* & S_{16}^* \\ S_{16}^* & S_{26}^* & S_{66}^* \end{pmatrix} = [R][T]^{-1}[R]^{-1}[S][T] \tag{3.58}$$

Because $[S]$ involves four independent elastic constants $(E_1, E_2, G_{12}$ and either v_{12} or $v_{21})$, $[S^*]$ only involves the same four elastic constants together with θ.

3.3.4 Apparent moduli for off-axis loading

The apparent elastic constants for off-axis loading under any one *single* stress are E_x, E_y, G_{xy} and v_{xy} (Figure 3.30). If a long strip is loaded only by σ_x, then measurements of axial strain ε_x and transverse strain ε_y will identify

$$\varepsilon_x / \sigma_x = S_{11}^* = 1/E_x \tag{3.59}$$

$$\varepsilon_y / \sigma_x = S_{12}^* = -v_{xy}/E_x \tag{3.60}$$

Similarly under σ_y acting alone

$$\varepsilon_y / \sigma_y = S_{22}^* = 1/E_y \tag{3.61}$$

It is more difficult to apply an in-plane shear stress τ_{xy}, but the strain response γ_{xy} enables us to determine

$$\gamma_{xy} / \tau_{xy} = S_{66}^* = 1/G_{xy} \tag{3.62}$$

More generally the elements of $[S^*]$ can be evaluated from Equation (3.58) as

$$S_{11}^* = S_{11} \cos^4 \theta + S_{22} \sin^4 \theta + (2S_{12} + S_{66}) \cos^2 \theta \sin^2 \theta$$

$$S_{12}^* = (S_{11} + S_{22} - S_{66}) \cos^2 \theta \sin^2 \theta + S_{12} (\cos^4 \theta + \sin^4 \theta)$$

$$S_{22}^* = S_{11} \sin^4 \theta + S_{22} \cos^4 \theta + (2S_{12} + S_{66}) \cos^2 \theta \sin^2 \theta$$

$$S_{66}^* = 4(S_{11} - 2S_{12} + S_{22}) \cos^2 \theta \sin^2 \theta + S_{66} (\cos^2 \theta - \sin^2 \theta)^2$$

$$S_{16}^* = (2S_{11} - 2S_{12} - S_{66}) \cos^3 \theta \sin \theta - (2S_{22} - 2S_{12} - S_{66}) \cos \theta \sin^3 \theta)$$

$$S_{26}^* = (2S_{11} - 2S_{12} - S_{66}) \cos \theta \sin^3 \theta - (2S_{22} - 2S_{12} - S_{66}) \cos^3 \theta \sin \theta) \tag{3.63}$$

Expressing each of the compliances S_{ij} in terms of E_1, E_2, G_{12} and v_{12} now

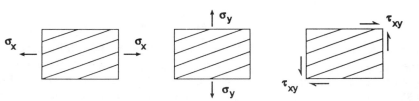

Figure 3.30 Single stress loading of unidirectional ply off-axis.

gives the required expressions for the off-axis elastic constants:

$$S^*_{11} = (1/E_x) = (\cos^4\theta/E_1) + (\sin^4\theta/E_2) + (1/G_{12} - 2v_{12}/E_1)\cos^2\theta\sin^2\theta$$
$$S^*_{22} = (1/E_y) = (\sin^4\theta/E_1) + (\cos^4\theta/E_2) + (1/G_{12} - 2v_{12}/E_1)\cos^2\theta\sin^2\theta$$
$$S^*_{66} = (1/G_{xy}) = 4(1/E_1 + 2v_{12}/E_1 + 1/E_2)\cos^2\theta\sin^2\theta + (1/G_{12})(\cos^2\theta - \sin^2\theta)^2$$
$$-E_x S^*_{12} = v_{xy} = E_x[(v_{12}/E_1)(\sin^4\theta + \cos^4\theta) - (1/E_1 + 1/E_2 - 1/G_{12})\cos^2\theta\sin^2\theta]$$

$$(3.64)$$

For plies which have been properly made and carefully tested, there is close agreement between theoretical prediction and experimental results. It is now possible to calculate the dependence of stiffness on angle for a unidirectional ply, given the properties in the principal directions. Some authors plot modulus vs θ, some plot normalized modulus, i.e. E_x/E_1, E_y/E_1 and G_{xy}/G_{12}. Representative plots are shown in Figures 3.31 and 3.32.

It is apparent that E_x/E_1 changes smoothly from $E_x = E_1$ to $E_x = E_2$ over the range $0° < \theta < 90°$. The variation is large if the anisotropy ratio E_1/E_2 is large. This is particularly so for cord reinforced elastomers, and for carbon fibre reinforced epoxy or polyester at high volume fractions of fibres.

The shear modulus G_{xy} reaches its maximum value at $\theta = 45°$. This can be deduced by setting $dS^*_{66}/d\theta = 0$ and finding one minimum.

Values of Poisson's ratio v_{xy} can vary widely, especially where the anisotropy ratio is large. It is not uncommon for the major Poisson's ratio v_{xy} to exceed unity, which is quite permissible for unidirectionally reinforced

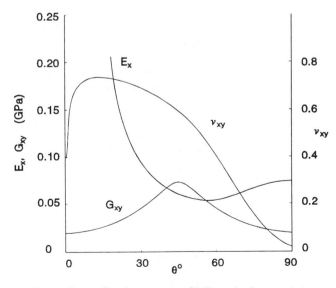

Figure 3.31 In-plane off-axis response of NE to single stress(es).

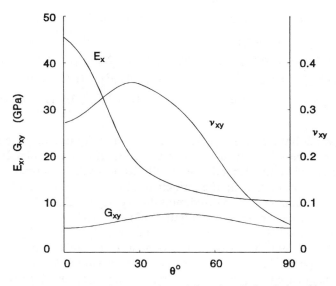

Figure 3.32 In-plane off-axis response of a unidirectional ply of glass fibres in epoxy.

materials (even though the maximum legitimate value for an isotropic material is 0.5).

The physical meanings of S_{16}^* and S_{26}^* are evident from Equation (3.57). Under off-axis loading, a single direct stress σ_x induces not only ε_x and ε_y, but also $\gamma_{xy} = S_{16}^* \sigma_x$. We can infer this visually by resolving σ_x along the diagonal AC in Figure 3.33, which extends readily, as it is dominated by matrix behaviour, and along DB which deforms little because it is dominated by the stiff fibres. Similarly a shear stress acting alone will induce $\varepsilon_x = S_{16}^* \tau_{xy}$ and $\varepsilon_y = S_{26}^* \tau_{xy}$.

We can see therefore that S_{12}^* (and indeed S_{12}), and S_{16}^* and S_{26}^* have the role of coupling coefficients between one type of strain and another. To relate direct strains, we use the familiar concept of Poisson's ratio, v_{12}, which is a coupling coefficient. To relate shear to a direct strain we use S_{16}^* or S_{26}^*, which are not associated with the name of their discoverers.

The reader will note that there is no counterpart to S_{16}^* and S_{26}^* in isotropic materials whatsoever. Equally the reader will see (from Equation (3.63) or from Section 3.2 above) that for a unidirectional material stressed in the plane in the principal directions that $S_{16}^* = S_{26}^* = 0$.

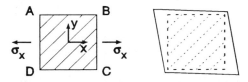

Figure 3.33 Negative shear as a part response to applied stress σ_x.

Problem 3.19: A thin flat sheet made from unidirectional fibres in a polymer matrix has the elastic constants $E_1 = 14\,\text{GPa}$, $E_2 = 3.5\,\text{GPa}$, $G_{12} = 4.2\,\text{GPa}$ and $v_{12} = 0.4$. In-plane loads are applied in the x, y directions which make an angle of $60°$ to the principal directions. A compressive stress of $3.5\,\text{MPa}$ is applied in the x direction, a tensile stress of $7\,\text{MPa}$ in the y direction and a negative shear stress of $1.4\,\text{MPa}$. Calculate the stresses in the principal directions and the strains in the x, y directions.

Problem 3.20: A unidirectional ply is made from high modulus carbon fibre reinforced epoxy resin, and has the following elastic constants: $E_1 = 180\,\text{GPa}$, $E_2 = 8\,\text{GPa}$, $G_{12} = 5\,\text{GPa}$, $v_{12} = 0.25$. Tests are to be made in uniaxial tension. What angles of misalignment to the principal (1) direction would cause a 1% and a 5% reduction in the value of E_1? What conclusion do you draw about the need for precision of alignment of fibres to the test direction?

Problem 3.21: For the ply used in Problem 3.20, what angle of misalignment would cause a 5% error in the value of the in-plane shear modulus G_{12}?

Problem 3.22: At what angle off-axis will a single uniaxial load give the maximum value of the major Poisson's ratio v_{xy} for a unidirectional ply?

Problem 3.23: Tests made on specimens cut from a unidirectional fibre-reinforced PEEK at $0°$, $90°$ and $45°$ to the principal directions give the following values of tensile modulus: $E_1 = 134\,\text{GPa}$, $E_2 = 8.9\,\text{GPa}$ and $E_{45} = 12.83\,\text{GPa}$. What value of modulus E_{60} would you expect for specimens tested at $60°$ to the principal direction?

3.3.5 Stress–strain behaviour under in-plane off-axis strain

The argument detailed above has focused on the strain response to single applied stresses. In order to calculate the more general extensional stiffnesses [A] and the bending stiffnesses [D], we need to pursue a similar argument. This time we apply successive modifications to stresses in order to obtain the strains.

$$
\begin{pmatrix} \sigma_x \\ \sigma_y \\ \tau_{xy} \end{pmatrix} = [T]^{-1} \begin{pmatrix} \sigma_1 \\ \sigma_2 \\ \tau_{12} \end{pmatrix} = [T]^{-1}[Q] \begin{pmatrix} \varepsilon_1 \\ \varepsilon_2 \\ \gamma_{12} \end{pmatrix}
$$

$$
= [T]^{-1}[Q][R] \begin{pmatrix} \varepsilon_1 \\ \varepsilon_2 \\ \gamma_{12}/2 \end{pmatrix} = [T]^{-1}[Q][R][T] \begin{pmatrix} \varepsilon_x \\ \varepsilon_y \\ \gamma_{xy}/2 \end{pmatrix}
$$

$$
= [T]^{-1}[Q][R][T] \begin{pmatrix} \varepsilon_x \\ \varepsilon_y \\ \gamma_{xy}/2 \end{pmatrix} = [T]^{-1}[Q][R][T][R]^{-1} \begin{pmatrix} \varepsilon_x \\ \varepsilon_y \\ \gamma_{xy} \end{pmatrix} = [Q^*] \begin{pmatrix} \varepsilon_x \\ \varepsilon_y \\ \gamma_{xy} \end{pmatrix}
$$

$$
(3.65)
$$

Evaluating the multiplication of the five matrices $T^{-1}QRTR^{-1}$ will give expressions for elements of the transformed reduced stiffness matrix $[Q^*]$ in terms of the reduced stiffnesses (and ultimately the elastic constants E_1, E_2, v_{12} and G_{12}). The intermediate step is:

$$Q_{11}^* = Q_{11}\cos^4\theta + Q_{22}\sin^4\theta + (2Q_{12} + 4Q_{66})\cos^2\theta\sin^2\theta$$
$$Q_{12}^* = Q_{12}(\cos^4\theta + \sin^4\theta) + (Q_{11} + Q_{22} - 4Q_{66})\cos^2\theta\sin^2\theta$$
$$Q_{22}^* = Q_{11}\sin^4\theta + Q_{22}\cos^4\theta + (2Q_{12} + 4Q_{66})\cos^2\theta\sin^2\theta$$
$$Q_{66}^* = (Q_{11} + Q_{22} - 2Q_{12} - 2Q_{66})\cos^2\theta\sin^2\theta + Q_{66}(\cos^4\theta + \sin^4\theta)$$
$$Q_{16}^* = (Q_{11} - Q_{22} - 2Q_{66})\cos^3\theta\sin\theta - (Q_{22} - Q_{12} - 2Q_{66})\cos\theta\sin^3\theta$$
$$Q_{26}^* = (Q_{11} - Q_{12} - 2Q_{66})\cos\theta\sin^3\theta - (Q_{22} - Q_{12} - 2Q_{66})\cos^3\theta\sin\theta \quad\quad (3.66)$$

We see that $[Q^*]$ is a fully-populated matrix. Q_{11}^*, Q_{22}^* and Q_{66}^* relate direct *applied* stresses to similar *induced* strains. Q_{12}^* describes lateral contraction. Q_{16}^* and Q_{26}^* relate direct *applied* shear stresses to *induced* shear strains, and these coefficients have no counterpart for isotropic materials (nor for an orthotropic material stressed in its principal directions). It is perhaps helpful to think of Q_{16}^* and Q_{26}^* as representing an unnamed ratio analogous to Poisson's ratio which relates direct stress σ_j to direct strain ε_1.

The graphs of Q_{ij}^* versus θ are plotted in Figure 3.34 for a high modulus carbon fibre epoxy composite having reduced stiffnesses $Q_{11} = 180.7\,\text{GPa}$, $Q_{12} = 2.41\,\text{GPa}$, $Q_{22} = 8.032\,\text{GPa}$, $Q_{66} = 5\,\text{GPa}$.

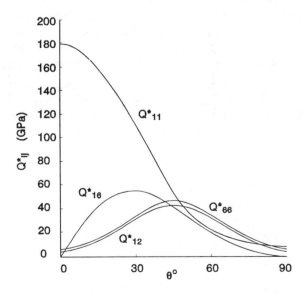

Figure 3.34 Transformed reduced stiffnesses for a high modulus carbon fibre reinforced epoxy unidirectional ply.

Q_{11}^*, Q_{12}^*, Q_{22}^* and Q_{66}^* are even functions of θ, so $Q^*(-\theta) = Q^*(+\theta)$, but Q_{16}^* and Q_{26}^* are odd functions and hence $Q_{16}^*(-\theta) = -Q_{16}^*(+\theta)$ and $Q_{26}^*(-\theta) = -Q_{26}^*(+\theta)$.

Problem 3.24: What value of θ gives the maximum value of Q_{16}^*?

3.3.6 Extensional stiffness matrix for in-plane off-axis loading

Given the transformed reduced stiffness matrix $[Q^*]$ for a given angle θ, we calculate the extensional stiffness matrix $[A]$ using $[Q^*]$ to transform applied strains: the procedure is analogous to that used previously:

$$[N] = [Q^*]h[\varepsilon^\circ] = [A][\varepsilon^\circ] \qquad (3.67)$$

and by inversion we obtain

$$[\varepsilon^\circ] = [A]^{-1}[N] = [a][N] \qquad (3.68)$$

3.3.7 Application of single off-axis ply behaviour

Although as we shall see later, there are several well-known applications of balanced symmetric laminates where in-plane coupling effects are exploited, it is difficult to identify applications for a single lamina, not least because the single lamina is so often flexible or fragile when loaded in the transverse principal direction.

The nearest to reality is to consider a tube wound with stiff fibres at some angle θ to the hoop direction embedded in and well-bonded to a flexible matrix. For demonstration purposes it is possible to make one by coating a tube about 50 to 100 mm diameter, 200 to 300 mm long, with rubber latex, letting the water evaporate, winding on some unwaxed light string or cotton at $+\theta$ (*not* at $-\theta$ as well), then applying a layer of latex to hold the fibres in place. When the rubber has cured, the reinforced rubber tube can (with care) be removed from the former and dipped in dilute sodium hypochlorite to make the rubber less sticky.

Under axial tension of the vertical tube (with the top clamped and the bottom end free) the tube will extend and also rotate. It is the off-axis loading component via a_{16} which causes the shear, and hence the rotation. Applying an oscillating load will induce oscillating extension and oscillating rotation, which can be quite clearly seen even under modest loads, because of the thinness and extensibility of the rubber.

A similar behaviour is seen when open-coiled helical springs are loaded axially in tension: this is most readily seen in large-diameter thin springs having a large number of turns, which provide good amplification of extension under self-weight loading. Small thick springs do the same, but the effect is very small.

The leading dimensions of an open-coiled helical spring are shown in

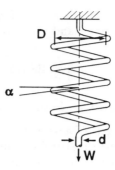

Figure 3.35 Leading dimensions of an open-coiled helical spring.

Figure 3.35. The coil of mean diameter D, of wire diameter d, with n coils wound at a helix angle α is wound from an isotropic material having a Young's modulus E and a Poisson's ratio v. Under axial load F_z, the axial extension z is given by

$$z = 16\,F_z D^3 n(1 + v\cos^2\alpha)/[Ed^4\cos\alpha]$$

and the twist ϕ of the free end

$$\phi = 32\,F_z D^2 nv\sin\alpha/[Ed^4]$$

With some care in construction to avoid pinholes it will also be possible to apply a small internal pressure to the reinforced rubber tube. In addition to extension (readily observable) and an increase in diameter (usually less easy to discern by eye alone), the tube will also rotate (readily observable) as caused by

$$\gamma_{xy}^\circ = a_{16}N_x + a_{26}N_y$$

With some imagination it is possible to model many of the outer muscles in fish having bony skeletons as a helical array of 'fibres'. The muscles themselves are bundles of microfibres with the property that selected segments of their length can be caused to contract. This contraction acts against the stiffness of the vertebral column, so the body bends, and in so doing it exerts a force on the surrounding water, thus propelling the fish in the desired direction. This helical patterns of muscles was reported by Alexander (1969); more accessible summaries are given by Blake (1983) and by Moyle (1982).

To avoid wrinkling in muscles on the concave side of the fish, i.e. to avoid buckling, the helical muscles are always in a state of sufficient latent tensile strain that superposed compressive bending strains always result in a (reduced) net tension.

It will be appreciated that the relevant muscles of fish are wound with both positive and negative helical angles (otherwise the fish would only be able to turn in one plane). We use just one of the windings in the discussion above because only one set of fibres is strained to change the direction of motion. Contracting positive and negative angle windings to the same amount would make the fish feel uncomfortable without changing its direction of motion. What is clever is that the fish can select which muscles to contract to achieve the desired change of direction; in most synthetic composites under in-plane loads at constant temperature we normally strain all the fibres in a given area rather than a selection of them.

3.3.8 General off-axis direct stress response

Previous discussion of the result of applying an off-axis direct in-plane stress to a lamina oriented at $+\theta$ to the principal direction (Figure 3.36) has indicated that the shear response can be large. When we calculate strain responses to a uniaxial stress over the complete range $0° \leqslant \theta \leqslant 90°$, we find some interesting results. Although direct strains are much as expected, shear strains may cause some surprise. For the NE lamina the direct and shear strain responses to a direct stress are presented in Table 3.5.

These data show that at $\theta = 0°$ or $90°$ (the principal directions) there is no shear response (as expected). At about $\theta = 54.63°$ there is also a zero shear response (corresponding to maximum longitudinal strain ε_x). Within

Figure 3.36 Off-axis loading of unidirectional ply.

Table 3.5 Values of strain in lamina NE stressed at θ to the principal direction for $N_x = 1$ N/mm. (Thickness 2 mm)

θ	$\varepsilon_x(\%)$	$\varepsilon_y(\%)$	$\gamma_{xy}(\%)$
0	0.05495	−0.01978	0
10	0.896	−0.6274	−4.51
20	3.003	−2.166	−7.32
30	5.66	−3.916	−7.465
40	7.923	−5.057	−5.15
50	9.113	−5.057	−1.597
54.6	9.24	−4.652	−0.0086
60	9.089	−3.916	+1.532
70	8.251	−2.16	+2.913
80	7.307	−0.6274	+2.167
90	6.906	−0.0198	0

Table 3.6 Stiffnesses and compliances of lamina NE for $N_x = 1$ N/mm, stressed at 54.63° to the principal direction (2 mm lamina)

A(kN/mm) B(kN)			B(kN) D(kNmm)		
0.2165	0.4088	0.2855	0	0	0
0.4088	0.8127	0.5676	0	0	0
0.2855	0.5676	0.4072	0	0	0
0	0	0	0.07216	0.1363	0.09518
0	0	0	0.1363	0.2709	0.1892
0	0	0	0.09518	0.1892	0.1357
a (mm/MN) *h* (/MN)			*b* (/MN) *d* (/MNmm)		
92400	−46490	11.04	0	0	0
−46490	69800	−64690	0	0	0
11.04	−64690	92610	0	0	0
0	0	0	277200	−139500	33.11
0	0	0	−139500	209400	−194100
0	0	0	33.11	−194100	277800

$0 < \theta < 54.6°$ the shear strain is negative, and within $54.64° < \theta < 90°$ the shear strain is positive. Even at $\theta = 54.63°$ all terms in the [A] matrix are substantially populated, as shown in Table 3.6. But in the inverse matrix [a], the term a_{16} is negligible compared with all other a_{ij} hence the minimal shear response to direct stress.

The condition that under a direct stress σ_x the shear strain can change sign as θ varies from 0° to 90° depends on the values of elastic constants of the plies. Either of the following expressions will cause a change of the sign of the shear strain: $G_{12} > E_1/(2(1 + v_{12}))$ or $G_{12} < E_2/(2(1 + v_{21}))$. The ply NE fulfils the second expression. In later chapters we shall examine a rayon/ rubber ply and a Kevlar/epoxy ply, both of which satisfy the second expression. In contrast the high modulus carbon/epoxy HM-CARB and the epoxy/glass ply EG satisfy neither expression and so the shear strain does not change sign under direct stress as the angle changes from 0° to 90°.

3.3.9 Bending stiffnesses for moments applied off-axis

Where the fibres are set in the plane of a flat rectangular sheet at some angle θ to the sides of the sheet, calculation of the response to applied moments at the edges of the sheet follows exactly the same principle as already discussed for isotropic panels in Section 3.1.3.

Using the appropriate values of transformed reduced stiffnesses [Q*], the relationship between applied mid-plane curvatures and moment resultants takes the form

$$[M] = [D][\kappa] = [Q^*]h^3[\kappa]/12 \qquad (3.69)$$

The inverse form is

$$[\kappa] = [D]^{-1}[M] = [d][M] \qquad (3.70)$$

Both [D] and [d] are fully populated matrices and hence the application of any one moment will induce not only anticlastic curvature (κ_x and κ_y) but also the less familiar twisting curvature κ_{xy}. The curvature κ_x and κ_y are quite straightforward. An additional explanation may help the reader to visualize why the twisting curvature arises.

Let us model the off-axis plate as two identical layers bonded at the midplane (Figure 3.37). Applying a positive bending moment M_x will cause a (net) compressive stress in the top layer and a (net) tension in the bottom layer. The consequence, for separated layers, would be positive shear in the top layer and negative shear of the bottom layer, arising from the element a_{16} in the extensional stiffness matrix. (We use the uniform 'net' stress in each ply merely to simplify the argument; it is a (rough) model of the actual varying stress present in each ply.)

The net effect of shear must be zero in the real single plate, so we imagine the forces needed to cancel these shears. We therefore impose a negative restoring shear to the top layers and a positive restoring shear to the lower layer: the layers are now rectangular so we bond them together. When the bond is secure

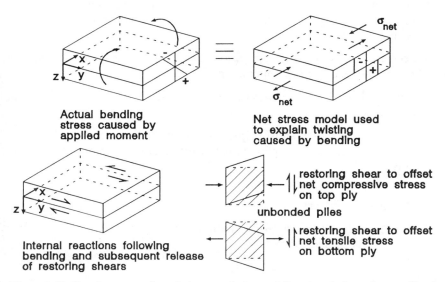

Figure 3.37 Development of a twisting couple in a unidirectional ply under an off-axis moment.

we now release the restoring shear forces because in the single plate we are only applying moment result M_x, with no applied torque. This leaves internal reacting shear forces of opposite sign locked within the real sheet. These internal reacting forces represent an internal couple which induces the negative twist.

3.3.10 Transformation of moments and curvatures

We have already seen that stresses in (x, y) at θ to the principal directions $(1, 2)$ can be related using the $[T]$ matrix, and that strains can be transformed using $[T]$ provided $[\gamma_{xy}/2]$ is used for shear strain.

We can also analogously relate curvatures and moments in (x, y) to corresponding curvatures and moments in $(1, 2)$.

$$\begin{pmatrix} \kappa_1 \\ \kappa_2 \\ \frac{1}{2}\kappa_{12} \end{pmatrix} = [T] \begin{pmatrix} \kappa_x \\ \kappa_y \\ \frac{1}{2}\kappa_{xy} \end{pmatrix} \quad \text{and} \quad \begin{pmatrix} M_1 \\ M_2 \\ M_{12} \end{pmatrix} = [T] \begin{pmatrix} M_x \\ M_y \\ M_{xy} \end{pmatrix} \tag{3.71}$$

Thus the character of curvatures depend dramatically on the directions of the co-ordinate systems used to describe them. The reader is invited to sketch each of the following and confirm visually the conclusions stated. A single curvature (e.g. $\kappa_x = 0$, $\kappa_y = \kappa_y$) looks like synclastic curvature in co-ordinate axes $(x + 45°, y + 45°)$. Equal synclastic curvatures $\kappa_x = \kappa_y$ will always be synclastic as the co-ordinates are rotated in the plane. For anticlastic curvatures $(\kappa_x + \text{ve}, \kappa_y - \text{ve})$ in (x, y), rotating the axis by a certain value of θ to x will suppress one of the curvatures completely, thus giving single curvature. Applying a twisting curvature κ_{12} in the $(1, 2)$ set will lead to anticlastic curvatures (say, $\kappa_x + \text{ve}, \kappa_y - \text{ve}$) in (x, y).

Problem 3.25: Given the application of κ_{12}, what is the value of $\kappa_x(\theta)$, and at what angle θ is $\kappa_x(\theta)$ a maximum?

Problem 3.26: Given the application of κ_1 and κ_2, what is the maximum value of κ_x?

Problem 3.27: Given an isotropic plate where only anticlastic curvatures are developed, at what angles to the 1 direction are the curvatures κ_x and κ_y zero?

3.3.11 Formal summary

Stress and strain calculations for a unidirectional ply loaded in-plane off-axis may be summarized as follows:

From elastic constants calculate $[Q]$
Calculate $[Q*]$
Calculate $[A]$ and $[D]$

Calculate resultants [N] and [M] from applied loads and moments
Calculate midplane strains $[\varepsilon^\circ]$ and curvatures $[\kappa_x]$
Calculate strain profiles $[\varepsilon(z)]$
Calculate stress profiles $[\sigma(z)]$
Calculate midplane strains $[\varepsilon^\circ]$ and curvatures $[\kappa]$ in principal directions
Calculate strain profiles $[\varepsilon(z)]$ in principal directions
Calculate stress profiles $[\sigma(z)]$ in principal directions

3.3.12 Unidirectional lamina NE30 loaded in-plane off-axis

This is the same sheet of nylon cord reinforced elastomer material as discussed in Section 3.2.3, but the cords are now arranged at 30° to the x axis (Figure 3.38). The transformed reduced stiffness matrix Q^*_{ij} for the lamina oriented at 30° to the (x, y) directions is shown in Table 3.7. The effect on behaviour is substantial and dramatic, and it is now necessary to distinguish behaviour in the (x, y) directions and in the principal $(1, 2)$ directions. The ABD and abhd matrices for 2 mm thick NE30 are given in Table 3.8.

The A matrix is now fully populated and the non-zero terms A_{16} and A_{26} are responsible for an in-plane shear strain response to an applied direct stress (or a direct strain when a shear stress is applied). These effects are not coupled together for an isotropic material or for a unidirectional material stressed only in the principal directions. The D matrix is also fully populated and the non-zero terms D_{16} and D_{26} couple bending moments with twisting curvatures. The lamina is still symmetric however as indicated by [B] = 0, so there is no bending or twisting as the result of applying only in-plane forces.

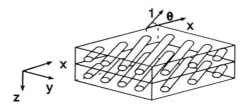

Figure 3.38 The unidirectional ply NE30.

Table 3.7 Properties of lamina NE30

$E_{11} = 0.91\,\text{GPa}$	$[Q^*(+30°)]$ in GPa		
$E_{22} = 0.00724\,\text{GPa}$	0.5152	0.1724	0.2937
$G_{12} = 0.0018\,\text{GPa}$	0.1724	0.6334	0.9760
$v_{12} = 0.36$	0.2937	0.9760	0.1716
$\theta = 30°$			

Table 3.8 Stiffnesses and compliances of lamina NE30 (total thickness 2 mm)

A(kN/mm) B(kN)			B(kN) D(kNmm)		
1.03	0.3449	0.5874	0	0	0
0.3449	0.1267	0.1952	0	0	0
0.5874	0.1952	0.3433	0	0	0
0	0	0	0.3435	0.115	0.1958
0	0	0	0.115	0.04223	0.06507
0	0	0	0.1958	0.06507	0.1144
a (mm/MN) h (/MN)			b (/MN) d (/MNmm)		
56630	− 39160	− 74650	0	0	0
− 39160	90890	15320	0	0	0
− 74650	15320	121900	0	0	0
0	0	0	169900	− 117500	− 224000
0	0	0	− 117500	272700	45950
0	0	0	− 224000	45950	365800

Application of single force resultant $N_x = 1$ N/mm to NE30

The strain responses $\varepsilon_x^\circ = a_{11}N_x = 5.663\%$, $\varepsilon_y^\circ = a_{12}N_x = -3.916\%$ and $\gamma_{xy}^\circ = a_{16}N_x = -7.465\%$. Now that the nylon cords are at 30° to the applied load N_x, their reinforcing effect is small in the (x, y) directions, and the in-plane direct strains are much larger than those observed in NEO. The deformations are shown schematically in Figure 3.39. The major Poisson's ratio $v_{xy} = -(-3.916)/5.663 = 0.6915$, is substantially larger than the value v_{12} in the principal direction.

The stress and strain distributions are all uniform through the thickness of the sheet and are $\sigma_x = 0.5$ MPa, $\sigma_y = 0$, $\tau_{xy} = 0$, $\varepsilon_x = 5.663\%$, $\varepsilon_y = -3.916\%$, $\gamma_{xy} = -7.47\%$.

The strains are of course the same as the midplane strains and the profiles may be formally calculated as $[\varepsilon(z)] = [\varepsilon^\circ]$ using Equation (3.11). Stresses may be calculated from the off-axis equivalent to Equation (3.12) e.g.

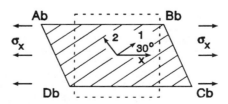

Figure 3.39 Deformation of NE30 under $N_x = 1$ N/mm.

$[\sigma(z)] = [Q^*][\varepsilon^\circ]$. Hence

$$\begin{aligned}
\sigma_x &= Q_{11}^* \varepsilon_x^\circ + Q_{12}^* \varepsilon_y^\circ + Q_{16}^* \gamma_{xy}^\circ \\
&= 0.5159 \times 10^9 \times 0.05663 - 0.1724 \times 10^9 \times 0.03916 - 0.2937 \times 10^9 \times 0.07465 \\
&= (29.1758 - 6.7511 - 21.9247) \times 10^6 = 0.5\,\text{MPa}
\end{aligned}$$

Note the small differences between large quantities. It is usually unrewarding to try and identify the dominant term. It is advisable to round off figures at the end of a chain of calculations, not in the middle.

Give the transformation matrix for 30° (from Equation (3.43))

$$[T(30^\circ)] = \begin{pmatrix} 0.75 & 0.25 & 0.866 \\ 0.25 & 0.75 & -0.866 \\ -0.433 & 0.433 & 0.5 \end{pmatrix}$$

we can readily find the stresses in the principal directions, e.g.:

$$\sigma_1 = T_{11}\sigma_x = 0.375\,\text{MPa}, \quad \sigma_2 = 0.125\,\text{MPa} \quad \text{and} \quad \tau_{12} = -0.21\,\text{MPa}$$

The strains in the principal directions can also be calculated, remembering to use Reuter's matrix to allow for the factor of (1/2) for shear strain transformations. Thus from Equation (3.65):

$$\begin{pmatrix} \varepsilon_1 \\ \varepsilon_2 \\ \gamma_{12} \end{pmatrix} = [R][T][R]^{-1} \begin{pmatrix} \varepsilon_x \\ \varepsilon_y \\ \gamma_{xy} \end{pmatrix}$$

We then obtain $\varepsilon_1 = 0.036\%$, $\varepsilon_2 = 1.712\%$ and $\gamma_{12} = -12\%$, which has a quite different character from the strains in (x, y).

Note that this equation gives a direct relation between off-axis strain and strains in the principal directions. This is therefore the direction relationship mentioned in Section 3.3.2, and eliminates the factor of (1/2) which is essential when using the stress transformation matrix [T] alone.

Application of single force resultant $N_y = 1\,N/mm$ to NE30

The strain responses are $\varepsilon_x^\circ = -3.916\%$, $\varepsilon_y^\circ = 9.089\%$ and $\gamma_{xy}^\circ = 1.532\%$ (as shown in Figure 3.40(a)), the minor Poisson's ratio is $v_{yx} = -(-3.916)/9.089 = 0.4309$, which is much larger than the value v_{21} in the principal direction, as expected.

The stresses in global co-ordinates are found formally from Equation (3.65):

$$\begin{aligned}
\sigma_x &= Q_{11}^* \varepsilon_x^\circ + Q_{12}^* \varepsilon_y^\circ + Q_{16}^* \gamma_{xy}^\circ = 0 \\
\sigma_y &= Q_{12}^* \varepsilon_x^\circ + Q_{22}^* \varepsilon_y^\circ + Q_{26}^* \gamma_{xy}^\circ = 0.5\,\text{MPa} \\
\tau_{xy} &= Q_{16}^* \varepsilon_x^\circ + Q_{26}^* \varepsilon_y^\circ + Q_{66}^* \gamma_{xy}^\circ = 0
\end{aligned}$$

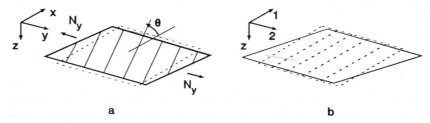

Figure 3.40 Deformations of NE30 in global and principal directions under $N_y = 1$ N/mm.

The stresses in the principal directions are found from Equation (3.43):

$$\sigma_1 = T_{11}\sigma_x + T_{12}\sigma_y + T_{16}\tau_{xy} = 0.125\,\text{MPa}$$
$$\sigma_2 = T_{12}\sigma_x + T_{22}\sigma_y + T_{26}\tau_{xy} = 0.375\,\text{MPa}$$
$$\tau_{12} = T_{16}\sigma_x + T_{26}\sigma_y + T_{66}\tau_{xy} = 0.2165\,\text{MPa}$$

The strains in the principal directions are $\varepsilon_1 = -0.001099\%$, $\varepsilon_2 = 5.175\%$ and $\gamma_{12} = 12.03\%$ as sketched in Figure 3.40(b).

Application of unidirectional strain $\varepsilon_x = 5\%$ to NE30

The force resultants needed to give the single strain response are $N_x = A_{11}\varepsilon_x^{\circ} = 51.52$ N/mm, $N_y = A_{12}\varepsilon_x^{\circ} = 17.24$ N/mm, and $N_{xy} = A_{16}\gamma_{xy}^{\circ} = 29.37$ N/mm. The corresponding stresses are $\sigma_x = 25.76$ MPa, $\sigma_y = 8.622$ MPa and $\tau_{xy} = 14.69$ MPa. The stresses in the principal directions are $\sigma_1 = 34.19$ MPa, $\sigma_2 = 0.1884$ MPa, and $\tau_{12} = -0.07994$ MPa.

Application of single moment resultant $M_x = 0.1$ N to NE30

The resulting curvatures are $\kappa_x = d_{11}M_x = 16.99$/m, $\kappa_y = d_{12}M_x = -11.75$/m and $\kappa_{xy} = d_{16}M_x = -22.4$/m. As illustrated schematically in Figure 3.41 there is substantial anticlastic curvature with superposed twist. The major Poisson's

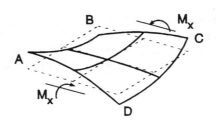

Figure 3.41 Deformation of NE30 under $M_x = 0.1$ N.

ratio is $v_{xy} = -\kappa_y/\kappa_x = 11.75/16.99 = 0.6916$, which corresponds to the value under in-plane load N_x, as it should for a single ply.

The twisting response can be visualized as follows. The midplane when twisted alone (with no bending) will deform out of plane as shown in Figure 3.42. Where the midplane adopts anticlastic curvature, under superposed twist it will deform as indicated in Figure 3.41. Under the action of the compressive bending stress, a top surface layer of the block would in isolation deform as shown in Figure 3.43 and an isolated lower surface layer would wish to deform as shown in Figure 3.40(a). In the real sheet there is a linear variation of strain from the lower surface to the upper surface and all 'layers' are bonded together. The effect is the bending and twisting shown in Figure 3.41. The implication from Figure 3.43 is that in the bonded sheet the top surface diagonal AtCt is shorter than BtDt and indeed the curvature in Figure 3.41 is positive along AtCt and negative along BtDt. The undersurface consistently shows a longer diagonal AbCb. Note that the twisting curvature is opposite in sign to the positive direction of the twisting moment by the sign convention used in the theory (Section 2.5.9).

The linear variations in stress and strain in the (x, y) directions and in the principal directions are shown in Figure 3.44. The stress profiles can be calculated from the curvatures and the transformed reduced stiffness using equations such as

$$\sigma_y(z) = Q_{12}^* z\kappa_x + Q_{22}^* z\kappa_y + Q_{26}^* z\kappa_{xy}$$

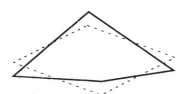

Figure 3.42 Twisting of midplane of NE30 under $M_x = 0.1$ N.

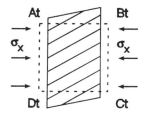

Figure 3.43 Isolated behaviour of top surface of NE30 under $M_x = 0.1$ N.

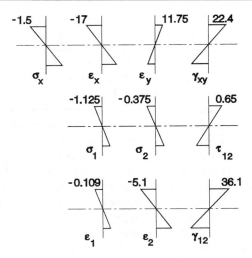

Figure 3.44 Stress and strain profiles in NE30 under $M_x = 0.1$ N.

Suppression of one curvature

If we apply $M_x = 0.1$ N then to suppress the curvature κ_y we need to apply an additional $M_y = 0.04308$ N: this changes all three curvatures to $\kappa_x = d_{11}M_x + d_{12}M_y = 11.93/$m, $\kappa_y = 0$, $\kappa_{xy} = -20.42/$m as shown schematically in Figure 3.45.

If we apply $M_x = 0.1$ N, we can suppress the twist by applying an additional torque $M_{xy} = 0.06121$ N, giving $\kappa_x = d_{11}M_x + d_{16}M_{xy} = 3.281/$m, $\kappa_y = d_{12}M_x + d_{26}M_{xy} = -8.934/$m and $\kappa_{xy} = 0$, as shown schematically in Figure 3.46.

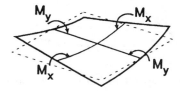

Figure 3.45 Deformation of NE30 under $M_x = 0.1$ N with κ_y suppressed.

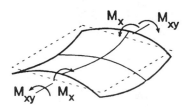

Figure 3.46 Deformation of NE30 under $M_x = 0.1$ N with κ_{xy} suppressed.

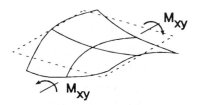

Figure 3.47 Deformation of NE30 under $M_{xy} = 0.1$ N.

Application of single moment resultant $M_{xy} = 0.1$ N to NE30

The resulting curvatures are $\kappa_x = d_{16} M_{xy} = -24.4/m$, $\kappa_y = d_{26} M_{xy} = 4.595/m$ and $\kappa_{xy} = d_{66} M_{xy} = 36.58/m$ as shown in Figure 3.47.

Application of $\kappa_x = 10/m$ to NE30

Assuming that only $\kappa_x = 10/m$ is applied, we set $\kappa_y = \kappa_{xy} = 0$. Hence $M_x = D_{11}\kappa_x = 3.435$ N, $M_y = D_{12}\kappa_x = 1.15$ N, $M_{xy} = D_{16}\kappa_x = 1.958$ N. The bending and twisting stresses in global and principal coordinates are shown in Figure 3.48.

Problem 3.28: What are the values of the apparent elastic constants E_x, ν_{xy}, E_y, and G_{xy} for NE30 when under the appropriate single stress? How would you try and measure the longitudinal modulus E_x?

Problem 3.29: Calculate the midplane strains and curvatures in a 2 mm ply of NE30 under the application of single force resultant $N_{xy} = 1$ N/mm.

Problem 3.30: A unidirectional strain $\varepsilon_x = 5\%$ is applied to NE30, and other strains are allowed to develop by suitable design of grips and by using a long specimen. Compare your estimate of in-plane force resultants with those discussed for a sole strain of $\varepsilon_x = 5\%$ above.

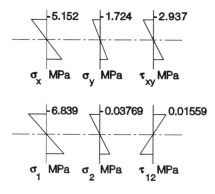

Figure 3.48 Stress profiles in NE30 under $\kappa_x = 10/m$.

Problem 3.31: A uniform shear strain $\gamma_{xy} = 5\%$ is applied to NE30. Calculate the stresses in the principal directions.

Problem 3.32: Calculate the curvatures in a single ply of NE30 under the application of a single moment resultant $M_y = 0.1$ N.

Problem 3.33: A moment $M_y = 0.1$ N and a force $N_x = 1$ N/mm are applied to the ply NE30. What are the midplane strains and curvatures if: (a) the twisting curvature is suppressed; or (b) if the twisting curvature and the curvature κ_x are suppressed by careful choice of edge conditions?

Problem 3.34: A panel of NE30 develops a curvature $\kappa_x = 10/\text{m}$; transverse curvature κ_y is prevented by suitable grips, but twisting curvature is allowed. Compare the moments and stress profiles with those where only κ_x is allowed.

3.3.13 Glass fibre epoxy lamina loaded in-plane off-axis

A single ply EG45 made from a 0.5 mm thin square sheet of unidirectional glass fibres in epoxy resin has the fibres aligned at 45° to the sides of the square. The ABD matrix and its inverse are given in Table 3.9.

Problem 3.35: This sheet EG45 (Table 3.9) is subject to in-plane force resultants N_x, N_y, N_{xy}. These resultants each can have the value $+1$ N/mm or -1 N/mm, and may be applied singly or in any combination (but positive and negative values of the same force are not allowed). Sketch the deformation responses.

Table 3.9 Stiffnesses and compliances of 0.5 mm ply of EG with fibres at 45° to the principal directions

A(kN/mm) B(kN)			B(kN) D(kNmm)		
10.49	5.346	4.437	0	0	0
5.346	10.49	4.437	0	0	0
4.437	4.437	6.42	0	0	0
0	0	0	0.2185	0.1114	0.09244
0	0	0	0.1114	0.2185	0.09244
0	0	0	0.09244	0.09244	0.1337

a (mm/MN) h (/MN)			b (/MN) d (/MNmm)		
148.8	−45.72	−71.27	0	0	0
−45.72	148.8	−71.27	0	0	0
−71.27	−71.27	254.3	0	0	0
0	0	0	7144	−2195	−3421
0	0	0	−2195	7144	−3421
0	0	0	−3421	−3421	12210

Problem 3.36: The sheet EG45 (Table 3.9) is subject to moment resultants M_x, M_y, M_{xy}. These resultants each can have the value $+0.1$ N/mm or -0.1 N/mm, and may be applied singly or in any combination (but positive and negative values of the same force are not allowed). Sketch the deformation responses.

Problem 3.37: A long weightless thin-walled pipe, sealed at each end, is to be wound with polymer-impregnated fibres at $+30°$ to the hoop direction. The unidirectional composite (NE30) may be assumed linear elastic with these values of the extensional compliance matrix:

$$[a] = \begin{pmatrix} 5.66 & -3.92 & -7.47 \\ -3.92 & 9.09 & 1.53 \\ -7.47 & 1.53 & 12.2 \end{pmatrix} \text{mm/kN}$$

The pipe is clamped vertically at the upper end and the lower end is free to move in response to any applied load.

Explain, with annotated sketches and reference to components of the extensional stiffness matrix, the patterns of deformation you would expect to see in the pipe wall well away from the ends, and comment briefly on the displacements at the lower end, for each of the following loading conditions:

(a) axial tension applied at the free end;
(b) internal pressure;
(c) opposing torques at each end.

3.3.14 Aspects of the stiffness of wood

In this section we shall examine a simple rods-in-matrix model of the behaviour of wood under uniaxial tension, in which we relate the macroscopic stiffness of wood to the stiffness of cellulose microfibrils. This provides an example of the use of arguments involving volume fraction and off-axis loading to estimate the behaviour of a complicated structure from fundamental properties. Further detail is given in Jeronimidis (1980), Vincent (1982), Wainwright *et al.* (1976), and Dinwoodie (1989).

Wood from trees contains much cellulose. Cellulose is a long chain polymer consisting of repeating saccharide units. Cellulose can exist in both crystalline and amorphous forms. The modulus, E_c of the crystalline form is about 250 GPa, and of the amorphous form, E_a, about 50 GPa. Wood also contains a flexible matrix, having a modulus, E_m, of about 1 GPa, consisting of lignin, hemicelluloses and low molecular mass sugars.

In terms of composite mechanics, the basic building block (called a microfibril), has three constituents, which may be pictured as a long series of bundles of short parallel crystalline rods ($V_c = 0.4$, some 30 to 60 nm long) joined end to end by amorphous material ($V_a = 0.1$), encased in the matrix ($V_m = 0.5$). It can be shown that the longitudinal modulus of the microfibril, E_f, is about 70 GPa.

The simplest model of wood consists of bundles of tubes called tracheids, which permit the transport of nutrient and other fluids from roots along the trunk and branches to transpire through the leaves. These tracheids are stiff and strong, but the matrix which holds them together is much more flexible and is weaker, and it is this rods-in-matrix structure which gives the familiar characteristics that wood is stiffer and stronger along the axis of the tracheids ('the grain') than across the grain.

Each typical tracheid is formed as four concentric tubes. the thin inner tube has the cellulose-based microfibrils spirally wound at about $-80°$ to the longitudinal axis. The next tube, representing about 80% of the total wall thickness and usually called the secondary S2 wall, is spirally wound in the opposite sense at about $+10°$ to the axis. Next is a thin layer which behaves rather like an angleply wound at about $\pm 70°$ to the axis. The thin outer layer looks rather like a random mat of fibres. We shall concentrate just on the thickest layer.

With microfibrils at $10°$ to the axis of the tracheid, we can now estimate the modulus of the S2 tube along its axis as $E_{S2} \approx E_f \cos^4 \theta = 65$ GPa. Neglecting the stiffness of the other tubes in the tracheid, and knowing that the volume fraction of the S2 is about $V_f = 0.8$, the longitudinal modulus of the tracheid, E_t is about $E_t \approx V_f E_{S2} = 52$ GPa.

The volume fraction of cell material in wood is about 25%, so the modulus of wood along the grain E_1 is about $E_1 \approx 0.25 E_t = 13$ GPa. It must be emphasized that this is an extremely rough calculation, but the longitudinal modulus of most softwoods and hardwoods is in the range 11 to 16 GPa.

FURTHER READING

Alexander, R. McN. (1967) *Functional Design in Fishes*, Hutchinson.

Alexander, R. McN. (1969) The orientation of muscle fibres in the myomeres of fishes, *J. Marine Biological Association of the UK*, **49**, 263–90.

Blake, R.W. (1983) *Fish Locomotion*, Cambridge University Press.

Dinwoodie, J. M. (1989) *Wood, Nature's Polymeric Fibre-Composite*, The Institute of Metals, London.

Jeronimidis, G. (1980) Wood, one of nature's challenging composites; in J.F.V. Vincent and J.D. Currey (eds), *The Mechanical Properties of Biological Materials,* Cambridge University Press.

Moyle, P.B. and Cech, J.J. (1982) *Fishes, an Introduction to Ichthyology*, Prentice Hall.

Vincent, J.F.V. (1982) The mechanical design of grass, *J. Mat. Sci.* **17**, 856–60.

Vincent, J.F.V. (1990) *Structural Biomaterials*, Macmillan.

Wainwright, S.A., Biggs, W.D., Currey, J.D. and Gosline, J.M. (1976) *Mechanical Design in Organisms*, Arnold.

Laminates based on isotropic plies

4

4.1 INTRODUCTION

The discussion of the behaviour of plates under in-plane or bending loads has so far focused on a single sheet of an isotropic linear elastic material, or a single unidirectional ply. Before we discuss the similar behaviour for a laminate based on a bonded stack of unidirectional plies orientated at different angles to a reference direction, it will be helpful to examine the mechanics for laminates based on a bonded stack of isotropic linear elastic plies. This will provide the helpful stepping stone of enabling the reader to appreciate the essential concepts of laminates before tackling the more complicated effects of varying the directionality through the thickness.

The essential steps in the arguments in this chapter, and the special cases, are as follows:

1. We describe the stacking sequence for the laminate, from the top down, using shorthand if necessary and appropriate.
2. We define the externally applied force and moment resultants per unit width for the laminate as a whole in terms of the resultants for each ply.
3. We define stiffnesses and compliances for the laminate as a whole, in order to relate midplanes strains and curvatures to the force and moment resultants. We not only use the stiffness matrices [A] and [D], but also the new bending coupling matrix [B] when dealing with non-symmetric laminates.
4. Given the relationships for the laminate-as-a-whole between midplane strains and curvatures and the force and moment resultants, we can use geometry to calculate strain profiles within the laminates and from these, together with ply stiffnesses, we can calculate the stress profile within each ply and hence through the laminate thickness.

4.2 POSITIONS AND THICKNESSES IN A BONDED LAMINATE

A laminate has its own midplane on which midplane strains and curvatures can be described.

It is necessary to describe in a consistent way the positions and thicknesses of each ply within the stack. We number the plies from the top down, as shown in Figure 4.1. The total number of plies is F. The lower surface of the fth ply is assigned the co-ordinate h_f, so that the thickness of this ply is

$$h(f) = h_f - h_{f-1} \tag{4.1}$$

In this co-ordinate system numerical values of h_f above the midplane are negative. The total thickness of the laminate is

$$h = \Sigma h(f) = -h_0 + h_F \tag{4.2}$$

The midplane of the laminate is the geometrical plane dividing the laminate into two halves of equal thickness. The midplane may be within a ply, or at an interface between two plies. Strains are measured from the midplane ($z = 0$). In laminates the midplane is not necessarily the neutral plane of bending. In laminates of complicated construction, especially when non-symmetric or composed of plies of different types of material, it is not always obvious where the neutral plane(s) is (are). In some laminates the neutral plane has three different locations: in the two in-plane reference directions for the two direct bending stresses, and in a third parallel plane for shear stresses caused by twisting. We therefore use the unambiguous midplane throughout, and the calculated strain profiles through the thickness will disclose the location of the neutral plane(s).

If the force resultants per unit width in the fth ply are $[N]_f$, then the total force resultant in the laminate $[N]$ of F plies about the midplane of the laminate is

$$[N] = \sum_{f=1}^{F} [N]_f \tag{4.3}$$

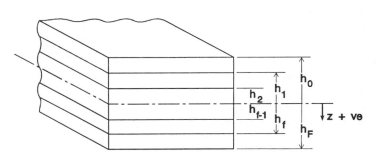

Figure 4.1 Laminate co-ordinates.

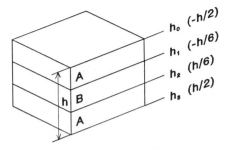

Figure 4.2 Laminate co-ordinates.

Similarly the moment resultant of the laminate [M] about its midplane is related to the moment resultants of each ply, $[M]_f$, by

$$[M] = \sum_{f=1}^{F} [M]_f \qquad (4.4)$$

So for the three layer symmetric laminate $h = 2h_A + h_B$ (Figure 4.2), we have

$$[N] = 2N_A + N_B \qquad (4.5)$$

4.3 RELATING MIDPLANE STRAINS AND CURVATURES TO FORCE AND MOMENT RESULTANTS PER UNIT WIDTH

The approach for a laminate closely follows that for a single isotropic sheet. We express for any ply the force resultant in terms of the stress within that ply, and within that ply the stress is related by the reduced stiffness of that ply to the midplane strain and curvature of that ply. The force resultants for the laminate are therefore given by the sum of the force resultants for each ply:

$$[N]_L = \sum_{f=1}^{F} [N]_f = \sum_{f=1}^{F} \int_{z_{f-1}}^{z_f} [\sigma]_f \, dz = \sum_{f=1}^{F} \left(\int_{z_{f-1}}^{z_f} [Q]_f [\varepsilon^\circ] \, dz + \int_{z_{f-1}}^{z_f} [Q]_f [\kappa] z \, dz \right)$$

$$(4.6)$$

For a thin laminate, $[\varepsilon^\circ]$ and $[\kappa]$ relate to the midplane of the laminate and do not depend on z, and within a given ply $[Q]_f$ is constant (although $[Q]_f$ may vary from ply to ply). As we saw in Chapter 3, this enables us to simplify Equation (4.6) by bringing terms in Q, ε° and κ in front of the integrals, to give

$$[N]_L = \left(\sum_{f=1}^{F} [Q]_f [\varepsilon^\circ] \int_{z_{f-1}}^{z_f} dz + [Q]_f [\kappa] \int_{z_{f-1}}^{z_f} z \, dz \right) = [A][\varepsilon^\circ]_L + [B][\kappa]_L \quad (4.7)$$

where

$$[A] = \sum_{f=1}^{F} [Q]_f (h_f - h_{f-1}) \qquad (4.8)$$

and

$$[B] = \left(\frac{1}{2}\right) \sum_{f=1}^{F} [Q]_f (h_f^2 - h_{f-1}^2) \tag{4.9}$$

For the three ply symmetric laminate of Figure 4.2, the co-ordinates of ply surfaces may be written from top to bottom as h_0, h_1, h_2 and h_3, which gives

$$[A] = [Q]_A (h_1 - h_0) + [Q]_B (h_2 - h_1) + [Q]_A (h_3 - h_2)$$
$$= 2[Q]_A h_A + [Q]_B h_B \tag{4.10}$$

It can be seen then that in a laminate the extensional stiffness matrix $[A]$ has the character of a thickness-weighted stiffness.

If the laminate is regular, i.e. all the plies are of the same thickness, then in a three-ply stack the thickness co-ordinates can be designated (starting at the top surface) $- h/2, - h/6, + h/6$, and $+ h/2$. On this basis we calculate

$$[A] = [Q]_A ((- h/6) - (- h/2)) + [Q]_B (h/6 - (- h/6)) + [Q]_A (h/2 - (h/6))$$
$$= (\tfrac{1}{3}) [2 [Q]_A h + [Q]_B h] \tag{4.11}$$

which is the same as we obtained previously, if $h_A = h_B$.

The matrix $[B] = 0$ for a symmetrical laminate.

The moment resultant per unit width can be derived in a similar fashion to give

$$[M]_L = \sum_{f=1}^{F} [M]_f = \sum_{f=1}^{F} \int_{z_{f-1}}^{z_f} [\sigma]_f z \, dz = [B][\varepsilon^\circ]_L + [D][\kappa]_L \tag{4.12}$$

where

$$[D] = \left(\frac{1}{3}\right) \sum_{f=1}^{F} [Q]_f (h_f^3 - h_{f-1}^3) \tag{4.13}$$

Using the same three-ply regular example as before

$$[D] = (\tfrac{1}{3}) [[Q]_a ((- h/6)^3 - (- h/2)^3) + [Q]_B ((h/6)^3 - (- h/6)^3)$$
$$+ [Q]_A ((h/2)^3 - (h/6)^3)]$$
$$= (\tfrac{1}{3}) [(\tfrac{13}{54}) [Q]_A + (\tfrac{1}{108}) [Q]_B] h^3 \tag{4.14}$$

The bending stiffnesses of the laminate are clearly the bending stiffnesses of each ply weighted by the second moment of the thickness about the midplane of the laminate. In this three-ply example the outer plies obviously contribute far more to the bending stiffnesses of the laminate than the middle ply. This conforms with the elementary principle of the I beam, where the flanges contribute most and the web serves largely to separate the webs and transfer shear between them.

We now need to be able to calculate the strain and stress profiles. For the special case of a symmetric laminate made from isotropic plies we have the stiffness relationships $[N]_L = [A][\varepsilon^\circ]_L$ and $[M]_L = [D][\kappa]_L$, which are not coupled, so by matrix inversion $[\varepsilon^\circ]_L = [a][N]_L$ and $[\kappa]_L = [d][M]_L$. The essential point is that the ABD matrix relates laminate inputs and outputs as a whole; $[\varepsilon°]$ and $[\kappa]$ refer to the midplane and are related to the force and moment resultants per unit width.

We are now therefore able to find the strain profiles $[\varepsilon(z)]$ using $[\varepsilon(z)] = [\varepsilon°]_L + z[\kappa]_L$. For the fth ply we can calculate the stress profiles $[\sigma(z)] = [Q]_f([\varepsilon°]_L + z[\kappa]_L)$. There will be a discontinuity in the stress profile at the interface between dissimilar materials because the modulus (E_f) or the reduced stiffnesses $(Q_{ij})_f$ will be different for different materials.

Problem 4.1: Distinguish when it is appropriate to use $[\varepsilon°]_L$ and $[\kappa]_L$, and when to use $[\varepsilon]_f$ and $[\kappa]_f$.

Problem 4.2: Show that for an isotropic linear elastic material the bending stiffness coefficient D_{66} reduces to $(\frac{1}{2})D(1 - v)$, which was derived in Chapter 2.

Problem 4.3: Evaluate the extensional and bending stiffness matrices for a regular symmetric laminate consisting of four isotropic plies, two each of materials a and b, arranged as $(a/b)_s$, giving a total thickness h. The properties of each material are given by $[Q_A]$ and $[Q_B]$.

4.4 BEHAVIOUR OF SYMMETRIC STEEL/ALUMINIUM LAMINATE SAL4S

The two constituent materials are isotropic and have the elastic constants at 20 °C given in Tables 4.1 and 4.2. The reduced stiffnesses for individual layers of steel and aluminium are also given in Table 4.1 and 4.2.

The symmetric laminate SAL4S consists of four plies each 0.25 mm thick, arranged as S/Al/Al/S. The A, B and D matrices for this 1 mm thick laminate in Table 4.3 show all the characteristics of a material isotropic in the plane and specially orthotropic through the thickness. All the plies are perfectly bonded together.

Table 4.1 Properties of steel

$E_{11} = 208$ GPa	[Q] in GPa		
$E_{22} = 208$ GPa	225.7	63.19	0
$G_{12} = 81.3$ GPa	63.19	225.7	0
$v_{12} = 0.28$	0	0	81.3

Table 4.2 Properties of aluminium

$E_{11} = 70$ GPa	[Q] in GPa		
$E_{22} = 70$ GPa	78.55	25.92	0
$G_{12} = 26$ GPa	25.92	78.55	0
$v_{12} = 0.33$	0	0	26

Table 4.3 Stiffnesses and compliances of regular symmetric steel/aluminium laminate $(S/Al)_s$. Each ply 0.25 mm

A (kN/mm) B (kN)					B (kN) D (kNmm)
152.1	44.56	0	0	0	0
44.56	152.1	0	0	0	0
0	0	53.65	0	0	0
0	0	0	17.28	4.878	0
0	0	0	4.878	17.28	0
0	0	0	0	0	6.199
a (mm/MN) h (/MN)					b (/MN) d (/MNmm)
7.19	−2.106	0	0	0	0
−2.106	7.19	0	0	0	0
0	0	18.64	0	0	0
0	0	0	62.9	−17.76	0
0	0	0	−17.76	62.9	0
0	0	0	0	0	161.3

4.4.1 Application of force resultant $N_1 = 20$ N/mm to SAL4S

The strain responses are uniform through the thickness of the symmetrical laminate: $\varepsilon_1^\circ = a_{11}N_1 = 0.01438\%$ and $\varepsilon_2^\circ = -0.004212\%$. The total applied force N_1 acting at the midplane causes a uniform extension, but because the tensile modulus of the aluminium is lower than that of the mild steel, the direct stress in the aluminium is lower, as shown in Figure 4.3.

It may be tempting to calculate the average stress in the laminate: in the reference direction this would give 'σ_{1L}' $= N_1/h_L = 20/1 = 20$ MPa, but this has no physical meaning; stresses in each ply should be used.

Note that if the plies were not bonded at their interfaces, but still subjected to a total force N_1, the stresses in the separately (but equally strained) plies would be given by

$$\sigma_A = N_1/(2h_A + (E_S/E_A)2h_S) = 20/0.5(1 + (208/70)) = 10.0715 \text{ MPa}$$
$$\sigma_S = (E_S/E_A)\sigma_A = 29.928 \text{ MPa}$$

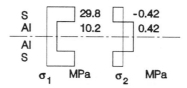

Figure 4.3 Stress profiles in SAL4S under $N_1 = 20$ N/mm.

and the strain in both plies would be

$$\varepsilon_1 = \varepsilon_A = \varepsilon_S = \sigma_A/E_a = 0.01438\%$$

The transverse stress in the bonded laminate, σ_2, has a quite different character from that in a single isotropic sheet, and can be visualized as follows.

If the plies were unbonded but strained equally in the longitudinal direction, the mild steel (with the smaller Poisson's ratio) will contract laterally less than the aluminium: $\varepsilon_{2S} = 0.28 \times 0.01438\% = 0.0040286\%$ and $\varepsilon_{2A} = 0.33 \times 0.01438 = 0.004748\%$, as shown in Figure 4.4(a). Each unbonded ply will carry no transverse stress.

But when bonded together, all the plies must contract the same (Figure 4.4(b), and hence the steel is compressed by the aluminium, and the aluminium is stretched by the steel. The strain in the bonded laminate may be calculated as

$$\varepsilon_2^\circ = a_{12}N_1 = -2.106\,\text{mm/MN} \times 20\,\text{N/mm} = -0.004212\%$$

Figure 4.3 not only shows the expected internal transverse stress profile σ_2, but it also shows that the net *force* in the transverse direction, F_2, is zero, as equilibrium requires.

The stress responses in the bonded laminate can be calculated from the midplane strains using values of [Q] for the material concerned. For example

$$\sigma_1(\text{Al}) = Q_{11}(\text{Al})\varepsilon_1 + Q_{12}(\text{Al})\varepsilon_2 = 10.2\,\text{MPa}$$
$$\sigma_2(\text{S}) = Q_{12}(\text{S})\varepsilon_1 + Q_{22}(\text{S})\varepsilon_2 = -0.42\,\text{MPa}$$

The difference in stresses and strains between unbonded and bonded plies due to one longitudinal strain is a consequence of the lack of restraint caused by lack of interaction between layers in the transverse direction, or the restraint caused by the interaction between layers.

Problem 4.4: A single force resultant $N_{12} = 20\,\text{N/mm}$ is applied to SAL4S. Calculate the stress profiles.

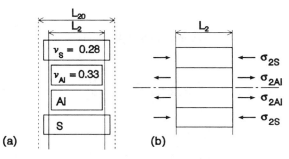

(a)   (b)

Figure 4.4 Response of $(\text{S/Al})_s$ under uniformly distributed load N_1: (a) separate layers; (b) bonded layers.

Problem 4.5: What are the apparent elastic constants E_1, E_2, v_{12} and G_{12} for SAL4S?

Problem 4.6: A single strain $\varepsilon_y = 0.1\%$ is applied to laminate SAL4S. Calculate the induced force resultants, the average stresses in the laminate, and the stress profiles.

Problem 4.7: A thin-walled pipe with sealed ends is 40 mm mean diameter, 950 mm long, and made from the laminate SAL4S. The pipe carries a torque of 100 Nm applied at each end and an internal pressure of 2.5 MPa. Ignoring end effects, calculate the increase in enclosed volume of the pipe, and the stress profiles through the wall.

4.4.2 Application of single moment $M_1 = 2\,N$ to SAL4S

The resulting curvatures are $\kappa_1 = 0.125/\text{m}$ and $\kappa_2 = -0.0355/\text{m}$, with anticlastic curvature as expected and indicated in Figure 4.5. The bending strains are linear and continuous through the laminate, and zero at the midplane. There is a change of modulus at the interface between the two materials, so there must be a discontinuity in stress at the interface, and that for the applied bending moment is as expected.

The bending stress and strain profiles are shown in Figure 4.6. Note that the value of transverse stress in the aluminium layer at the interface with the steel has the opposite sign to the value of the transverse strain, and the slope of the stress profile is opposite to the slope of the strain profile in the aluminium. The

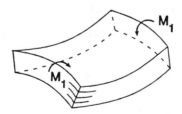

Figure 4.5 Anticlastic curvature in SAL4S under a single moment M_1.

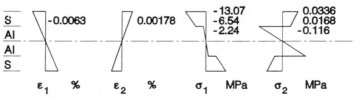

Figure 4.6 Stress profiles for SAL4S under $M_1 = 2\,\text{N}$.

net transverse force is zero, as required, because there is no applied transverse moment.

The stress profiles can be calculated in accordance with

$$\sigma_1 = Q_{11}z\kappa_1 + Q_{12}z\kappa_2$$
$$\sigma_2 = Q_{12}z\kappa_1 + Q_{22}z\kappa_2$$

using the values of [Q] for the material of the ply being considered.

The transverse stress distribution can be explained as follows. It is not possible to try and use the analogy of bending of a separate unbonded stack of plies – bending theory breaks down because of the shear between unbonded plies. But at the bonded interface between dissimilar materials there is the same tensile longitudinal strain in the aluminium and the steel. Using the fact that the Poisson's ratio for the two materials are different (as before), the aluminium will be in transverse tension at the interface, and the steel in compression. The linear variation of stress through each ply is a consequence of the bending.

For an initially-flat isotropic panel under bending moment M_1, $v = -\kappa_2/\kappa_1$ gives the same result as for the same sheet under a uniform force N_1: $v = -\varepsilon_2/\varepsilon_1$. For a laminated sheet consisting of different isotropic plies (or differently aligned similar plies), under bending the plies strain differently, and the weighting they have depends on their distance from the midplane. There is therefore no justification for calculating Poisson's ratio for laminates under bending loads.

Problem 4.8: What are the stresses, strains, and curvatures in laminate SAL4S under the application of single twisting moment $M_{12} = 2\,\text{N}$.

Problem 4.9: Calculate the response of SAL4S to application of $\kappa_y = 0.1/\text{m}$.

4.5 SYMMETRIC STEEL/FOAM LAMINATE MSF23S

This laminate consists of two skins of mild steel enclosing a core of 'stiff' isotropic low-density polymer foam. The elastic constants and reduced stiffnesses of the mild steel are given in Table 4.1, and comparable data for the foam in Table 4.4. The mild steel sheets are 0.2 mm thick and the two foam layers are 0.3 mm thick; the arrangement is (MS/F/F/MS), with a total thickness of 1 mm.

Table 4.4 Properties of foam

$E_{11} = 0.07\,\text{GPa}$	[Q] in GPa		
$E_{22} = 0.07\,\text{GPa}$	0.08777	0.0395	0
$G_{12} = 0.0241\,\text{GPa}$	0.0395	0.08777	0
$v_{12} = 0.45$	0	0	0.0241

The A, B and D matrices for this laminate are shown in Table 4.5. A comparison with data for the steel/aluminium sandwich panel in Table 4.3 shows that comparable values of elements of the ABD matrix are somewhat reduced, and that elements of [a] and [d] are increased. This is entirely consistent with the steel taking almost all of the load, because of the negligible modulus of the foam. A detailed examination of the behaviour of MSF23S supports this commonsense view: it becomes clear that the steel/foam sandwich is a special case. In summary, the foam contributes virtually nothing to the in-plane stiffnesses, but does contribute strongly to bending stiffnesses by spacing the steel from the midplane.

4.5.1 Application of single force resultant $N_1 = 20$ N/mm to MSF232S

The strains are $\varepsilon_1^\circ = a_{11}N_1 = 0.02403\%$, and $\varepsilon_2^\circ = a_{12}N_1 = -0.00673\%$. The stresses in the panel are shown in Figure 4.7 and confirm that the steel takes

$$\sigma_1 \quad \text{MPa} \qquad \sigma_2 \quad \text{kPa}$$

Figure 4.7 Stress profiles for MSF232S under $N_1 = 20$ N/mm.

Table 4.5 Stiffnesses and compliances of symmetric mild steel/foam laminate MSF23S. Steel ply 0.2 mm, foam 0.3 mm; total thickness 1 mm

A (kN/mm) B (kN)					B (kN) D (kNmm)
90.33	25.3	0	0	0	0
25.3	90.33	0	0	0	0
0	0	32.53	0	0	0
0	0	0	14.75	4.129	0
0	0	0	4.129	14.75	0
0	0	0	0	0	5.312

a (mm/MN) h (/MN)					b (/MN) d (/MNmm)
12.02	−3.365	0	0	0	0
−3.365	12.02	0	0	0	0
0	0	30.74	0	0	0
0	0	0	73.59	−20.6	0
0	0	0	−20.6	73.59	0
0	0	0	0	0	188.3

the longitudinal load. The stresses in the transverse direction are minute, though they have the same character as those for the symmetric steel/aluminium laminate. The small transverse stresses reflect the dominance of the steel.

From the definition of the extensional stiffness matrix in Equation (4.8), we have for the bonded sandwich

$$[A] = \Sigma[Q]_f(h_f - h_{f-1}) = 2[Q](S)h_S + [Q](F)h_F \qquad (4.16)$$

and the second term is negligible because $[Q](F) \ll [Q](S)$. Hence, ignoring the foam we can approximate to

$$A_{11} \approx 2Q_{11}(S)h_S = 2 \times 225.7\,\text{GPa} \times 2 \times 10^{-4}\,\text{m} = 90.28\,\text{kN/mm}$$

which is very close to the value for the real laminate, 90.33 kN/mm, in Table 4.5, which includes the foam.

As expected, we obtain almost exactly the same results for in-plane stress by taking just the steel alone. Based on 0.4 mm of steel, the stress is $\sigma_1 = N_1/h = 20/0.4 = 50\,\text{MPa}$. The strains are $\varepsilon_1 = \sigma_1/E = 50/208000 = 0.024038\%$, and $\varepsilon_2 = -v\varepsilon_1 = -0.0067308\%$.

4.5.2 Application of single moment resultant $M_{12} = 2\,\text{N}$ to MSF232S

The midplane twisting curvature is $\kappa_{12} = d_{66}M_{12} = 0.3765/\text{m}$. This results in shear stress and shear strain profiles as indicated in Figure 4.8: the strain profile is linear, and almost all the shear stress is taken by the steel, as expected.

Symmetric foam sandwich structures are widely used in practice because in bending they offer high bending stiffness for low weight and cost. Using fibre-reinforced polymer skins increases the advantage. For optimum advantages it is necessary to consider modes of failure such as debonding at the interface, and the risk of local buckling of the outer load-bearing plies under compression. These matters lie outside the scope of this introduction, but are covered in some detail in Allen (1969) (for isotropic skins) and in Vinson and Sierakowski (1986) (for orthotropic skins).

Problem 4.10: What are the stress and strain responses in laminate MSF23S under the application of single force resultant $N_{12} = 20\,\text{N/mm}$?

Problem 4.11: Calculate the curvatures and the stress and strain profiles

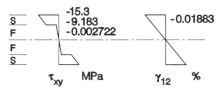

Figure 4.8 Response of MSF232S under $M_{12} = 2\,\text{N}$.

in laminate MSF23S under the application of single bending moment $M_1 = 2\,N$.

Problem 4.12: The laminate MSF232S is to be replaced by one with the same thickness of steel skins but a foam core 5 mm thick. Calculate the stiffnesses of the new laminate and compare them with MSF232S.

4.6 NON-SYMMETRIC LAMINATES BASED ON ISOTROPIC PLIES

For a non-symmetric laminate it is necessary and important to realize that there now *is* coupling between applied moments and resulting in-plane strains, or between applied in-plane force (resultants) and resulting curvatures.

An important everyday application involving isotropic materials is the bi-metallic strip which is made from two metals of differing thermal expansion coefficients, where a change of temperature induces bending. Subjecting just one strip along to a temperature change induces direct thermal strain, as described in Section 2.3.8. We shall analyse the behaviour of the bimetallic strip in detail in Chapter 7.

Another common example is either slice of bread in the simple sandwich consisting of two layers of bread buttered on one side and enclosing a filling. If the sandwich is left exposed to the atmosphere, the exposed unbuttered surface of the bread dries out, and contracts biaxially: this induces synclastic curvature and the curl develops in the upper slice, so that the filling becomes more visible than intended; the sandwich lacks appeal.

Bending-extension coupling is described by the [B] matrix calculated as

$$[B] = (\tfrac{1}{2})\Sigma[Q]_f (h_f^2 - h_{f-1}^2) \tag{4.17}$$

Each non-zero coefficient B_{ij} has its greatest value for a two-ply laminate where each ply (e.g. material a and b) has the same thickness and the ratio of moduli $(E_a/E_b$ or $G_a/G_b)$ is very large. If the top layer is a, then the values of B_{ij} are negative when $E_a > E_b$ and positive when $E_a < E_b$.

Where plies are isotropic, the [B] matrix has the form

$$[B] = \begin{pmatrix} B_{11} & B_{12} & 0 \\ B_{12} & B_{22} & 0 \\ 0 & 0 & B_{66} \end{pmatrix} \tag{4.18}$$

The complete relationship between midplane strains and curvatures and the in-plane forces and the bending moments per unit width involve the 6×6 stiffness matrix

$$\begin{pmatrix} N_x \\ N_y \\ N_{xy} \\ M_x \\ M_y \\ M_{xy} \end{pmatrix} = \begin{pmatrix} A_{11} & A_{12} & 0 & B_{11} & B_{12} & 0 \\ A_{12} & A_{22} & 0 & B_{12} & B_{22} & 0 \\ 0 & 0 & A_{66} & 0 & 0 & B_{66} \\ B_{11} & B_{12} & 0 & D_{11} & D_{12} & 0 \\ B_{12} & B_{22} & 0 & D_{12} & D_{22} & 0 \\ 0 & 0 & B_{66} & 0 & 0 & D_{66} \end{pmatrix} \begin{pmatrix} \varepsilon_x^\circ \\ \varepsilon_y^\circ \\ \gamma_{xy}^\circ \\ \kappa_x \\ \kappa_y \\ \kappa_{xy} \end{pmatrix} \tag{4.19}$$

Equation (4.19) shows the significance of the coefficients B_{ij}. If a single mid-plane strain ε_x° is present, then in addition to the in-plane forces N_x and N_y, there are also developed moment resultants $M_x = B_{11}\varepsilon_x^\circ$ and $M_y = B_{12}\varepsilon_x^\circ$. If a single curvature κ_y is applied (alone) then, as well as moments $M_x = D_{12}\kappa_y$ and $M_y = D_{22}\kappa_y$, we also encounter force resultants $N_x = B_{11}\kappa_y$ and $N_y = B_{12}\kappa_y$. In this laminate made from isotropic plies the coupling between shear modes is less complicated: applying an in-plane shear strain induces $N_{xy} = A_{66}\gamma_{xy}^\circ$ and a twisting moment $M_{xy} = B_{66}\gamma_{xy}^\circ$.

In short, the [B] matrix links applied direct strains with internal moments and applied bending curvatures with internal force resultants, and it links in-plane shear strain with a twisting moment. (It will be noticed that, for the laminates discussed in this chapter, $B_{16} = B_{26} = 0$, and hence there is no coupling between twisting deformations and direct stresses, or direct strains and twisting moments. Such coupling does occur in angle-ply laminates, and we shall discuss this in Chapter 5).

Now that [B] can have finite coefficients, the result of applying loads and moments can only be obtained by formally inverting the 6×6 matrix

$$\begin{pmatrix} A & B \\ B & D \end{pmatrix}^{-1} = \begin{pmatrix} a & b \\ h & d \end{pmatrix} \tag{4.20}$$

i.e.

$$\begin{pmatrix} \varepsilon^\circ \\ \kappa \end{pmatrix} = \begin{pmatrix} a & b \\ h & d \end{pmatrix} \begin{pmatrix} N \\ M \end{pmatrix} \tag{4.21}$$

Thus for a non-symmetric laminate based on isotropic plies, we can calculate the midplane strains and curvatures using the inverted form of the ABD matrix:

$$\begin{pmatrix} \varepsilon_x^\circ \\ \varepsilon_y^\circ \\ \gamma_{xy}^\circ \\ \kappa_x \\ \kappa_y \\ \kappa_{xy} \end{pmatrix} = \begin{pmatrix} a_{11} & a_{12} & 0 & b_{11} & b_{12} & 0 \\ a_{12} & a_{22} & 0 & b_{12} & b_{22} & 0 \\ 0 & 0 & a_{66} & 0 & 0 & b_{66} \\ h_{11} & h_{12} & 0 & d_{11} & d_{12} & 0 \\ h_{12} & h_{22} & 0 & d_{12} & d_{22} & 0 \\ 0 & 0 & h_{66} & 0 & 0 & d_{66} \end{pmatrix} \begin{pmatrix} N_x \\ N_y \\ N_{xy} \\ M_x \\ M_y \\ M_{xy} \end{pmatrix} \tag{4.22}$$

The use of the [h] sub-matrix may seem strange, and would seem to be unnecessary in this chapter. The sub-matrix [h] has the same set of coefficients as $[b]^T$, and for most simple laminates $[b]^T = [b]$. Only where [B] is fully populated is [h] different from [b]: this does not arise for laminates made from a range of isotropic materials, but we shall see (in Section 5.6.2) that the non-symmetric quasi-isotropic laminate is an example giving a fully populated [B] matrix.

It should be carefully noted (again) that any of the coefficients in the inverse 6×6 ABD matrix *cannot* be obtained simply by inverting the apparently appropriate 3×3 matrix such as [A] alone, expect in the special case where [B] = 0, which we mentioned in Section 3.1.3.

We can see, from the [abhd] matrix, the role of lack of symmetry in a

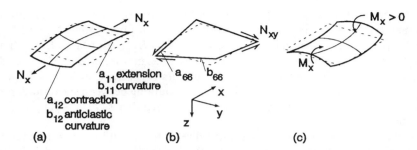

Figure 4.9 Deformation responses of SAL2NS to: (a) direct load; (b) in-plane shear load; and (c) bending moment.

laminate based on isotropic plies. Applying direct in-plane force resultants induces direct mid-plane strains and bending curvatures, but does not induce shear stress or twisting curvature. Applying a shear force resultant induces both shear stress and twisting curvature, but not direct strain or simple bending. A graphical representation of deformations for several types of applied force or moment resultants is presented and annotated in Figure 4.9.

Problem 4.13: Calculate the value of B_{11} for a laminate of one flat 0.5 mm thick sheet of steel bonded to a flat 0.5 mm sheet of aluminium. The physical properties are given in Tables 4.1 and 4.2.

Problem 4.14: Show that non-zero coefficients of [B] are greatest for a two layer non-symmetric laminate.

Problem 4.15: Show the condition when [B] will be negative.

4.7 NON-SYMMETRIC STEEL/ALUMINIUM LAMINATE SAL2NS

This laminate consists of one sheet of mild steel and one of aluminium bonded together. The reduced stiffnesses for individual layers of steel and aluminium are given in Table 4.1 and 4.2. Both sheets are 0.5 mm thick, so the total thickness of the laminate is the same as for the previous problem. The behaviour patterns are rather different because of the lack of symmetry.

Inspection of the A and D matrices in Table 4.6 reveals no surprises. The A matrix is identical to that for SAL4S because the thicknesses of steel and aluminium are the same. The D matrix is different from SAL4S because of the different special arrangement in SAL2NS, but the population is the same. But the B matrix is no longer zero because there is no symmetry: the non-zero values of B_{ij} lead to coupling between direct force resultants and curvatures and are negative because $Q_S > Q_{A1}$.

Table 4.6 Stiffnesses and compliances of regular non-symmetric steel/aluminium laminate SAL2NS – each ply 0.5 mm

A (kN/mm) B (kN)					B (kN) D (kNmm)
152.1	44.56	0	−18.39	−4.659	0
44.56	152.1	0	−4.659	−18.39	0
0	0	53.65	0	0	−6.912
−18.39	−4.659	0	12.68	3.713	0
−4.659	−18.39	0	3.713	12.68	0
0	0	−6.912	0	0	4.471

a (mm/MN) h (/MN)					b (/MN) d (/MNmm)
8.823	−2.735	0	13.13	−4.573	0
−2.735	8.823	0	−4.573	13.13	0
0	0	23.28	0	0	35.99
13.13	−4.573	0	105.9	−32.82	0
−4.573	13.13	0	−32.82	105.9	0
0	0	35.99	0	0	279.3

4.7.1 Application of single force resultant $N_1 = 20\,N/mm$ to SAL2NS

The strain and curvature responses are given by $\varepsilon_1^\circ = a_{11}N_1 = 0.01765\%$, $\varepsilon_2^\circ = a_{12}N_1 = -0.00547\%$, $\kappa_1 = b_{11}N_1 = 0.2627/m$ and $\kappa_2 = -0.09145/m$. This deformation of the midplane is shown diagrammatically in Figure 4.10.

If the two isotropic sheets had the same elastic properties, there would be no curvature under an in-plane load. For dissimilar isotropic materials it is easy to find an intuitive reason for the bending behaviour by taking the extreme case where the modulus of one ply is very small compared with the other. If an in-plane load is now applied and acting at the midplane of the two-ply composite, the effect is almost as if the load were only applied to the stiff ply at the interface surface (Figure 4.11). This is equivalent to the same direct stress applied in the midplane of the stiff ply, together with a bending moment M, and it is the latter which is responsible for the curvature. the anticlastic curvature follows from similar reasoning.

The strain and stress profiles reflect the combination of strains and curvatures, as shown in Figure 4.12. the strains vary linearly through the thickness,

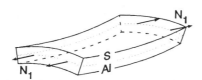

Figure 4.10 Bending curvatures in SAL2NS under N_1.

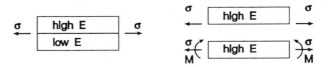

Figure 4.11 Development of moment when non-symmetric laminates is loaded at its midplane.

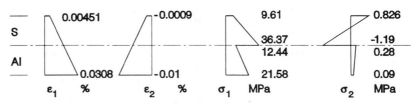

Figure 4.12 Responses of SAL2NS to $N_1 = 20$ N/mm.

and the neutral plane is outside the sheet. There is a discontinuity in the stress profile. For σ_1 this is simply the result of the difference in modulus at the midplane interface: for a strain of $\varepsilon_1 = 0.0177\%$, the more flexible aluminium is stressed less than the steel.

The strain profile ε_2 transverse to the applied load can be visualized as follows. At the bonded interface, the strain in the loaded direction is 0.0177%. If the two layers were unbonded but carrying the same strain, the steel would not contract laterally as much as the aluminium because of the different Poisson's ratios. But because the layers are bonded, the steel is compressed by the aluminium, and the aluminium is extended by the steel. The sloping stress profile from the interface to the outer surface is caused by the bending.

The stress profiles can be calculated in detail using appropriate values of Q_{ij} for the material and ply, from

$$\sigma_1 = Q_{11}(\varepsilon_1^\circ + z\kappa_1) + Q_{12}(\varepsilon_2^\circ + z\kappa_2)$$
$$\sigma_2 = Q_{12}(\varepsilon_1^\circ + z\kappa_1) + Q_{22}(\varepsilon_2^\circ + z\kappa_2) \tag{4.23}$$

4.7.2 Application of shear force resultant $N_{12} = 20$ N/mm to SAL2NS

The response is the simple one: $\gamma_{12} = a_{66}N_{12} = 0.04655\%$ together with $\kappa_{12} = b_{66}N_{12} = 0.7198$/m. This is shown schematically in Figure 4.13.

4.7.3 Application of twisting moment $M_{12} = 2$ N to SAL2NS

The response is the midplane shear strain $\gamma_{12}^\circ = b_{66}M_{12} = 0.007198\%$ and $\kappa_{12} = d_{66}M_{12} = 0.5586$/m with a character qualitatively similar to that in Figure 4.13.

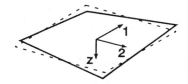

Figure 4.13 Response of SAL2NS to $N_{12} = 20\,\text{N/mm}$.

Problem 4.16: What are the deformation responses and stress profiles in laminate SAL2NS under the application of single bending moment resultant $M_1 = 2N$?

Problem 4.17: How does [B] change in a regular balanced non-symmetric laminate consisting of alternate plies a and b as the number of plies increases, assuming the total thickness of the laminate is kept constant?

Problem 4.18: Determine the response of SAL2NS to strain $\varepsilon_y^\circ = 0.1\%$.

4.8 NON-SYMMETRIC STEEL/ALUMINIUM LAMINATE SAL4NS

Laminate SAL4NS has the same quantity of each metal as SAL2NS but arranged as (S/Al/S/Al), with each ply now only 0.25 mm thick.

The A and D matrices (Table 4.7) are the same for both SAL2NS and

Table 4.7 Stiffnesses and compliances of regular non-symmetric steel/aluminium laminate S/Al/S/Al. Each ply 0.25 mm

A(kN/mm) B(kN)			B(kN) D(kNmm)		
152.1	44.56	0	−9.196	−2.329	0
44.56	152.1	0	−2.329	−9.196	0
0	0	53.65	0	0	−3.456
−9.196	−2.329	0	12.68	3.713	0
−2.329	−9.196	0	3.713	12.68	0
0	0	−3.456	0	0	4.471
a(mm/MN) h(/MN)			b(/MN) d(/MNmm)		
7.539	−2.236	0	5.608	−1.879	0
−2.236	7.539	0	−1.879	5.608	0
0	0	19.62	0	0	15.16
5.608	−1.879	0	90.46	−26.83	0
−1.879	5.608	0	−26.83	90.46	0
0	0	15.16	0	0	235.4

SAL4NS, but the components of the B matrix are smaller, thus reducing the effect of non-symmetry compared with SAL2NS.

4.8.1 Application of single force resultant $N_1 = 20\,\text{N/mm}$ to SAL4NS

The deformations are caused by $\varepsilon_1^\circ = 0.01508\%$, $\varepsilon_2^\circ = -0.00447\%$, $\kappa_1 = 0.11/\text{m}$ and $\kappa_2 = -0.0376/\text{m}$, and the curvatures are much reduced because values of [B] are much reduced. The deformed sheet has a similar shape to that of Figure 4.10, but less pronounced. The stress and strain profiles in SAL4NS look more complicated (Figure 4.14) because there are now four plies to consider, but the principles are exactly the same as those for Figure 4.12.

It will be noted that the direct strains ε_1 and ε_2 in response to $N_1 = 20\,\text{N/mm}$ are different for the three laminates SAL4S, SAL2NS and SAL4NS, even thought the total thickness of steel and of the aluminium are identical and the components of the extensional stiffness matrix A are the same for each laminate. When the ABD matrix is inverted, the terms of the [B] and [D] matrix contribute to the inverted forms [a], [b], [h] and [d]. It is of course the values of [a] and [b] which are used to calculate the direct strain responses to the applied force resultants N_1 and N_2. i.e.:

$$\begin{pmatrix} \varepsilon^\circ \\ \kappa \end{pmatrix} = \begin{pmatrix} a & b \\ h & d \end{pmatrix}\begin{pmatrix} N \\ M \end{pmatrix}$$

4.8.2 Application of single moment resultant $M_1 = 2\,\text{N}$ to SAL4NS

The deformation responses are $\varepsilon_1^\circ = 0.001122\%$, $\varepsilon_2^\circ = -0.000376\%$, $\kappa_1 = 0.181/\text{m}$ $\kappa_2 = -0.0537/\text{m}$, and these have the same general character as

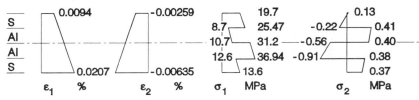

Figure 4.14 Stress and strain profiles for SAL4NS under $N_1 = 20\,\text{N/mm}$.

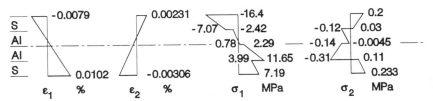

Figure 4.15 Stress and strain profiles for SAL4NS under $M_1 = 2\,\text{N}$.

those for the two-ply unsymmetric laminate having the same amounts of each material. The stress and strain profiles are illustrated in Figure 4.15.

Problem 4.19: Why is [D] the same for laminates SAL2NS and SAL4NS?

4.9 COMPARISONS

It is interesting to compare the profiles for the three laminates SAL4S, SAL2NS and SAL4NS, under the same force resultant N_1.

For the symmetric laminate SAL4S, the strain is uniform throughout the thickness at 0.0426%, and there is no curvature; the maximum difference in stress at the interface between the steel and the aluminium is 55.2 MPa.

There is a linear variation in strain in SAL2NS, with a midplane strain of 0.0485%, and the difference in stress at the interface is now 66.2 MPa.

For SAL2NS the midplane curvature is 0.505/m. For the four-ply non-symmetric laminate SAL4NS, the curvature has reduced to 0.22/m as expected, but the maximum difference in stress between the steel and the aluminium is now 67.6 MPa.

For an eight-ply regular non-symmetric laminate of total thickness 1 mm, the midplane curvature reduces to 0.106/m, but the maximum difference in stress between different materials at an interface is still high at 63.2 MPa.

The reader is invited to explore what happens with further increase in the number of plies for the same total laminate thickness, in a non-symmetric bonded stack.

FURTHER READING

Allen, H.S. (1969) *Analysis and design of Structural Sandwich Panels*, Pergamon.
Vinson, J.R. and Sierakowski, R.L. (1986) *The behaviour of structures composed of composite materials*, Nijhoff.

Laminates based on unidirectional plies

5.1 INTRODUCTION

We noticed in Chapter 3 that one major limitation of unidirectional plies is the difference in stiffnesses in the two principal directions of up to one or two orders of magnitude. In Chapter 4 we looked at the stiffness properties of laminates based on stacks of different isotropic plies. We are now in the position where we can explore (and confirm) the usefulness of laminates based on unidirectional plies for solving many defined stiffness problems.

One simple approach to reducing the directionality of unidirectional plies is to use the crossply arrangement. Although the simplest such crossply consists of two plies laid up at $0°$ and $90°$ to a reference direction (Figure 5.1(a)), it is obviously not symmetric, and its behaviour is rather complicated. We have already discussed some aspects of non-symmetric laminates in Sections 4.5 to 4.7, and we shall explore further aspects in this chapter. More usefully representative of usable crossply laminates is the symmetric form $(0°/90°/90°/0°)$ (Figure 5.1(b)), for which the variation of modulus with angle of rotation to the reference direction is shown in Figure 5.1(c). Balanced bidirectional woven fabrics used to reinforce polymer matrices can also be represented approximately by the symmetric crossply $(0°/90°/90°/0°)$, even if only one layer of cloth is used. This approach of using a bonded stack of (quasi) unidirectional plies, admittedly in a multisandwich form, is encountered in plywood, polyethylene bags, and (in effect) in most of the body of a cylindrical plastics carbonated-drinks bottle, as well as in papyrus, the precursor to paper.

Another simple approach to reducing the directionality of unidirectional plies is to make up an angleply laminate. Again, although the simplest arrangement is to stack two plies at $+\theta$ and $-\theta$ to the reference direction (Figure 5.2(a)), the lack of symmetry raises usually unwelcome aspects of performance. Much more useful is the symmetric angleply consisting of four plies laid up as $(\theta/-\theta)_s$, as shown in Figure 5.2(b). Flat (and nearly flat) angle

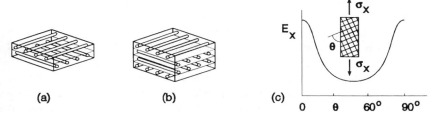

Figure 5.1 Two crossply arrangements and variation of longitudinal modulus of (b) with direction of load to the principal direction.

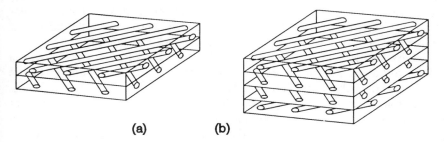

Figure 5.2 Two angleply arrangements.

ply laminates can be readily laid up by hand with some care, and thin-walled tubular forms can be made with varying degrees of approximation by filament winding or braiding. Woven cloth laminates stressed 'on the bias' (i.e. at 45° to the principal directions) also behave as angleply laminates.

This chapter explores a wide range of simple laminate constructions, with the emphasis on understanding and illustration of general and special features rather than exhaustive coverage of all possibilities.

The chapter provides examples of a wide range of patterns of behaviour relating not only to applied loads and moments, but also to applied deformations and curvatures. The stress and strain profiles are also given, where non-trivial, not only in global co-ordinates but also in the principal directions (if different). These stress and strain profiles (particularly in ply principal directions) are given some attention because of their importance in strength calculations, which provides the subject matter of Chapter 6.

5.1.1 A shorthand laminate stacking code

Even with simple laminate constructions, there is merit in achieving a concise description of the stacking sequence. We make many laminates from a stack of identical plies aligned at different angles to the reference direction. Such a laminate is 'regular'. We can describe such a stack by simple listing the ply orientations from the top down. Thus the notation $[90°/0°/90°]_T$ uniquely describes a three ply laminate. The subscript T refers to the total number of layers in the arrangement . The angle states the orientation of the principal direction in the ply to the reference direction in the laminate.

Any laminate in which the ply stacking sequence above the mid-plane is a mirror image of the stacking sequence below the midplane is described as a symmetric laminate. The laminate $[0°/90°/90°/0°]_T$ (also written as $[0°/90_2°/0°]_T$) can be abbreviated as $(0°/90°)_s$. The subscript s describes that the stacking sequence is symmetric. The subscript 2 in the total stacking sequence notation shows that the 90° ply is repeated. Angle ply laminates can be described in a similar way, for example $(+\theta/-\theta/-\theta/+\theta)_T = (+\theta/-\theta)_s$. The reader should be aware that there is a wide range of shorthand conventions for describing laminate stacking sequences, showing subtle differences and every opportunity for ambiguity: please check the definitions.

Many engineering structures deform in the same sorts of ways as woven cloth, although the movements in composites are usually minute. Before looking at the behaviour of crossply laminates, it is therefore helpful to discuss in Section 5.2 some simple aspects of fabric behaviour under in-plane tensile or shear loads. Similarly, before looking at the behaviour of angleply laminates, it is helpful to discuss in Section 5.4 the behaviour of fabrics loaded off-axis.

Budding and practising engineers and technologists can learn a great deal about engineering by studying the behaviour of cloth in garments – shear, extension, and buckling are readily evident. Observation of these phenomena is fascinating even if the underlying physics and mechanics are left unpondered.

5.2. WOVEN, FLEXIBLE AND KNITTED FABRICS

Woven, flexible and knitted fabrics are widely used to reinforce polymer matrices, because they are easy and convenient to handle. The volume fractions of fibres are not high, because of the interlacing of fibres, but the composite products have adequate properties for a range of markets. We shall discuss here aspects of the stiffnesses of each of these fabrics when stressed in their principal directions. We shall discuss off-axis behaviour in Section 5.4.

The basic building block is the fibre, which by definition is long and thin. Fibres can be continuous ('filaments') or short ('staples'). Naturally occurring filaments may be represented by silk or the web of a spider; synthetic filaments include polyamide. Natural staple fibres include cotton and wool, and it is practicable to make synthetic staple fibres by cutting glass or carbon filaments.

Yarns consist of bundles of filaments or staples. Filaments or bundles can be held together by twisting, which is quite normal in many used for garments. But in composites many yarns are used untwisted (rovings), and unless a light binder is used, the untwisted rovings will spread when under diametral loading. Staple fibres must be twisted together to make a yarn: friction between individual staples holds them together under axial load.

5.2.1 Woven cloth stressed in principal directions

Conventional woven fabrics consist of two sets of yarns interlaced at right angles to each other. Warp yarns run along the length and weft yarns run across the width of the fabric. Fabrics may be woven in a variety of patterns – e.g. plain weave, twill weave, and satin weave. These basic weaves are shown schematically in Figure 5.3. From these different weaves a range of mechanical properties, surface finishes and drapabilities can be obtained.

The simplest interlacing produces a plain weave. Closely spaced yarns prevent sideways movement of yarns in the fabric, so it is a tight ungiving structure. The stiffness of the fabric in the warp and weft directions depends on the tightness of the weave and the properties of the yarn. A square of wire gauze is quite stiff because the wire is stiff. A fine cotton plain weave is flexible and shows cloth-like behaviour because of the low stiffness of the cotton yarn.

From a composites viewpoint, the openness of the wire gauze provides many large-scale interstices where resin (if applied) would form unreinforced pockets. On the other hand, the closest interlacing makes it difficult to impregnate the fabric and fully wet out the fibres in the yarn with any but the most runny of polymer compositions. Compared with the use of twisted yarns, a plain weave based on untwisted yarns (or rovings) will give a flatter fabric which has a denser appearance, because yarns will spread out in the plane under the action of forces between touching interlacing yarns through the thickness of the fabric.

For use in composites, one major disadvantage of conventional woven fabrics in the degree of crimp imparted to the fibres, i.e. the extent of waviness of yarn out-of-plane and its period. This leads to a reduction in properties of the composite laminate caused by misalignment of fibres relative to the direction of the applied load. One measure of crimp, c, relates the length of straight

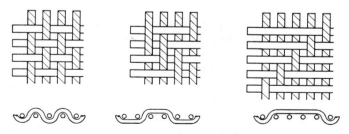

Figure 5.3 Plain, twill and satin weave.

yarn L_y to the length of fabric made from that yarn, L_{fab}: $c = (L_y - L_{fab})/L_{fab}$. The values of crimp may be different in the warp and weft directions, depending on the weave pattern. For plain weave, crimp is typically 8 to 12%.

The closest spacing of yarns in a parallel array one yarn thick is touching. The plain weave ('under one over one') requires that every alternate yarn is omitted to make space for interlacing; this gives a stable fabric. If the weft yarn interlaces over two under two, then more warp yarns can be used: each pair of yarns touch so the closeness increases, but yarns are only held in place in pairs so the fabric is slightly less stable. As the interlacings become less frequent, the importance of closeness of yarns becomes dominated by the lack of holding them in place. This leads to a progressively looser fabric which drapes well not least because of its lack of stability. Satin weave (e.g. one over eight) is an extremely flexible (but damage-prone) fabric.

Under tensile loads applied in the direction of the weave, plain weave cloth is not very extensible, as axially most fibres are quite stiff – loose weaves extend more than tight weaves because of some straightening of fibres; satin weaves are inherently stiffer and stronger in the warp and weft directions because most of the fibres are straight anyway.

The deformability of bidirectional woven cloth under shear forces applied in the plane in the warp and weft directions depends on the friction between touching yarns, and the bending stiffness of the yarns. If the friction is very small and the yarn is stiff in bending, then modest shear forces will cause rotation at the nodes and permanent shear deformation in the fabric. If the friction at the nodes is high, and the yarns are flexible in bending, then the nodes stay fixed, and the shear in the fabric is solely caused by elastic bending of the fibres. There is a spectrum of combinations of behaviour between these two extremes.

Shear deformation is an important basic property which influences whether a fabric will drape readily over a body having convex curvatures. If you wet out the woven cloth with resin which wets and sticks to the fibres and then solidifies, you prevent movement between fibres so the result is a much stiffer structure, which is much more difficult to drape over complicated shapes compared with the original cloth. Thus it is customary to drape to shape, then fix the shape by cooling a thermoplastic resin or crosslinking a crosslinkable one.

To greatly reduce the misalignment effects of crimp without suffering the instability of satin or twill weaves, fabrics destined for reinforcement of polymeric resins can now be woven which almost eliminate crimp. Parallel arrays of the main reinforcing fibres are laid down in one or more reference directions, and a fine binder yarn is woven to hold the main reinforcing fibres in place. The fine binder yarns are, of course, crimped, but they have minimal influence on the in-plane properties of the cloth and composite based on it, and they actually help to confer some shear strength and damage tolerance.

It would be unreasonable to suggest too close a correspondence between woven (or knitted) cloth mechanics and the behaviour of (even symmetric) cross-ply laminates based on unidirectional plies. There are many variables in cloth which we have not discussed in detail. There are features of structure at every scale of dimension which would need to be taken into account. At the fibre level, fibres are very slender and buckle readily under axial compression, the surface finish dictates the friction and hence load transfer between parallel or interlaced fibres.

Bundles of fibres are often twisted before weaving or knitting – the twist holds fibres together and suppresses the tendency to buckle, but the twist complicates the mechanics, because pulling causes extension and twist. The behaviour of the fabric depends on the weave, as we have already mentioned, and of course on the direction of stresses relative to the warp and weft direction. In addition, cloth is thin and so it bends readily to large deformations, and buckles all-too-readily under in-plane compression or in-plane shear. The large deformations especially in bending lie well outside the scope of the engineer's theory of bending (which assumes deflections are small), and we would have to use non-linear theories and numerical techniques which are outside the scope of this book. The ability of cloth to shear readily is an attractive feature in garment behaviour, but it introduces complications which are greatly suppressed in the behaviour of conventional continuous materials, whether isotropic or anisotropic. For a more detailed review of fabric mechanics, the interested reader may consult accessible texts such as Hearle *et al.* (1969), the NATO 1979 conference, and Postle *et al.* (1988). As with many topics in this book, there is also a rich research literature.

5.2.2 Flexible fabrics loaded in their principal directions

Sometimes we want a really flexible structure. One way is to weave a fabric from flexible fibres such as the elastomeric polyurethanes. Another way of achieving a flexible structure relies on the principle that it is easier to deform an elastic rod by bending or twisting it rather than trying to stretch it along its length. The starting point, for woven fabrics based on stiff polymers, is to curl or twist quite stiff fibres: on loading in tension, these rods then deform readily by uncurling or untwisting. The rod does not itself elongate much but the structure does. This is the basis of bulked or textured fibres such as Crimplene. We can use a weave based on these bulked fibres to make a cloth which is readily extensible (and the spring action ensures that this deformation is recoverable on removing the load).

Problem 5.1: An element of coiled yarn is modelled as an open coil spring having a mean diameter which is ten times the diameter of the yarn. There are five turns, with a helix angle of 30°. Compare the axial stiffness (load per unit extension) of the coiled yarn with that of the same length of straight yarn.

5.2.3 Knitted fabrics

We can also achieve flexibility by knitting a structure from stiff fibres. Knitted fabric elongates under tension by straightening the loop yarns and then pulling through in the direction of stretch, as shown in Figure 5.4. You get a lot of extension (and much lateral contraction too) when you pull on the top of a sock. Snug fitting knitted sweaters can look terrific and they owe much to Poisson's ratio – we shall discuss woven fabric and laminates stressed 'on the bias' in Section 5.4.5: the result is the same but the mechanism is different.

> **Problem 5.2(a):** Measure the Poisson's ratio of an old knitted garment such as a sock, a stocking, or a pullover.

There is no reason why we should not knit with a flexible yarn: at the simplest level of concept, we can model these fabrics quite well – but by no means perfectly – as a crossply structure. It will be appreciated that the flexibility of the yarn can derive from two sources. Either the yarn is made from a stiff polymer and is based on thin bulked fibres or yarns; or the yarn is made from a flexible polymer such as an elastomeric polyurethane; or a mixture of both polymer yarn types.

Stretch garments are deliberately made undersize. When the wearer puts one on, the garment stretches, and the tensions in the fibres reach equilibrium. The tension depends on the 'power' of the fabric, and in our crossply model this power comes from the stiffness of the rods, and their diameter and spacing. The tensioned fabric compresses the body underneath. The pressure on the body depends on the local curvature according to the expression

internal pressure × radius = fabric tension (per unit width)

This expression was derived for stress in Section 2.3.3, and is here recast using force resultant per unit width from Section 3.1.3.

Consider as a potential pressure vessel a pair of ordinary ladies' fashion tights weighing about 40 grams and typically described as 20 denier. The leg of this pressure vessel is knitted as a tube with (typically) 401 stitches round the circumference and about 1000 rows from toe to waistband. The waist end – a

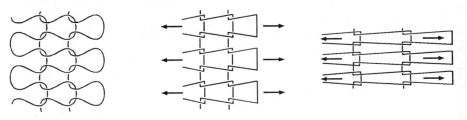

Figure 5.4 Mechanism for extension of a plain-knit fabric under a unidirectional uniform load.

heavier knit – is split with a hot wire, and pairs of split tubes are stitched together to make the upper part of the garment. Each leg takes typically about half a minute to knit, so the knitting process is rapid. The garment as sold is grossly undersize but readily stretchable with very little tension, to closely fit the leg. On the leg the pressure is at most about 1 mm of mercury, about 0.25 kPa, more so at the ankle than at the thigh (why do you think this is so?).

In contrast, power-stretch knitted support hose are worn for comfort or to prevent excessive distension of varicose veins in the leg. The garments are tailored during knitting so that the tension in the thigh region is comparable with that near the ankle even though there are the same numbers of stitches round the circumference. So if the tension is similar all over the leg (and if it is not, then the tensions adjust themselves during leg movement until they are similar), the product of body pressure and leg radius is constant. This means that the greatest pressure is developed at the least radius, namely the ankle – and this is just where the hydrostatic head of blood is a maximum in a standing person. As a rough guide, support tights can exert a pressure of up to 16 mm of mercury, 2 kPa, at the ankle, but only about a quarter of this value at the upper thigh. The body does not feel particularly uncomfortable under these sorts of pressures.

> **Problem 5.2(b):** What is the diameter and total length of filament in a pair of 20 denier tights having a mass of 40 g? The yarn is twisted from five separate filaments. The density of polyester may be taken as 1350 kg/m³. One denier is defined as the mass in grams of a continuous filament 9 kilometres long.

The prospect for using conventional knitted structures based on relatively inextensible glass, carbon, or aromatic polyamide fibres to reinforce polymers in load-bearing engineering structures would seem to be unrewarding. In such a knitted structure the spacing of yarns is large (even though the yarns themselves consist of dense bundles of fibres, twisted or untwisted), so the overall volume fraction of fibres is modest, and there are resin-rich pockets of unreinforced resin in the interstices of the fabric. The looped structure means that virtually none of the length of the fibres is aligned in the direction of applied tension, so that the stiffness of the fibre material is greatly degraded by the off-axis loading. Putting it another way, a knitted fabric stretches readily in any direction in the plane.

Yet there is a niche in fibre reinforced polymer technology which exploits knitted fabrics because of the ease with which the knitted fabric can be placed over a preform of complicated convex shape, which it would be difficult to cover using woven fabrics or other methods of fibre placement. Thus the basic idea of a knitted sock without the sewn toe lends itself to the manufacture of fibre reinforcement for pipe bends. It is not too difficult to knit a sweater or pullover in one piece. Now imagine knitting a sweater where the neck is completely sealed. Not much use as a sweater, but with some imagination this shape is what is wanted in order to reinforce a Tee-piece connector for a pressure pipeline system. Extending this basic one-piece sweater to include a hood but no aperture for the face, and adjusting proportions appropriately, leads to the imaginative knitter rolling up a preform for a small sailplane into a suitcase... but however appealing, such a resin-impregnated plain-knit structure on a suitable lightweight former could not be stiff enough in service, and must remain a flight of fancy.

Recent developments in technical knitting technology use parallel arrays of straight reinforcing fibres in one or more directions, which are then sown together by a knitting process using fine yarns. This opens up the possibility of good mechanical properties in-the-plane (because of the straight reinforcing fibres and their alignment in the directions of applied loads), and the stitching through the thickness provides improved shear strength and damage tolerance.

5.3 CROSSPLY LAMINATES

Crossply laminates can be either symmetric or non-symmetric, can be balanced or unbalanced, and can be regular or non-regular. A wide range of structures encountered in engineering or other technologies can be represented usefully by the concept and formal properties of a crossply laminate. The reader will therefore encounter many familiar examples from his or her experience to supplement this formal discussion.

Crossply laminates are usually based on a stack of unidirectional plies arranged at $0°$ and $90°$ to a chosen reference direction. Analytical discussion usually concentrates on the special circumstance where the principal directions in each ply correspond to the 'principal directions' in the laminate, as followed in Section 5.3.2. Other reference directions in the laminate can be chosen to suit the design problem in hand, as we shall explore in Section 5.4.5.

The calculation of the [ABD] matrix, or its inverse [abhd], follows the general principles outlined in Chapter 4. There are some simplifications which apply to all crossply laminates, and then some special features which depend on the arrangement of plies. To keep this book to a reasonable length, we shall just discuss the behaviour of regular crossply laminates, but the principles can be readily extended to embrace non-regular laminates.

For a unidirectional ply stressed in-plane in its principal directions $Q_{16} = Q_{26} = 0$. It is not therefore surprising that no crossply laminates stressed in the principal directions exhibit shear-extension coupling or bending-twisting coupling: $A_{16} = A_{26} = 0 = D_{16} = D_{26}$, and similarly for the inverse coefficients.

Balanced regular symmetric crossply laminates will have identical stiffnesses $A_{11} = A_{22}$, but D_{11} and D_{22} have different values because the weighting of each ply under bending depends on its distance from the midplane.

There are many examples of this kind of laminate (or structure which can be reasonably so represented) in a wide range of technologies. Most of the non-laminate structures usually involve only in-plane direct stresses, and a brief look at them, in the following section, provides useful insights on which to build a grasp of the more complex possibilities of deliberately designed laminates.

5.3.1 Examples of crossplies from other technology

Consider the film from which a polyethylene bag is made. This starts life as a thin-walled tube of molten plastic having the consistency of treacle or chewing gum. If the tube is pulled upwards faster than it emerges from the extrusion die, it is stretched along its length. So some of the long-chain molecules in the polymer melt line up along the length of the tube. If the tube is pressurized internally the diameter increases, so some other molecules will line up round the circumference. A few molecules get rather confused and do not line up in either direction – we will ignore them here. But there is a substantial commitment to lining up along the tube or around the circumference. The stretched tube – a pressure vessel – is thin and cools quickly to a solid with the molecular alignment frozen-in in this rather unnatural way: this makes the tube much stiffer and stronger than it would otherwise be. Usually the tube is stiffer along the length than round the circumference – this is advantageous in a carrier bag. The lower end of the bag is heat welded together to provide the container form, and a handle is provided to help carry the contents.

But consider stretching a polyethylene bag at 45° to either the longitudinal or transverse edge: this is the equivalent of stretching cloth in the bias direction. Compare the uniaxial stiffness response in the 45° direction with that in the longitudinal direction by cutting out parallel-sided strips at 0° and 45° to the length of the bag.

We have already said that the frozen-in alignment of molecules in tubular film is unnatural – the molecules much prefer to be all coiled up, and tangled up for good measure. So if you warm aligned molecules too much, they become restless and curl up again. If you put plastics film on a bonfire (which is *not* recommended), the film shrivels up as it gets warm due to the curling up process. If the film is wrapped round a few boxes and then warmed up, it shrinks and holds the boxes tightly together – this is the basis of shrink-wrapping packages behind the commercial and retail scene.

Many packaging films are laminates consisting of, say, a polymer layer chosen for its strength, toughness and durability superbly bonded in the melt state to another layer which is more impermeable to the gases which would otherwise diffuse into a package and cause the contents to spoil, e.g. the fat in biscuits and pastry might oxidize or go rancid, or moisture may make crisp confectionery become soggy, sticky or otherwise unappetizing.

Let us now examine a sample of plastics netting of the kind often containing oranges or sprouts in the supermarket. How is this netting made? It is tempting to imagine lots of people with soldering irons putting blobs of plastic on the criss-crossed strands of polyethylene filaments. But the mesh is actually extruded in one operation.

Making the strands themselves is easy – just extrude through strand-shaped holes in a die – but this only gives many parallel unconnected strands. If you rotate a die with the holes arranged in a circle about the centre of rotation you produce an advancing helix of unconnected strands Figure 5.5(a). If you drill another slightly bigger circle of holes on the same centre and counter-rotate both sets of holes, you produce opposing helices which are still not joined up. But if you have one set of holes with the die split along the pitch line of the holes, and counter-rotate both parts of the die, you can still make two opposing helices (b), and where the holes momentarily coincide (c), a blob emerges which joins the strands. One piece mesh in tubular form looks suspiciously like a crossply structure without a resin matrix, although there is little orientation in the strands. Under suitable conditions it is possible to stretch the net to align the molecules unidirectionally in the strands: the alignment does much to strengthen the enlarged mesh. The net used in the supermarket for packaging nuts has a highly oriented form along the strand length, which gives good dimensional stability. A damage-resistant version is made from foamed but not oriented strands, and is slipped over a bottle in many airport duty-free shops to protect the bottle in transit.

What happens when you stretch rectangular mesh plastics netting in the diagonal direction? How does this compare with the stiffness when stretched in the direction of one strand?

A heavier duty mesh has also been developed, by taking a sheet of polymer, punching holes in it and under suitable conditions stretching the sheet along its length and width. This can produce a mesh with tremendous orientation, and hence fantastic stiffness and strength. A lot of this material was used recently to stabilize marshy soil and extend the runway at Port Stanley in the Falkland Islands at very short notice, so that military aircraft could land safely. Applications of geotextiles are growing rapidly.

A quite different example of the crossply model is a sandwich of balanced woven cloth encased in and bonded to a synthetic weatherproof and abuse-

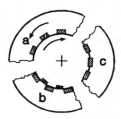

Figure 5.5 A counter-rotating extrusion die system for making plastics netting.

resistant rubber. This is used to make basic tube shapes which are assembled to make inflatable boats which are well-known for their performance and reliability. The yarns in the cloth are lined up along the tube and round the circumference. Inflation pressures are only a tenth of an atmosphere but this is sufficient to give good stiffness and prevent local buckling even under localized high compressive loads. More generally there is a wide range of structural applications of coated fabrics in inflatable buildings.

Pastry for tarts and pies often has a crossply laminated structure due to the many operations of rolling out the pastry and folding it over for re-rolling. In short pastry the bond between layers is quite good but because short pastry is not rolled out much, the cooked pastry has a more or less uniform, slightly crumbly texture and only modest stiffness and strength. In flaky pastry (millefeuille) the construction is a very large number of extremely thin layers with hardly any bonding between them – they are deliberately separated by fat during manufacture – and the properties of the cooked flaky pastry vary markedly with direction. The individual foils have considerable stiffness and strength in their plane (because of the molecular orientation of the gluten, which is a polymer), so that they are quite tricky to cut through with a knife or pastry fork, but the difficulty arises not least from the negligible interlaminar bond: if the knife is pressed on to the top surface other than perpendicularly, the horizontal component of the applied force causes adjacent layers underneath the blade to slide past each other. Any cream (in a cream slice) helps this process and the result can be dramatic and messy. Another problem is that the cream squeezes out so that there is not much reaction against the applied load so the cutting efficiency drops.

In synthetic composites we usually apply external loads and the strain response of the structure is then fixed by the size of the load and the way the load is distributed within the structure. Stresses and strains may be uniform across a cross-section, as in a thin-walled pipe under internal pressure, and will vary along the length of the pipe only if, for example, there is a change in wall thickness or a change in diameter. A plate under an applied bending moment has stresses and strains which vary through the thickness. In nature's living systems it is possible, and often essential, deliberately to vary the loading patterns at will *locally* within the structure as a function of time: this provides changes of shape necessary for movement and survival.

We can model the muscle systems in the outer layers of an earthworm most simply as a sort of crossply laminate, even though this is a tremendous oversimplification. There are longitudinal muscles and belts of circumferential muscles: these, with the skin, encase the viscera which we shall regard here as essentially an incompressible fluid under internal pressure at constant volume.

If the earthworm contracts its circumferential muscles and relaxes its longitudinal muscles, then the internal pressure forces the worm to elongate, perhaps by a factor of two or so if necessary. If the worm relaxes its circumferential muscles and contracts its longitudinal muscles, then the worm increases its diameter and becomes shorter. Alternating the two activities, differently along its length as necessary, allows the worm to travel. Contraction and extension of muscles on opposite ends of the diameter of a short length of its body induces a bend, so the worm can also change direction of travel. Because of the internal pressure, the muscles are always in some tension even in their relaxed state, so the worm does not wrinkle and develop buckling collapse when it bends.

In practice some muscles are also aligned at an angle to the length direction in a balanced crossed helical pattern in order to provide the possibility of a change of cross-sectional shape, but we shall not discuss this here – Wainwright *et al.* (1976) summarize further details.

The cylindrical wall of a sea anemone such as *Metridium senile* has highly extensible circumferential and longitudinal muscles. The volume fraction of muscles (fibres) is very small. The anemone is fixed to the sea bed at one end and open at the other, so there is no question of development of an internal pressure on the enclosed volume of water. By contracting or relaxing different muscles, the sea anemone can change its body size and shape through an enormous range, from a wrinkled blob to a relatively large stiff vertical column some four times taller than its diameter. By local muscular activity the cylindrical form can bend over and sweep the sea floor in search of food.

5.3.2 Symmetrical regular crossply laminate

In a balanced regular symmetric crossply laminate, the simplest arrangement consists of four identical plies arranged in the sequence $(0°/90°)_s$, i.e. $0°/90°/90°/0°$. Each ply has the same properties E_1, E_2, G_{12}, and v_{12}. We shall choose the co-ordinate axes (x, y, z) to align with the top $0°$ principal directions $(1, 2, z)$ as shown in Figure 5.1(b).

As each ply has the properties[Q], with $Q_{16} = Q_{26} = 0$, it follows that the extensional stiffness matrix has a similar population, with $A_{16} = A_{26} = 0$, because $[A] = \Sigma[Q]_f(h_f - h_{f-1})$. There is therefore no coupling between applied shear stress and direct strain, because the plies are each stressed only in their principal directions. The reader will need to ensure that the appropriate values of Q_{ij} are used to calculate A_{ij}. In the particular example (only), where the same number of plies is used in each direction, then $A_{11} = A_{22}$.

For a symmetric laminate, $[B] = 0$, so there is no coupling between bending and in-plane stresses and strains.

For bending we calculate the bending stiffnesses as

$$[D] = \Sigma[Q]_f(h_f^3 - h_{f-1}^3)$$

and as expected $D_{16} = D_{26} = 0$. But note that even when there are the same number of plies in each direction, D_{11} is not the same as D_{22} because D_{ij} depends on both the orientation of each ply and its distance from the midplane.

It becomes apparent therefore that the symmetric regular crossply laminate is well-behaved when stressed in its principal directions, and the data in the following subsection confirm this.

Problem 5.3: Show that as the number of plies increases in a regular symmetric crossply laminate of given total thickness, $D_{11} \to D_{22}$.

Balanced regular symmetric crossply laminate stressed in principal directions (C + 4S)

This laminate consists of four plies of unidirectional high modulus carbon fibres in epoxy resin, each 0.125 mm thick arranged as $(0/90)_s$. The properties and reduced stiffnesses for the single ply stressed in its principal directions are given in Table 5.1. For this 0.5 mm thick laminate the ABD and abhd matrices are given in Table 5.2.

Table 5.1 Properties of lamina HM-CARB

$E_{11} = 180\,\text{GPa}$	$Q*(0°)$ in GPa	$Q(90°)$ in GPa
$E_{22} = 8\,\text{GPa}$	(180.7 2.41 0)	(8.032 2.41 0)
$G_{12} = 5\,\text{GPa}$	(2.41 8.032 0)	(2.41 180.7 0)
$v_{12} = 0.3$	(0 0 5)	(0 0 5)
$\theta = 0°$		

Table 5.2 Stiffnesses and compliances of regular symmetric crossply $(0°/90°)_s$ C + 4S stressed in principal directions. Each ply 0.125 mm

A (kN/mm) B (kN)			B (kN) D (kNmm)		
47.19	1.205	0	0	0	0
1.205	47.19	0	0	0	0
0	0	2.5	0	0	0
0	0	0	1.658	0.0251	0
0	0	0	0.0251	0.3085	0
0	0	0	0	0	0.05208

a (mm/MN) h (/MN)			b (/MN) d (/MNmm)		
21.21	− 0.5414	0	0	0	0
− 0.5414	21.21	0	0	0	0
0	0	400	0	0	0
0	0	0	604	− 49.14	0
0	0	0	− 49.14	3245	0
0	0	0	0	0	19200

Application of single force resultant $N_x = 10\,N/mm$ *to C + 4S*
The strain responses, uniform through the thickness of the laminate, are $\varepsilon_x^\circ = 0.0212\%$ and $\varepsilon_y^\circ = -0.00054\%$, as shown diagrammatically in Figure 5.6(a). The stress profiles in Figure 5.6(b) are entirely as expected, and for any fth ply can be calculated from equations such as

$$\sigma_{xf} = Q_{11f}(\varepsilon_x^\circ + z\kappa_x)$$

It is instructive to compare the stress profiles in Figure 5.6 with those for a foam sandwich panel faced with stiff isotropic skins in Figure 4.7.

Application of single twisting moment $M_{xy} = 1\,N$ *to C + 4S*
The resulting curvature $\kappa_{xy} = 19.2/m$ and the linear shear stress and shear strain profiles are shown in Figure 5.7.

Problem 5.4: Check the numerical values of A_{11} and D_{12} in Table 5.1 by calculation.

Problem 5.5: Explain, without detailed calculations, the shape of the transverse stress profile $\sigma_y(z)$ which arises when laminate C + 4S carries: (a) N_x alone; and (b) M_x alone.

Problem 5.6: Calculate the apparent elastic constants for C + 4S under single in-plane loads.

Problem 5.7: Calculate the strain response of C + 4S to the application of single shear force resultant $N_{xy} = 10\,N/mm$. Explain why the profile $\tau_{xy}(z)$ is uniform whereas under direct stress the stress profiles in Figure 5.7 are stepped.

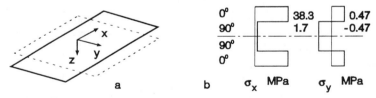

Figure 5.6 Responses of C + 4S to $N_x = 10\,N/mm$.

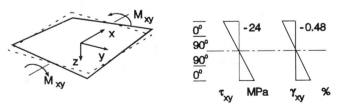

Figure 5.7 Responses of C + 4S to $M_{xy} = 1\,N$.

Problem 5.8: What happens when you apply a single moment resultant $M_x = 1\,N$ to $C + 4S$?

Problem 5.9: Construct the stress profiles for $C + 4S$ when subjected to the strain $\varepsilon_x^{\circ} = 0.02\%$.

Problem 5.10: A wide sheet of $C + 4S$ is wrapped round a drum of radius 2 m. Calculate the stress profiles in the laminate.

Approximate methods of calculation

Although in practice the designer would use computers to make the necessary calculations, it is informative to review approximate methods for calculations, and their laminations, for crossply laminates where $E_1 \gg E_2$.

The in-plane extensional stiffnesses can be calculated as

$$[A] = \Sigma [Q]_f (h_f - h_{f-1})$$

If $Q_{11} \gg Q_{22}$, A_{11} dominates in the x direction and A_{22} dominates in the y direction. This applies even if the number of plies in the x and y directions is not the same, because A_{ij} are thickness-weighted stiffnesses. So for direct stresses we can ignore the transverse layers and use the approximate expressions

$$A_{11} = Q_{11}(0°)(-h/4 - (-h/2) + h/2 - h/4) + Q_{11}(90°)(h/4 - (-h/4))$$
$$= Q_{11}(h/2) + Q_{22}(h/2) \approx Q_{11}h/2, \text{ provided } Q_{11} \gg Q_{22} \cdot A_{22} = A_{11}.$$

This approximation does not work for applied shear because the shear modulus is the same in each principal direction in the ply; we must therefore make the full calculation, and fortunately this is straightforward.

For pure bending we can follow similar simplifications, but they are less accurate and convincing than those for in-plane direct loads.

Taking bending stiffnesses as proportional to $E_f h_f^3$ in a regular four-ply $(0°/90°)_s$ laminate of thickness h, we have

$$\kappa_x \alpha E_1 h^3 - E_1 (h/2)^3 + E_2 (h/2)^3$$
$$\kappa_y \alpha E_2 h^3 - E_2 (h/2)^3 + E_1 (h/2)^3$$

Ignoring the small contributions from E_2, we find

$$D_{11}/D_{22} \sim \kappa_x/\kappa_y = 7$$

Using $E_1 = 180\,GPa$ and $E_2 = 8\,GPa$ the full calculation gives $D_{11}/D_{22} = (180 \times 7 + 8)/(8 \times 7 + 180) = 1268/236 = 5.373$ which checks with the data in Table 5.2.

As was observed for in-plane stiffnesses, the approximate method of neglecting transverse plies does not work for applied torque, because the shear contributions are the same for plies in both orientations. Under torque the full method should be used.

5.3.3 Unbalanced regular symmetric crossply laminates

Unbalanced regular symmetric crossply laminates offer different extensional stiffnesses A_{11} and A_{22} and different bending stiffnesses D_{11} and D_{22}, however many plies there may be in the laminate. Many commercial plywoods consist of an odd number of nominally identical plies arranged in this way.

Data for plywood

The elastic constants for flat-sawn spruce are quoted by Stavsky and Hoff (1969) as $E_1 = 9.86\,\text{GPa}$, $E_2 = 0.355\,\text{GPa}$, $G_{12} = 0.364\,\text{GPa}$, $v_{12} = 0.539$, $v_{21} = 0.0194$ (Table 5.3).

Table 5.3 Unbalanced regular symmetric crossply $(0°/90°/0°)_T$ three-ply laminate FSP based on 1.5 mm plies of flat-sawn spruce

A (kN/mm) B (kN)			B (kN) D (kNmm)		
30.43	0.8702	0	0	0	0
0.8702	16.02	0	0	0	0
0	0	1.638	0	0	0
0	0	0	72.96	0.1468	0
0	0	0	0.1468	5.426	0
0	0	0	0	0	2.764
a (mm/MN) *h* (/MN)			*b* (/MN) *d* (/MNmm)		
32.91	−1.787	0	0	0	0
−1.787	62.51	0	0	0	0
0	0	610.5	0	0	0
0	0	0	13.78	−3.729	0
0	0	0	−3.729	185.3	0
0	0	0	0	0	361.8

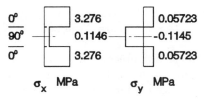

Figure 5.8 Response of FSP to $N_x = 10\,\text{N/mm}$.

Application of force resultant $N_x = 10\,N/mm$ to FSP
The deformations are represented by the uniform midplane strains $\varepsilon_x^\circ = 0.03291\%$ and $\varepsilon_y^\circ = -0.001787\%$, giving a Poisson's ratio of $v_{xy} = -\varepsilon_y^\circ/\varepsilon_x^\circ = 0.0543$ for this laminate. The direct stress profiles are given in Figure 5.8.

Application of force resultant $N_{xy} = 10\,N/mm$ to FSP
Under a uniform shear stress $\tau_{xy} = 2.222\,MPa$, the strain response is simply a uniform $\gamma_{xy} = 0.6105\%$.

Application of force resultant $M_y = 1\,N$ to FSP
The midplane curvatures are $\kappa_x = -0.003729/m$ and $\kappa_y = 0.1853/m$, with stress and strain profiles in Figure 5.9.

Application of torque $M_{xy} = 1\,N$ to FSP
The twisting curvature is $\kappa_{xy} = 0.3618/m$, with the shear stress and strain profiles in Figure 5.10.

Problem 5.11: Calculate the midplane strains and important stress profiles for FSP under $N_y = 10\,N/mm$, and compare your results with those responses for FSP under $N_x = 10\,N/mm$.

Problem 5.12: Calculate the midplane curvatures of FSP under $M_x = 1\,N$. Compare and contrast your results for $M_x = 1\,N$ with those given for $M_y = 1\,N$, and discuss any differences.

Problem 5.13: Based on the information given in Section 5.3.2 (text and problems), calculate the major and minor Poisson's ratios for laminate FSP under single stress. What are the apparent moduli for FSP?

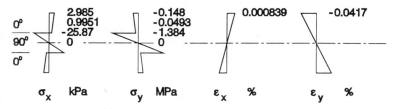

Figure 5.9 Response of FSP to $M_y = 1\,N$.

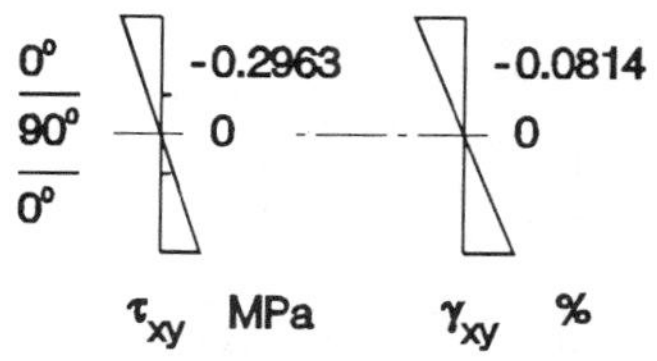

Figure 5.10 Response of FSP to $M_{xy} = 1\,N$.

5.3.4 Balanced non-symmetrical regular crossply laminates

A balanced regular non-symmetric crossply laminate will retain the simplification that $A_{11} = A_{22}$, but there is extension-bending coupling arising from the lack of symmetry: the terms $B_{11} = -B_{22}$ are non-zero. Under direct load in a principal direction the extension and lateral contraction are accompanied by curvature. Although we shall not discuss the reasons until Chapter 7, any non-symmetric lay-up cured above room temperature will show curvature on cooling, so it is good practice to avoid non-symmetric flat crossply laminates.

Problem 5.14: Show that $B_{11} = -B_{22}$ for a two-ply non-symmetric regular crossply laminate. How would the magnitude for B_{11} change with increasing numbers of pairs of plies in an alternating $0°/90°$ non-symmetric crossply laminate, assuming all laminates have the same thickness?

Balanced non-symmetric crossply laminate stressed in principal directions C + 4NS

This laminate consists of four plies of unidirectional high modulus carbon fibres in epoxy resin, each ply 0.125 mm thick arranged in the sequence $0°/0°/90°/90°$. The properties of each ply are given in Table 5.1. For this 0.5 mm thick regular non-symmetric laminate the ABD and abhd matrices are given in Table 5.4.

The [A] and [D] matrices are appropriate for a laminate where each ply is stressed in its principal directions; $A_{16} = A_{26} = D_{16} = D_{26} = 0$. The interest-

Table 5.4 Stiffnesses and compliances of regular non-symmetric crossply $0°/0°/90°/90°$ C + 4NS stressed in principal directions – each ply 0.125 mm

A (kN/mm) B (kN)			B (kN) D (kNmm)		
47.19	1.205	0	−5.397	0	0
1.205	47.19	0	0	5.397	0
0	0	2.5	0	0	0
−5.397	0	0	0.9831	0.0251	0
0	5.397	0	0.0251	0.9831	0
0	0	0	0	0	0.05208
a (mm/MN) h (/MN)			b (/MN) d (/MNmm)		
57.03	−1.456	0	313.1	0	0
−1.456	57.03	0	0	−313.1	0
0	0	400	0	0	0
313.1	0	0	2738	−69.89	0
0	−313.1	0	−69.89	2738	0
0	0	0	0	0	19 200

ing behaviour comes from the terms B_{11} and B_{22}, which offer substantial coupling between in-plane forces and simple curvatures.

Application of force resultant $N_x = 10$ N/mm to C + 4NS
The resulting deformations are $\varepsilon_x^\circ = 0.057\%$, $\varepsilon_y^\circ = -0.00146\%$ and $\kappa_x = 3.13$/m. There is therefore substantial curvature (in one plane only) superposed on the laminate becoming longer in the stressed direction and narrower transversely. The stress and strain profiles are shown in Figure 5.11. Note the substantial mismatch of stress σ_x at the midplane interface, which arises from the change of modulus across the interface between adjacent plies of different orientation.

Application of moment resultant $M_x = 1$ N to C + 4NS
The reponses are $\varepsilon_x^\circ = 0.0313\%$, $\varepsilon_y^\circ = 0$, $\kappa_x = 2.738$/m and $\kappa_y = -0.0699$/m. The stress and strain profiles shown in Figure 5.12 are as expected, there being a linear variation of strain through the thickness.

Problem 5.15: Why is the curvature zero in the y direction of laminate C + 4NS, when N_x is applied?

Problem 5.16: What moments would need to be applied at the grips to suppress the curvatures in C + 4NS when $N_x = 10$ N/mm is applied? What effect would this have on the strain ε_x in the sample?

Problem 5.17: Sketch the deformations expected when the laminate C + 4NS is under: (a) a balanced biaxial in-plane tensile stress; and (b) under a single shear force resultant N_{xy}.

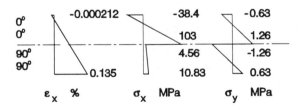

Figure 5.11 Response of C + 4NS to $N_x = 10$ N/mm.

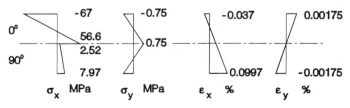

Figure 5.12 Response of C + 4NS to $M_x = 1$ N.

Problem 5.18: What midplane strains and curvatures are induced when a twisting moment $M_{xy} = 1$ N is applied to laminate C + 4NS?

Problem 5.19: A square plate made from C + 4NS carries a uniform moment $M_x = 1$ N, but the edges at which the moment is applied are reinforced with rigid grips which prevent curvature along their length. What are the midplane strains and curvatures in the plate?

Problem 5.20: (a) What forces are required to achieve imposition of $\varepsilon_x^\circ = 0$ and $\varepsilon_y^\circ = 0.005\%$ to C + 4NS? (b) Compare the result with that for an imposed $\varepsilon_y^\circ = 0.005\%$, and $\varepsilon_x^\circ = \kappa_x = \kappa_y = 0$.

5.3.5 Unbalanced regular non-symmetric crossply laminate

An unbalanced regular non-symmetric crossply laminate shows anticlastic curvatures under direct stress: the midplane strains and curvatures will differ according to the direction(s) in which loads are applied.

5.4 FABRICS STRESSED OFF-AXIS

Before discussing the behaviour of angle-ply laminates, it will be helpful to explore the characteristics of woven and braided cloth under in-plane stress. This will give an indication of the key features which the reader is encouraged to appreciate by observation of the behaviour of familiar structures. Other examples are also given of flexible structures which are not fabrics but which can be modelled as having similar mechanics of deformation.

5.4.1 Woven fabrics stressed off-axis

A woven fabric enables us to achieve a substantial measure of extensibility in the structure without much stretch in the fibres from which the structure is made. The fibres can be made from either a man-made polymer such as nylon or polyester, or a natural polymer such as cotton or wool.

Let us first take some non-stretch cotton cloth as used in a handkerchief. Along the fibre direction the handkerchief does not stretch much under unidirectional tension, and does not contract much in width either. But if we pull on diagonal corners – in the bias direction – the fabric extends quite a lot in the stretch direction, and contracts noticeably across the stretch. The cloth has sheared a lot too. The Poisson's ratio has a value of about 1. The stiffness of a handkerchief when stressed in different directions follows the pattern shown in Figure 5.1(c).

Indeed the dramatic increase in deformability when stressed at 45° to the warp or weft is exploited by dressmakers working with fabrics cut 'on the bias', and this is a phenomenon well-known to sailmakers and umbrella makers as well. So if you stretch a handkerchief on the diagonal it buckles and adopts a diamond shape. If you get a friend to help you by stretching the other diagonal at the same time and to the same extent, the handkerchief extends in area and becomes quite taut like the surface of a drum – the two perpendicular shears cancel out for quite legitimate technical reasons, the buckling has disappeared, and the sheet ends up quite flat. The effect is the same as would have occurred if you had applied a uniform tension at right angles to each edge of the square.

If you take a strip of cloth cut on the bias (at 45° to the warp and weft directions), it stretches readily and has a high Poisson's ratio so it gets quite narrow. If you make a tube cut on the bias, this behaves in quite the same way, provided you sew the seam so it can stretch as well. If you place a circle of card in one end to prevent a change in radius, and then pull on each end, the tube narrows down considerably at the free end (Figure 5.13(a)). Scraps of discarded domestic plain net curtain have the right sort of properties if you want to try this out. The experiment only works if the warp and weft are at 45° to the pull direction on the tube, and any seam is sown so that it is free to stretch in the same way as the net curtain, i.e. 'on the bias'. Stretch-fit garments have zig-zag stitches which permit considerable stretch along the line of the seam even though the fibre of the stitch stretches hardly at all.

If you put a ball of the same diameter as the unstretched tube inside, and then stretch, away from the ball the necking down occurs as before, but the ball prevents the cloth from elongating so it stays unsheared on the contact diameter (Figure 5.13(b)) and illustrates what engineers call 'restraint'.

We now have a basis for an effective design of snug-fit garments using suitable non-stretch fabrics. The hard way of achieving a good fit is to cut flat panels of cloth and stitch them to exact sizes so that the cloth is tensioned along warp and weft. The garment – a dress, say – may then be a good fit in one body position, but can be a poor fit and even very uncomfortable in any other position. Snug-fit ordinary cotton jeans may look good when the wearer is standing up, but are not necessarily comfortable to sit in for long periods

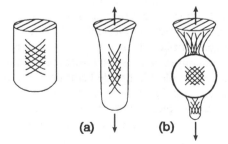

(a)   **(b)**

Figure 5.13 Stretching a tube of cloth cut 'on the bias'. Note the restraint at the upper end and that caused by the ball.

because the stiff cloth buckles and can form stiff creases which press hard, even mildly painfully on the skin.

But if the fabric is cut 'on the bias', so that tensions are applied at 45° to the fibres during wear, the fabric is much more flexible. String bags work on this principle to a considerable extent. In a dress with the bias in the vertical direction, gravity pulls the hem down and the Poisson effect ensures that the circumference of the tailored tube therefore decreases until prevented by some constraint like the body inside. Sound engineering principles exploit the fibres to ensure a snug fit where it is desired, a good drape in the skirt, a comfortable garment which stays comfortable in a range of body positions from standing to sitting or stretching upwards. Of course the seams must be sewn so they can stretch as well. All this is achieved without axially stretching the fibres much at all.

> **Problem 5.21:** The fabric top of an umbrella consists of many nearly triangular pieces (gores) of fabric, which are stretched when the frame is extended. Is it better to cut the fabric pieces so that the free edge is in the weave direction or in the bias direction?

5.4.2 Braided fabrics stressed off-axis

Braiding is a well-established process for textiles, which interweaves three or more strands to cross over one another in a diagonal formation. The technique is commonly used to make tubular fabrics and structures, but it can also produce flat tapes and fabrics with the fibres at a bias angle to the longitudinal direction. A variety of braid forms is possible, and there are two common forms: basket weave braid (over one, under one, Figure 1.44), and plain braid (over two, under two). Common examples of braided products include cotton bias binding, used domestically to protect the edges of fabrics in making garments, and the screening of television cable to prevent interference, using braided copper wire.

Most braids have fibres aligned at $+\theta$ and $-\theta$ to the braid axis. Triaxial braids are formed by introducing a third fibre parallel to the braid axis.

Polymer composites based on braided fabrics have reinforcement in the off-axis direction and, by suitable choice of braid angle, can be designed to give shear stiffness, and good torsional stiffness in tubular structures or, by using a different angle, can give pipes capable of withstanding high internal pressure.

The most common example of a braided composite is the reinforced rubber hose for water coolant supply to internal combustion engines for cars and lorries: in the simplest example, a rubber tube is extruded, the high tensile steel wires are braided over the tube, and the covered tube passes through a second extruder to provide a sheath of rubber which protects the braid and secures it in position. Other interesting examples include high-tech bicycle frames for professional riders and enthusiastic amateurs, where good torsional stiffness and bending stiffness per unit mass are required, and ski cores, which consist of a lightweight torsion box construction, which offers the torsional stiffness needed to transfer most of the skier's effort directly to the ski edge, to obtain maximum grip even on hard snow.

Braided structures usually deform axially by a lazy tongs mechanism of rotation at the nodes where fibres intersect. Such a mechanism gives good conformability to underlying surfaces provided that the radius of the former does not change too suddenly along the length. It must not be overlooked that when a braided tube is used to closely cover a former having different diameters along its length, the angle of the fibres also changes.

Problem 5.22: A braided tube is made from inextensible fibres and has a braid angle of $\pm 45°$, and a mean diameter of 25mm. A former consists of two straight sections having outside diameters of 20mm and 30mm, connected by a linear taper some 80mm long. Neglecting the thickness of the braid, estimate the angle the fibres make with the longitudinal axis of each of the straight tubes when the braid (under light axial tension) closely fits the former.

Problem 5.23: A braided tube made from inextensible fibres is 25mm diameter with a braid angle of 30°. Over what diameter former should it fit closely if, when impregnated with a suitable matrix, it is to be most efficient as a thin-walled torque tube?

5.5 ANGLE-PLY LAMINATES

The simplest angle ply laminate consists of plies laid-up and bonded at $+\theta$ and $-\theta$ to a chosen reference direction, x, in the plane. Angle-ply laminates can be either balanced or unbalanced, regular or non-regular, and symmetric or non-symmetric. We shall look at examples of regular angle-ply laminates in this section, although the principles can be readily extended to cover non-regular laminates.

5.5.1 Symmetrical angleply laminates

For a balanced symmetric angleply laminate, $A_{16} = A_{26} = 0$, $[B] = 0$, but D_{16} and D_{26} are non-zero. Such a laminate will certainly show bending-twisting coupling because of D_{16} and D_{26}. This principle is followed (but using a more complicated arrangement) to ensure that aircraft with forward-swept wings can be flown controllably even though its wings deflect considerably in extreme manoeuvres.

Fibre reinforced plastics pipes are widely used in the chemical and process industries. They can be made by helically winding fibres coated with a resin on to a rotating cylindrical former. The fibres form a criss-cross pattern which is quite well described by the angleply laminate. The fibres are usually glass, sometimes a special formula to achieve good chemical resistance, and the resin is often a polyester, epoxy or furan crosslinkable resin. Lampshades for domestic lighting, and some small Christmas tree decorations can also be filament wound in this way using either cylindrical or nearly spherical formers.

The cotton on bobbins and reels for sewing machines, and the coils of insulated wire for transformers and relays are wound in a similar way with the windings nearly in the circumferential direction, but of course without the resin. In these coils and reels the small angle means that filaments are wound like a closely coiled spring, with adjacent turns touching. The resulting structure is a good model for arrangement of fibres in an angleply construction, even if the angle between plies is very small. In bobbins and reels there is no interlacing of filaments, but in filament wound pipes and vessels there is usually some interlacing of filaments.

In a pipe only under internal pressure, the tension round the circumference is twice the tension along the length. Assuming that only the fibres take the load, the most efficient arrangement of fibres occurs when they are wound at $54.75°$ to the axial direction. A lot of glass fibre reinforced pipe is wound in this way with diameters from about 100mm up to at least 3m. Some textile reinforced rubber hose is wound as well. For example radiator hose for cars usually consist of a knitted braid sandwiched between the inner and outer layers of rubber: the braid withstands the pressure and the rubber stops leaks and copes with the environment and general abuse.

It is interesting to note that many structures in nature have the appearance of an angle-ply or as-if filament-wound laminate. The arrangement of many muscle fibres in the back of the human body provides a good example; and the way many sharks and whales navigate owes much to a helical pattern of winding of muscle fibres which can be caused to contract locally so the creature bends and heads in the desired direction (see also Section 3.3.7).

The arteries of mammals have a helical arrangement of aligned molecules to contain the pressure of blood inside – it also happens that our arteries are under considerable axial tension as well.

In living mammals, the variation in internal pressure in arteries caused by pumping blood causes mainly circumferential changes in dimension, with very little movement in the axial direction. For example, the change in diameter of the aorta in man during the cardiac cycle is typically about 9 to 12%. Wainwright *et al.* (1976), and Dobrin (1978), give information which suggests that the walls of arteries may be crudely modelled as helical windings of discontinuous collagen fibres in an elastic (elastin) matrix. In the body the artery is usually under considerable axial tensile strain: excised arteries retract to about two-thirds of their *in-situ* length, and this shortening causes thickening of the artery wall, and also causes a change in angle of the orientation of collagen. A change in diameter in a tube held at constant stretched length will cause fibres to change in length, and the angle of fibre orientation to increase with respect to the axis of the tube. This re-orientation is advantageous because it makes the tube progressively stiffer in the hoop direction and thus increasingly resistant to increasing pressure. The axial tension also suppresses any tendency to axial buckling: if the tubes buckled when we bent over, the blood flow would be substantially impeded leading to disturbing inconvenience – or, and very finally, worse.

We also find that Poisson's ratio can be tailored substantially by careful choice of orientation angle, albeit at the expense of other patterns of behaviour. With appropriate fibre-reinforced polymers it becomes feasible to achieve zero or even negative values of strain under biaxial stress. On this basis thin-walled tubes can be designed to decrease either their diameter or length (but not both) under internal pressure.

Certain types of rocket motor are driven by a solid fuel which takes the form of a solid cylinder with a small bore down its axis. When ignited, the solid fuel burns radially outwards from the hole, and generates tremendous pressures. Now, the propellant is very brittle and if allowed to expand under the pressure it would crack: the new surfaces of the crack would then burn as well and the rate of burning would rise dramatically and out of control. What is wanted therefore is a lightweight casing which will not expand in diameter under internal pressure. A casing made from a uniform material will always expand under internal pressure – making it thicker reduces but does not eliminate the radial swelling and this is at a severe weight penalty which is self-defeating. So how might we achieve our objective?

We have seen that a tensile force acting along a rubber band makes it thinner transversely. So let us consider separately the forces developed by the internal pressure in a cylinder with closed ends. The tension along the pipe causes a decrease in the circumference; the (bigger) tension round the circumference causes the girth to increase. We can choose the winding angle for the fibres in a filament wound pipe so that these changes in circumference just cancel each other out. There will then be no change in diameter under internal pressure, which is what is wanted for this type of rocket motor casing. The casing still gets longer under pressure but this is of minor significance. It must be admitted that there are some severe restrictions on the choice of fibres to achieve this effect but with rockets the price is worth paying.

It can be shown analytically that for a balanced symmetric angleply laminate at $(+45°/-45°)_s$, the shear modulus reaches a maximum. This is highly advantageous in principle in the design of stiff shafts to transmit torque, but we shall also see that we need a more complicated lay-up to meet other design requirements.

For angleply laminates the approach to calculating the ABD matrix and its inverse is exactly the same as we applied to the crossply laminate. We have seen that crossply laminates in general follow a pattern of behaviour having a character closely related to that of unidirectional plies stressed in their principal directions. Angleply laminates more closely follow the character of unidirectional plies stressed off-axis.

The major difference from loading in the principal directions is that under off-axis loading of a unidirectional ply, the matrix terms denoting shear coupling, e.g. A_{16} and D_{26}, are non-zero. These terms arise from the definitions of A_{ij}, B_{ij} and D_{ij}, bearing in mind that $Q_{16}^*(+\theta) = -Q_{16}^*(-\theta)$ and $Q_{26}^*(+\theta) = -Q_{26}^*(-\theta)$. These terms offer the possibility of tailoring the laminate to deliberately avoid or deliberately exploit shear-extension coupling, shear-bending coupling, extension-twist coupling or twist-bending coupling, or any combination of these effects.

These coupling effects undoubtedly introduce complexity into the design process. The designer needs to specify carefully the boundary conditions before rushing into the detailed mechanics.

Problem 5.24: Show that for a symmetric angle ply laminate with plies at $+\theta/-\theta$, the shear modulus G_{xy} reaches a maximum value at $\theta = 45°$.

Symmetrical angleply rayon/rubber laminate $(30/-30)_s$ RRX 30S

The basic lamina RR is a unidirectional set of rayon cords embedded in a vulcanized rubber matrix, and it technically described (Walter, 1981) as a 1650/3 rayon rubber composite 1 mm thick used in the manufacture of a HR78–15 radial car type. The properties of the ply are given in Table 5.5. For the $(30/-30)_s$ laminate of total thickness 4 mm, the ABD and abhd matrices are given in Table 5.6.

With $A_{16} = A_{26} = 0$ this balanced regular laminate shows no coupling between tension and shear, and the symmetry shows that there is no coupling between direct forces and curvatures. It is interesting to note that the D and d matrices are fully populated, so there is some coupling between bending moments and twist, and between torque and bending curvatures.

Table 5.5 Properties of lamina RR

E_{11}	1.74 GPa		$Q^*(+30°)$		(GPa)
E_{22}	0.014 GPa	(0.9867	0.3326		0.5622)
G_{12}	0.0025 GPa	(0.3326	0.1217		0.187)
v_{12}	0.547	(0.5622	0.187		0.3274)
θ	$+30°$				
E_{11}	1.74 GPa		$Q^*(-30°)$		(GPa)
E_{22}	0.014 GPa	(0.9867	0.3326		-0.5622)
G_{12}	0.0025 GPa	(0.3326	0.1217		-0.187)
v_{12}	0.547	(-0.5622	-0.187		0.3274)
θ	$-30°$				

Table 5.6 Stiffnesses and compliances of regular angle-ply laminate $(30°/-30°)_s$ RRX30S – Each ply 1 mm

A (kN/mm) B (kN)			B (kN) D (kNmm)		
3.947	1.33	0	0	0	0
1.33	0.4866	0	0	0	0
0	0	1.31	0	0	0
0	0	0	5.263	1.774	2.249
0	0	0	1.774	0.6489	0.748
0	0	0	2.249	0.748	1.746

a (mm/MN) h (/MN)			b (/MN) d (/MNmm)		
3225	− 8817	0	0	0	0
− 8817	26160	0	0	0	0
0	0	763.6	0	0	0
0	0	0	2729	− 6736	− 629.4
0	0	0	− 6736	19670	249.1
0	0	0	− 629.4	249.1	1277

Application of force resultant $N_x = 1$ N/mm to RRX30S
The strain responses are $\varepsilon_x^\circ = a_{11}N_x = 0.323\%$ and $\varepsilon_y^\circ = a_{12}N_x = -0.882\%$, both uniform across the thickness.

For the fth ply we can therefore calculate the stresses using $[\sigma]_f = [Q]_f[\varepsilon^\circ]$, hence $\sigma_x = Q_{11}\varepsilon_x^\circ + Q_{12}\varepsilon_y^\circ = 0.25$ MPa, and $\sigma_y = Q_{12}\varepsilon_x^\circ + Q_{22}\varepsilon_y^\circ = 0$, and as $Q_{11}(+30°) = Q_{11}(-30°)$, the stresses are the same throughout the laminate. For shear stresses we must distinguish the angles, as $Q_{16}(-30°) = -Q_{16}(+30°)$, so the shear stress has the same sign as the angle of the ply: $\tau_{xy}(+30°) = Q_{16}(+30°)\varepsilon_x^\circ + Q_{26}(+30°)\varepsilon_y^\circ = 0.164$ MPa, and $\tau_{xy}(-30°) = Q_{16}(-30°)\varepsilon_x^\circ + Q_{26}(-30°)\varepsilon_y^\circ = -0.164$ MPa. The total shear force is of course zero, as there is no applied shear force.

The stress and strain profiles in the reference directions (x, y) are shown in Figure 5.14. The most interesting profile is that for the shear stresses τ_{xy} and, compared with the formal discussion in the previous paragraph, a non-analytical explanation is as follows.

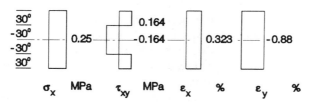

Figure 5.14 Responses in global directions of RRX30S to $N_x = 1$ N/mm.

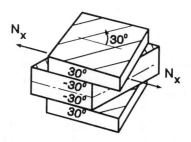

Figure 5.15 Deformation of RRX30S under $N_x = 1$ N/mm with unbonded plies.

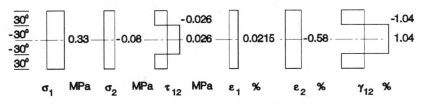

Figure 5.16 Responses in principal directions of RRX30S to $N_x = 1$ N/mm.

Imagine all four plies carry the same distributed load N_x, but that the plies are not bonded together. Because the load is applied at $+ 30°$ or $- 30°$ to the principal direction in the plies, each ply will become longer and thinner and will shear, and the four unbonded plies will behave as shown in Figure 5.15. The $+ 30°$ plies have developed a negative shear strain, and the $- 30°$ plies a positive shear.

The real laminate consists of bonded plies, and to prevent the shears shown in Figure 5.15, there must be an internal positive shear stress in the $+ 30°$ plies, and a negative internal shear stress in the $- 30°$ plies. This will result in a panel which does not shear and which carries no net shear force, because the two sets of shear stresses are equal and opposite in sign.

The stress and strain profiles in the principal directions of each lamina are shown in Figure 5.16, and are of course quite different from those in the (x, y) directions: note in particular the large shear strains in this example.

Application of moment resultant $M_x = 0.1$ N to RRX30S
The resulting curvatures are $\kappa_x = 0.273$/m, $\kappa_y = - 0.674$/m and $\kappa_{xy} = - 0.0629$/m. The laminate shows not only anticlastic curvature but also a small amount of negative twisting. The bending strain profiles vary linearly through the thickness as expected, as shown in Figure 5.17. These are found directly from the curvatures using $\varepsilon = \kappa z$.

The discontinuities in the stress profiles at the interface between $- 30°$ and $- 30°$ plies look unexpected until we recall that the stresses in the individual

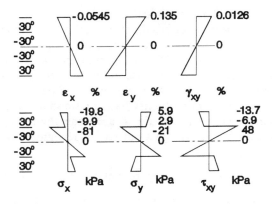

Figure 5.17 Responses in global co-ordinates of RRX30S to $M_x = 0.1$ N.

plies depend on Q^*, which is fully populated for off-axis loading, and that $Q_{16}^*(+\alpha) = -Q_{16}^*(-\alpha)$, and Q_{26}^* similarly. It is the shear-extension coupling for the two plies having the opposite sign which causes the discontinuity.

The strains at the interface A_1B_1 can be calculated from the curvatures as $\varepsilon_{x1} = 0.02729\%$, $\varepsilon_{y1} = -0.06736\%$ and $\gamma_{xy1} = -0.006249\%$. For a single ply under these strains, from Table 5.5, we find $Q_{11}^* = 0.9867$ GPa, $Q_{12}^* = 0.3326$ GPa, $Q_{16}^*(+30°) = 0.5622$ GPa, $Q_{16}(-30°) = -0.5622$ GPa. Hence

$$\sigma_x = Q_{11}^*\varepsilon_x + Q_{12}^*\varepsilon_y \pm Q_{16}^*\gamma_{xy}$$
$$= 0.0096 \text{ MPa } (+30°) \text{ or } 0.0804 \text{ MPa } (-30°)$$

Stress profiles σ_y and τ_{xy} can be explained in a similar manner, as can stress and strain profiles relating to the principal directions within the plies (Figure 5.18).

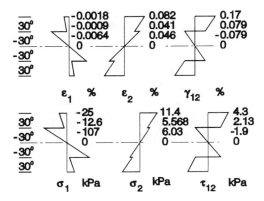

Figure 5.18 Responses in principal directions of RRX30S to $M_x = 0.1$ N.

Experience suggests that it is difficult to overemphasize that a stress profile in the principal directions describes the stress in that direction ply-by-ply. The physical direction of the stress (or strain) therefore changes (disconcertingly) at the interface between plies of different orientations. This therefore explains why there *seems* to be an apparent discontinuity in the strain profiles in the principal directions in Figure 5.18: in global co-ordinates (Figure 5.17) there is not of course a real discontinuity. In Chapter 6 we shall use the stress profiles in the principal directions as an aid to assessing the strength of a ply or laminates. Once the convention just outlined is recognized, the presentation becomes most useful.

Problem 5.25: What are the apparent elastic constants for RRX30S?

Problem 5.26: Calculate the stress profiles in global co-ordinates and principal directions when strains $\varepsilon_x^\circ = 0.3\%$ and $\varepsilon_y^\circ = 0$ are imposed on RRX30S.

Problem 5.27: A sheet of RRX30S is given a curvature $\kappa_y = 0.5/m$, but the curvature κ_x is prevented. Sketch the stress profiles in the sheet.

Problem 5.28: A sheet of RRX30S is wrapped round a drum of radius 2 m. What are the stress profiles if the axis of the drum corresponds to: (a) the x axis; and (b) the y axis of the RRX30S?

Introduction to behaviour of filament wound tubes

The symmetric angle-ply laminate can to a first approximation be used to examine the overall deformation behaviour of thin-walled filament-wound or braided tubes wound at a helix angle α to the hoop direction (Figure 5.19).

For a vertical tube clamped at the upper end and under axial tension, a small helix angle will give a large axial extension and minimal change in diameter. As the helix angle increases, the axial extension increases and at larger angles may

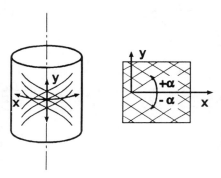

Figure 5.19 Modelling a thin-walled tube as an angle-ply laminate.

decrease again. The diametral contraction is greatest at 45°. Detailed results depend on the properties of the unidirectional lamina on which the laminate is based.

For example for a symmetric laminate $(+\alpha/-\alpha)_s$ using the rayon elastomer lamina RR in 1 mm plies, the result of applying an axial force $N_y = 1$ N/mm is shown in Table 5.7. The maximum axial elongation occurs at about 39°, whereas the maximum diametral contraction occurs at 45°.

For the same pipe under internal pressure such that $N_x = 2$ N/mm and $N_y = 1$ N/mm (corresponding to $\sigma_H = 0.5$ MPa and $\sigma_A = 0.25$ MPa), there are further curious dimensional changes as the helix angle varies, as shown in Table 5.8. For a helix angle in the range 6° to 34.5° this pipe actually contracts slightly in diameter but increases in length. At helix angles between about 34.6° and 35.2° this pipe will increase in both length and diameter. At helix angles above about 35.3° the pipe will increase its diameter but decrease its length.

The important practical conclusion is that where pipes to be made from these rayon/rubber undirectional plies are wound on a mandrel, the finished pipe may be eased off the mandrel by applying an internal pressure, provided the helix angle is greater than about 35.3°. Other removal techniques are required for smaller helix angles.

The data in Table 5.8 also prompt an interesting observation. The global strains at $\alpha = \pm 35°$ appear to suggest that such a pipe under internal pressure makes extremely efficient use of the material, there being hardly any change in either the hoop or axial directions. Are we almost getting something for nothing? But we must remember that small global strains $(\varepsilon_x, \varepsilon_y)$ do not imply small values of σ_x, σ_y and τ_{xy}. Closer investigation reveals that the stresses in

Table 5.7 Values of direct strain in regular symmetric angleply laminate $(\alpha/-\alpha)_s$ for $N_y = 1$ N/mm – each RR ply 1 mm

$\pm\alpha$	$\varepsilon_x(\%)$	$\varepsilon_y(\%)$
0	−0.00789	1.786
10	−0.0661	1.858
20	−0.2888	2.108
30	−0.8817	2.616
39	−1.992	3.041
40	−2.124	3.024
45	−2.486	2.514
50	−2.124	1.532
60	−0.8817	0.3225
70	−0.2888	0.0579
80	−0.0661	0.01758
90	−0.00789	0.01437

Table 5.8 Values of direct strain in regular symmetric angleply laminate $(+\alpha/-\alpha)_s$ for $N_x = 2\,N/mm$, $N_y = 1\,N/mm$ – each RR ply 1 mm

$\pm\alpha$	$\varepsilon_x\%$ (hoop)	$\varepsilon_y\%$ (axial)
0	0.0209	1.77
5	0.0079	1.76
6	0.00214	1.755
10	−0.0309	1.726
15	−0.0936	1.658
20	−0.173	1.531
30	−0.2367	0.8525
34.5	−0.00795	0.1465
35	0.04126	0.04436
40	0.941	−1.223
45	2.542	−2.458
50	3.924	−2.715
60	4.35	−1.441
70	3.927	−0.5196
80	3.651	−0.1146
90	3.564	−0.00135

global co-ordinates in the pipe are $\sigma_x = 0.5\,MPa$, $\sigma_y = 0.25\,MPa$ and $\tau_{xy} = 0.344\,MPa$ ($+35°$ ply) or $-0.344\,MPa$ ($-35°$ ply). In the principal directions (so important for strength) we see that the behaviour is dominated by $\sigma_1 = 0.741\,MPa$ (hence the stiff response), with σ_2 only 0.0093 MPa, and zero shear stress τ_{12}.

Problem 5.29: A laminate consists of an even number of thin identical laminae laid up alternatively at $+\theta$ and $-\theta$ to a reference direction. Show that if n is large, this laminate can be sensibly regarded as symmetric.

Problem 5.30: The propellant for a rocket motor is a brittle solid having the form of a cylinder with a small-diameter hole along its axis. An essential feature of the thin-walled case which fits closely round the propellant is that its diameter should not increase when the propellant is ignited and develops the required thrust. The case is to be wound from four layers of a carbon fibre-epoxy lamina having the reduced stiffnesses $Q_{11} = 207.6\,GPa$, $Q_{12} = 2.28\,GPa$, $Q_{22} = 7.6\,GPa$, and $Q_{66} = 4.8\,GPa$. If the laminate is regarded as symmetrical with two layers $+\theta$ and two layers at $-\theta$ to the axial direction, can a suitable casing be wound to meet this criterion? What constraints are there on materials properties for this problem?

Problem 5.31: For the four-ply laminate $(\alpha/-\alpha)_s$ based on 1 mm plies of RR, what is the angle which gives the maximum value of Poisson's ratio ν_{yx}?

5.5.2 Unbalanced symmetric angleply laminate

For an unbalanced symmetric angleply laminate, terms in Q_{16}^* and Q_{26}^* do not cancel out, and hence this laminate shows shear-extension coupling (caused by non-zero A_{16} and A_{26}) as well as twisting-bending coupling.

This can be seen for the laminate 3PLY45, which is a three-ply laminate based on 1.5 mm plies of flat sawn spruce arranged as $45°/-45°/45°$. Data for the individual ply are given in Section 5.3.3, Table 5.3. The stiffnesses of laminate 3PLY45 are given in Table 5.9.

For the regular unbalanced angleply laminate 3PLY30, three 1.5 mm plies of FSP are laid up as $30°/-30°/30°$, for which stiffness data are given in Table 5.10. Here the effect of different values of Q_{16}^* and Q_{26}^* can be seen.

Problem 5.32: Compare and contrast the behaviour of laminates 3PLY30 and 3PLY45 under in-plane or bending loads in the (x, y) directions.

Problem 5.33: Discuss the differences in stiffness characteristics between non-symmetric and symmetric angleply laminates using the data given in Tables 5.7 and 5.8 and Tables 5.9 and 5.10.

5.5.3 Balanced non-symmetric angleply laminate

For a balanced non-symmetric angleply laminate, $A_{16} = A_{26} = 0$, but B_{16} and B_{26} are non-zero, so this laminate will show twist-extension coupling, as well as twisting-bending coupling. Fan blades for wind power generation of

Table 5.9 Stiffnesses and compliances of regular symmetric angle-ply 3PLY45 based on three plies of flat sawn spruce each 1.5 mm thick laid up as $45°/-45°/45°$

A (kN/mm) B (kN)			B (kN) D (kNmm)		
13.69	10.41	3.602	0	0	0
10.41	13.69	3.602	0	0	0
3.602	3.602	11.18	0	0	0
0	0	0	23.1	17.57	16.88
0	0	0	17.57	23.1	16.88
0	0	0	16.88	16.88	18.86

a (mm/MN) h (/MN)			b (/MN) d (/MNmm)		
175.5	-129.7	-14.8	0	0	0
-129.7	175.6	-14.8	0	0	0
-14.8	-14.8	99	0	0	0
0	0	0	138.4	-42.54	-85.77
0	0	0	-42.54	138.4	-85.77
0	0	0	-85.77	-85.77	206.6

Table 5.10 Stiffnesses and compliances of regular symmetric angle-ply laminate 3PLY30 based on three 1.5 mm plies of flat sawn spruce laid up as $30°/-30°/30°$

A (kN/mm) B (kN)			B (kN) D (kNmm)		
26.88	8.025	4.496	0	0	0
8.025	5.265	1.742	0	0	0
4.496	1.742	8.793	0	0	0
0	0	0	45.36	13.54	21.08
0	0	0	13.54	8.885	8.168
0	0	0	21.08	8.168	14.84
a (mm/MN) h (/MN)			b (/MN) d (/MNmm)		
70.26	−101.9	−15.74	0	0	0
−101.9	351	−17.45	0	0	0
−15.74	−17.45	125.2	0	0	0
0	0	0	68.69	−30.37	−80.85
0	0	0	−30.37	241.3	−89.65
0	0	0	−80.85	−89.85	231.6

electricity can be designed to exploit twist-extension coupling. The faster the blade rotates, the greater the centrifugal stress along the blade. This can induce a twist in the blade, which alters its efficiency. The change in twist can be designed so that the drop in mechanical efficiency is sufficient to avoid a catastrophic run-away condition at abnormally high wind speeds without the use of any additional moving parts or control machinery.

A radial tyre on a car provides a good example of an embedded balanced non-symmetric angleply laminate. The business of containing the pressure is entrusted to plies of textile fibres and rubber in which the fibres are aligned in the radial direction. But the provision of steering and directionality is achieved by a pair of stiff tread-bracing plies which can be made from twisted high-tensile steel wires and rubber and having a form not very different in principle from the calendered conveyor belt we described earlier. These two plies are then laid up at about $\pm 20°$ to the crown angle of the tyre, as shown in Figure 5.20, and are located under the tread of the tyre. It can be seen that there are many other components in a car tyre, but we will not discuss them here.

Balanced regular non-symmetric angleply laminate RRX30NS2

This laminate is based on the unidirectional lamina RR having the properties given in Table 5.5. For a $30°/-30°$ regular non-symmetric laminate consisting of two plies each 2 mm thick, the ABD matrix and its inverted form are given in

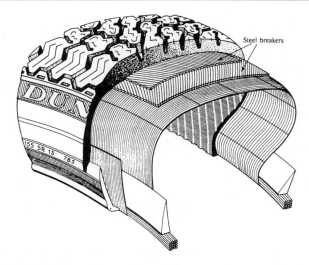

Figure 5.20 The anatomy of a radial car tyre (courtesy Dunlop).

Table 5.11 Stiffnesses and compliances of regular non-symmetric angleply laminate $(30°/-30°)_T$ RRX30NS2 – each ply 2 mm thick

A (kN/mm) B (kN)				B (kN) D (kNmm)	
3.947	1.33	0	0	0	−2.249
1.33	0.4866	0	0	0	−0.748
0	0	1.31	−2.249	−0.748	0
0	0	−2.249	5.263	1.774	0
0	0	−0.748	1.774	0.6489	0
−2.249	−0.748	0	0	0	1.746

a (mm/MN) h (/MN)				b (/MN) d (/MNmm)	
4160	−9187	0	0	0	1422
−9187	26310	0	0	0	−562.7
0	0	2883	1422	−562.7	0
0	0	1422	3120	−6890	0
0	0	−562.7	−6890	19730	0
1422	−562.7	0	0	0	2162

Table 5.11. the amount of each ply and orientation in this laminate is the same as that in RRX30S but the stacking sequence is quite different.

Comparison between Tables 5.5 and 5.11 show that the [A] matrix is the same for each laminate.

The major differences are in the terms B_{16}, B_{26}, D_{16}, and D_{26}. In the non-symmetric laminate $D_{16} = D_{26} = 0$ because the two layers are identical but laid at positive and negative angles to the reference x direction, and $Q_{16}^*(+\theta) = -Q_{16}^*(-\theta)$, and Q_{26}^* similarly. The terms B_{16} and B_{26} are non-zero, and are responsible for coupling between direct stress and twisting curvature, as we shall see below.

An angleply laminate is termed *antisymmetric* when all pairs of similar plies at identical locations above the below the midplane are oriented at the same absolute value of angle but of opposite sign: e.g. $+\theta$ and $-\theta$. It is clear that laminate RRX30NS2 is the simplest example of an antisymmetric angleply laminate, $(+\theta/-\theta)_T$.

Application of single force resultant $N_x = 1$ N/mm to RRX30NS2

Under $N_x = 1$ N/mm the response of RRX30NS2 is $\varepsilon_x^\circ = a_{11} N_x = 0.416\%$, $\varepsilon_y^\circ = a_{12} N_x = -0.919\%$ and $\kappa_{xy} = h_{16} N_x = 1.422$/m, and the character of this deformation is shown in Figure 5.21.

The stress and strain profiles in the global (x, y) co-ordinates are shown in Figure 5.22. The direct strain distributions are uniform, but the shear strain varies linearly through the thickness consistent with the twisting curvature caused by the applied load. This behaviour stems from the general form of the strains as a function of thickness: $\varepsilon = \varepsilon^\circ + z\kappa$: $\varepsilon_x = \varepsilon_x^\circ, \varepsilon_y = \varepsilon_y^\circ$, and $\gamma_{xy} = z\kappa_{xy}$.

The stress $\sigma_x(z)$ derives from $\sigma_x(z) = Q_{11}^* \varepsilon_x + Q_{12}^* \varepsilon_y + Q_{16}^* z\kappa_{xy}$. For the $+30°$ ply at the top surface, $z = -0.002$ m, and taking values of transformed reduced stiffness from Table 5.5, we have $\sigma_x(+30°, -2\,\text{mm}) = 0.987 \times 0.00416 + 0.333 \times (-0.00919) + 0.562 \times (-0.002) \times 1.422\,\text{GPa} = -0.522$ MPa.

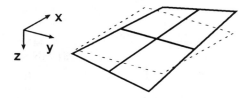

Figure 5.21 Response of RRX30NS2 to $N_x = 1$ N/mm.

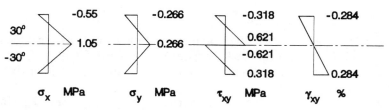

Figure 5.22 Responses in global directions of RRX30NS2 to $N_x = 1$ N/mm.

The value of σ_x increases to the midplane as the contribution from twisting reduces to zero. Below the midplace $Q^*_{16}(30°)$ is negative, and z is positive, so σ_x reduces with increase in co-ordinate below the midplane.

The profiles σ_y and τ_{xy} may be calculated in a similar way even though their net values must be zero because there is no external applied load N_y or N_{xy}. There is a discontinuity in the shear stress profile at the interface because $Q_{16}(-30°) = -Q_{16}(+30°)$.

Stress and strain profiles in the principal directions $(1, 2)$ for each ply are shown in Figure 5.23.

Application of moment resultant $M_x = 0.1\ N$ to RRX30NS2
The resulting deformations are $\varepsilon_x^° = \varepsilon_y^° = 0$, $\gamma_{xy}^° = 0.01422\%$, $\kappa_x = 0.312/m$, $\kappa_y = -0.689/m$, and $\kappa_{xy} = 0$. These deformations have the character of Figure 5.24, and the stress and strain profiles are given in Figure 5.25. The bending strain profiles are linear, and the shear strain is uniform as expected from the general equation $[\varepsilon] = [\varepsilon^°] + z[\kappa]$. The large discontinuity in direct stress σ_x at the midplane arises from $\sigma_x = Q^*_{16}\gamma_{xy}^°$, which is positive for the $+30°$ ply and negative for the $-30°$ ply.

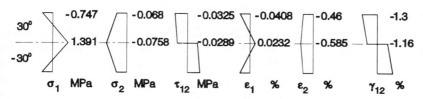

Figure 5.23 Responses in principal directions of RRX30NS2 to $N_x = 1$ N/mm.

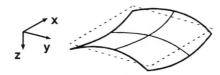

Figure 5.24 Response of RRX30NS2 to $M_x = 0.1$ N.

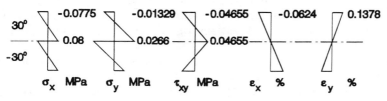

Figure 5.25 Responses in global directions of RRX30NS2 to $M_x = 0.1$ N.

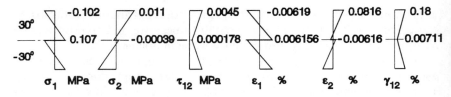

Figure 5.26 Responses in principal directions of RRX30NS2 to $M_x = 0.1$ N.

Stress and strain profiles in the principal directions of each ply are shown in Figure 5.26.

Problem 5.34: Identify which of the following laminates are anti-symmetric:

(a) $(+\theta/-\theta/+\theta/-\theta)_T$, (b) $(+\theta/-\theta/0°/-\theta/+\theta)_T$, (c) $(+\theta/-\theta/0°/+\theta/-\theta)_T$,
(d) $(+\theta_1/-\theta_2/-\theta_2/-\theta_1)_T$, (e) $(+\theta_1/-\theta_2/+\theta_2/-\theta_1)_T$.

Problem 5.35: Biaxial midplane strains $\varepsilon_x^\circ = \varepsilon_y^\circ = 0.5\%$ are imposed on RRX30NS2. Calculate the induced forces and sketch the stress profiles in global co-ordinates and in principal directions.

Problem 5.36: The proposed blade of a wind turbine for electricity genera-tion may be modelled initially as a long 'flat' blade for the purposes of this problem. The designer wishes to design a blade which will decrease in efficiency as the wind speed increases, to avoid destruction at very high wind speeds. Using unidirectional stiff fibre-polymer plies, the designer can investigate the feasibility of making a one-piece blade which will fulfil this criterion without the use of external parts or external control. How would you use lamination theory to give cost-effective protection from high speed steady winds? How would your outline proposal affect the axial and bending stiffnesses of the blade? What kinds of stacking sequence and orientations would you suggest, which would form the basis of later and more detailed investigations?

5.5.4 Unbalanced unsymmetric angleply laminate

In the unbalanced unsymmetric angleply laminate we find a fully populated ABD matrix and its inverse. There are therefore substantial coupling effects, and the response to combinations of applied loads or deformations is complicated.

Where plies of different thicknesses at two angles of the same value but of opposite sign are involved, the [A] and [D] matrices and the [abhd] matrices are fully populated, but the [B] matrix is not fully populated: terms B_{16} and B_{26} are non-zero, but $B_{11} = B_{12} = B_{22} = B_{66} = 0$. The 16 and 26 coefficients arise because $Q_{16}^*(-\theta) = -Q_{16}^*(+\theta)$, and Q_{26}^* similarly. Such a laminate gives full bending/extension coupling when loads or moments are applied.

It is quite possible to fill up the [B] matrix by choosing the two angles to be α and β, having different numerical values, so that terms contributing to each of the stiffness matrices do not cancel out. The unsymmetric crossply laminate rotated by any angle $\pm n\pi/4$, where n is any integer (including zero), will behave in this way.

Laminate APC44/−4

This laminate consists of three 0.125 mm plies of carbon fibre reinforced PEEK (APC2) bonded in the sequence $(+40°/+40°/-40°)_T$. The elastic constants of APC2 are given in Problem 3.8. The stiffnesses and compliances of the laminate APC44/−4 are given in Table 5.12.

Laminate APC44/−4 has the expected population of stiffness and compliance matrices. The response to any applied loads or deformations will normally be quite complicated, and the designer will approach problems with great care.

Application of single force resultant $N_x = 10$ N/mm to APC44/−4
Provided there are no restraints at the grips, the deformation responses will be $\varepsilon_x° = a_{11}N_x = 0.1477\%$, $\varepsilon_y° = a_{12}N_x = -0.07925\%$, $\gamma_{xy}° = -0.06202\%$, $\kappa_x = h_{11}N_x = -2.14$/m, $\kappa_y = h_{12}N_x = -1.683$/m, and $\kappa_{xy} = h_{16}N_x = +7.009$/m. The curvatures dictate that stresses vary linearly with thickness co-ordinate within and between the plies.

Table 5.12 Stiffnesses and compliances of regular unbalanced symmetric laminate APC44/−4; each ply 0.125 mm thick

A (kN/mm) B (kN)			B (kN) D (kNmm)		
20.28	11.69	4.502	0	0	−1.126
11.69	12.09	3.328	0	0	−0.8096
4.502	3.238	12.66	−1.126	−0.8096	0
0	0	−1.126	0.2376	0.137	0.005862
0	0	−0.8096	0.137	0.1417	0.004216
−1.126	−0.8096	0	0.005862	0.004216	0.4184

a (mm/MN) h (/MN)			b (/MN) d (/MNmm)		
147.7	−79.25	−62.02	−214	−168.3	700.9
−79.25	209.4	−48.79	−168.3	−132.5	551.5
−62.02	−48.79	211.1	700.9	551.5	−779.9
−214	−168.3	700.9	11840	−7361	−2800
−168.3	−132.5	551.5	−7361	17390	−2203
700.9	551.5	−779.9	−2800	−2203	15240

5.5.5 A special angleply laminate

The behaviour of the balanced $(+\theta/0/-\theta)$ angleply laminate depends on θ, and on the proportion of angle plies. three examples where this arrangement is used are the driveshaft of a car or lorry, the shaft of a golf club, and the wing of a conventional aircraft or glider.

In a driveshaft the obvious requirement is high stiffness in torsion, achieved by setting $\theta = 45°$. But at this angle alone the shaft would have too low a bending stiffness along its axis. The bending can be increased in principle by ensuring that some of the fibres are placed along the shaft. The practical arrangement is to wind the fibres at the lowest possible angle to the axis.

In the golf club, when the club hits the ball it is desirable that the shaft twists and bends to only a small and acceptable extent. The twisting is resisted by windings at $+45°$ and $-45°$, and longitudinal $0°$ fibres confer the bending stiffness.

In an aircraft wing the lift requires high longitudinal bending stiffness and calls for fibres along the wing. Resistance to torsion calls for a ply angle of $45°$.

Thus in all three examples the principle of $+45°/0/-45°$ is used, although with different angleply ratios and symmetric stacking sequences.

Tetlow (1973) provides a helpful pictorial view of the stiffness behaviour of the regular balanced symmetric laminate $(+\theta/0/-\theta)_s$. He uses a stylized building block of three plies in which the proportion of plies at an angle to the reference direction is $R/2$ at $+\theta$, $R/2$ at $-\theta$, and $(1-R)$ at $0°$. (Figure 5.27). Tetlow called R the crossply ratio, but in this book it would seem more useful to call R the angleply ratio. He then produced carpet plots to show how the modulus (under a single stress) varied with θ and R. Three such plots are given in Figure 5.28 for a high modulus carbon fibre epoxy ply having $E_1 = 207.6\,\text{GPa}$, $E_2 = 7.59\,\text{GPa}$, $G_{11} = 4.83\,\text{GPa}$ and $v_{12} = 0.3$.

The carpet plots confirm the behaviour of regular symmetric angleply laminates $(R = 1)$ already described in Section 5.5.1. In particular they emphasize the large shear modulus which can be achieved by lay-ups at $\pm 45°$ with high angleply ratios: these are of considerable technical interest in transmitting torque, and can be made by filament winding. The compromise introduced by

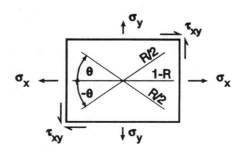

Figure 5.27 The Tetlow special angleply laminate.

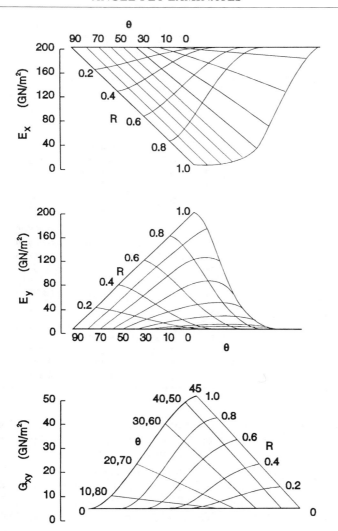

Figure 5.28 Carpet plots of modulus for laminate $(+\theta/0/-\theta)_s$ for different angleply ratios R. Carbon fibre epoxy unidirectional plies.

the need for a small proportion of plies at 0° is clear. Comments on other special laminates are made in Section 5.6 and 5.8.

5.5.6 Symmetric cross-ply laminates stressed off-axis

In practice crossply laminates are often made from woven cloth reinforced with a matrix. But the crossply laminate is loaded off-axis in service: the purpose of this subsection is to explore briefly the stiffnesses of the symmetric

Table 5.13 Stiffnesses and compliances for C + 4S stressed at $-\theta$ to the 0° direction

Coeff.	$\theta = 0°$	15°	30°	45°	60°	75°
A_{11} kN/mm	47.19	42.07	31.82	26.27	31.28	42.07
A_{12}	1.205	6.328	16.57	21.7	16.57	6.328
A_{22}	47.19	42.07	31.28	26.7	31.82	42.07
A_{66}	2.5	7.623	17.87	22.99	17.87	7.623
A_{16}	0	−8.873	−8.873	0	8.873	8.873
A_{26}	0	8.873	8.873	0	−8.873	−8.873
D_{11} kNmm	1.658	1.461	1	0.5562	0.3256	0.2922
D_{12}	0.0251	0.1318	0.3453	0.452	0.3453	0.1318
D_{22}	0.3085	0.2922	0.3256	0.5562	1	1.461
D_{66}	0.05208	0.1588	0.3723	0.4769	0.3723	0.1588
D_{16}	0	−0.3535	−0.477	−0.3373	−0.1072	−0.01622
D_{26}	0	−0.01622	−0.1072	−0.3373	−0.477	−0.3535
a_{11} mm/kN	0.02121	0.04349	0.08805	0.1103	0.08805	0.04349
a_{12}	−0.00541	−0.02282	−0.06739	−0.08967	−0.06739	−0.02282
a_{22}	0.02121	0.04349	0.08805	0.1103	0.08805	0.04349
a_{66}	0.4	0.3109	0.1326	0.04349	0.1326	0.3109
a_{16}	0	0.07719	0.07719	0	−0.07719	−0.07719
a_{26}	0	−0.07719	−0.07719	0	0.07719	0.07719
d_{11} /kNmm	0.604	1.734	4.124	5.738	5.445	4.022
d_{12}	−0.04914	−1.002	−2.909	−3.862	−2.909	−1.002
d_{22}	3.245	4.022	5.445	5.738	4.124	1.734
d_{66}	19.2	15.39	7.761	3.947	7.761	15.39
d_{16}	0	3.963	4.446	1.321	−2.159	−2.642
d_{26}	0	−2.642	−2.159	1.321	4.446	3.963

crossply laminate stressed in-plane and off-axis. Table 5.13 provides a source of stiffness and compliance coefficients for the crossply laminate C + 4S stressed at an angle $-\theta$ to the 0° direction.

The following points are apparent from perusal of these data. A_{12} and A_{66} peak at 45°, D_{16} peaks at 30°. a_{12} and v_{xy} reach maximum values at 45°. From other calculations not summarized in the table we find that a_{16} reaches its peak value of +0.08913 mm/kN at 22.5°, and d_{16} peaks at 4.767/kNmm at about 25°.

5.6 QUASI-ISOTROPIC LAMINATES

It is often convenient to have both isotropic behaviour in the plane and a high volume fraction of long fibres. As we shall in the next subsection, the (conventional short-fibre) random mat laminate usually shows isotropic behaviour in the plane but has a small volume fraction of fibres. We have seen that the simple laminates described earlier can readily have high volume fractions, but show marked directionality of properties.

The quasi-isotropic laminate attempts to achieve a good compromise, namely high volume fractions of fibres (often using prepreg forms or filament winding), and a *symmetric* sequence of ply orientations chosen such that when stressed in any direction *in the plane* the elastic constants E_x, E_y, G_{xy} and v_{xy} are independent of angle of loading to the chosen reference direction. The simplest and most economical example of a quasi-isotropic laminate involves the basic three-ply sublaminate $(0°/60°/-60°)$. We use 'quasi-isotropic' because these laminates are not isotropic under bending or twisting loads.

5.6.1 Symmetric quasi-isotropic laminates

The symmetric quasi-isotropic laminate we shall first discuss, QI60S is based on the ply of cold-cured epoxy with undirectional glass fibres EG having the elastic constants $E_1 = 45.6\,\text{GPa}$, $E_2 = 10.73\,\text{GPa}$, $G_{12} = 5.14\,\text{GPa}$, and $v_{12} = 0.274$. The stacking sequence in QI60S is $(0°/60°/-60°)_s$ and the laminate stiffnesses and compliances are shown in Table 5.14.

The population of the ABD and abhd matrices is as expected. $A_{16} = A_{26} = 0$ (and $a_{16} = a_{26} = 0$) because terms in $+\theta$ and $-\theta$ cancel out; $[B] = 0$ because of symmetry. $[D]$ is a fully populated matrix because of the weighting of the thickness of plies which prevents cancelling out of terms in $+\theta$ and $-\theta$; similarly $D_{11} \neq D_{22}$. The reader will recognize that this laminate is a particular example of the special angleply laminate described in the previous section.

Table 5.14 Symmetric balanced regular quasi-isotropic laminate $(0°/60°/-60°)_s$, QI60S, based on 0.25 mm plies of EG

A (kN/mm) B (kN)			B (kN) D (kNmm)		
37.23	10.26	0	0	0	0
10.26	37.23	0	0	0	0
0	0	13.48	0	0	0
0	0	0	10.36	1.323	0.272
0	0	0	1.323	4.809	0.6887
0	0	0	0.272	0.6887	1.927
a (mm/MN) h (/MN)			b (/MN) d (/MNmm)		
29.07	−8.013	0	0	0	0
−8.013	29.07	0	0	0	0
0	0	74.16	0	0	0
0	0	0	100.1	−26.89	−4.519
0	0	0	−26.89	226.4	−77.11
0	0	0	−4.519	−77.11	547.2

Table 5.15 Regular balanced symmetric quasi-isotropic laminate $(-30°/30°/-90°)_s$ QI60Sa, based on six 0.25 mm plies of EG

A (kN/mm) B (kN)			B (kN) D (kNmm)		
37.23	10.26	0	0	0	0
10.26	37.23	0	0	0	0
0	0	13.48	0	0	0
0	0	0	8.719	2.406	−1.377
0	0	0	2.406	4.281	−0.5439
0	0	0	−1.377	−0.5439	3.01

a (mm/MN) h (/MN)			b (/MN) d (/MNmm)		
29.07	−8.013	0	0	0	0
−8.013	29.07	0	0	0	0
0	0	74.16	0	0	0
0	0	0	143.3	−73.91	52.25
0	0	0	−73.91	277.2	16.27
0	0	0	52.25	16.27	359.1

Problem 5.37: Calculate the Poisson's ratios for QI60S.

Problem 5.38: The symmetric laminate QI60S is rotated by $-30°$ to the reference axis (previously 0°), to become QI60Sa. The elastic constants for this laminate $(-30°/30°/90°)_s$ are given in Table 5.15. Comment on the similarities and differences in laminate stiffness behaviour under the two reference directions.

Problem 5.39: Compare the twisting of QI60S, QI60Sa, and a sheet of isotropic material under the action of a single applied bending moment resultant $M_x = 20$ N.

Problem 5.40: Calculate the stress profiles in QI60S under an applied load $N_x = 10$ N/mm.

Problem 5.41: A sheet of QI60S is to be bent round a drum 4 m diameter. Calculate the stress profiles in the bent sheet.

5.6.2 Non-symmetric quasi-isotropic laminates

In this subsection we shall briefly examine the features of a non-symmetric laminate QI60NS based on three plies of EG each 0.5 mm thick laid up as $(0°/60°/-60°)_T$. The amount and total thickness of the EG is the same as for laminate QI60S. The stiffnesses and compliances of QI60NS are presented in Table 5.16.

Table 5.16 Regular unbalanced non-symmetric quasi-isotropic laminate $(0°/60°/-60°)_T$ QI60NS based on three plies each 0.5 mm thick EG

A (kN/mm) B (kN)			B (kN) D (kNmm)		
37.23	10.26	0	−8.099	1.444	−1.088
10.26	37.23	0	1.444	5.212	−2.755
0	0	13.48	−1.088	−2.755	1.444
−8.099	1.444	−1.088	8.331	1.684	−0.5439
1.444	5.212	−2.755	1.684	6.112	−1.377
−1.088	−2.755	1.444	−0.5439	−1.377	2.288

a (mm/MN) h (/MN)			b (/MN) d (/MNmm)		
39.86	−10.52	0.3623	43.3	−9.864	10.42
−10.52	35.28	−8.302	−11.18	−22.25	26.66
0.3623	−8.302	86.17	3.123	35.48	−42.09
43.3	−11.18	3.123	174.5	−42.62	20.97
−9.864	−22.25	35.48	−42.62	229.3	74.07
10.42	26.66	−42.09	20.97	74.07	550.3

Note that [h] is the transpose of [b], and that the matrices [b] and [h] are not symmetric. This is the reason why [b] and [h] have been kept separate up to this point in the book.

For QI60NS the [B] matrix is fully populated now, as expected, so there is substantial bending/stretching coupling. The [A] matrix for QI60NS is the same as for QI60S, because the same amounts of material are present in each of the ply directions. But because of the presence of the non-zero [B], the extensional compliance matrix [a] does not show isotropy in the plane: $a_{11} \neq a_{22}$.

Because the [abhd] matrix is fully populated, the response to even a single load or moment is as complicated as possible. For example, applying $N_x = 10 \, \text{N/mm}$ generates the responses $\varepsilon_x^\circ = a_{11} N_x = 0.03986\%$, $\varepsilon_y^\circ = -0.01052\%$, $\gamma_{xy}^\circ = 0.0003623\%$, $\kappa_x = 0.433/\text{m}$, $\kappa_y = -0.09864/\text{m}$ and $\kappa_{xy} = 0.1042/\text{m}$.

Table 5.17 Compliances of $(\theta/\theta + 60°/\theta - 60°)_T$ based on QI60NS for selected reference directions, θ

Coefficient	Maximum value	Minimum value
a_{11} mm/kN	0.3986 at 0°	
a_{12} mm/kN	−0.01177 at −15°	−0.006764 at +30°
a_{22} mm/kN	0.03986 at −30°	0.02923 at +30°
v_{xy} −	0.3068 at −20°	0.1761 at +30°

Rotating the reference direction by θ, i.e. treating QI60NS as $(\theta/\theta + 60°/\theta - 60°)_T$ demonstrates how anisotropic this non-symmetric laminate is under in-plane loads, and Table 5.17 gives maximum and minimum values of selected compliances.

Table 5.18 Regular unbalanced non-symmetric quasi-isotropic laminate $-15/45/-75$ QI60NSb based on three plies each 0.5 mm EG

A (kN/mm) B(kN)			B(kN) D(kNmm)		
37.23	10.26	0	−7.685	0	1.667
10.26	37.23	0	0	7.685	−1.667
0	0	13.48	1.667	−1.667	0
−7.685	0	1.667	7.462	1.443	−1.109
0	7.685	−1.667	1.443	7.462	−1.109
1.667	−1.667	0	−1.109	−1.109	2.047

a (mm/MN) h(/MN)			b(/MN) d(/MNmm)		
38.82	−11.77	−4.584	37.84	0.761	−20.27
−11.77	38.82	−4.584	−0.761	−37.84	20.27
−4.584	−4.584	81.16	−28.33	28.33	0
37.84	−0.761	−28.33	194.2	−34.95	54.87
0.761	−37.84	28.33	−34.95	194.2	54.87
−20.27	20.27	0	54.87	54.87	581

Table 5.19 Regular unbalanced non-symmetric quasi-isotropic laminate $(30°/90°/-30°)_T$ QI60NSc based on 0.5 mm plies of EG

A (kN/mm) B(kN)			B(kN) D(kNmm)		
37.23	10.26	0	0	0	−5.51
10.26	37.23	0	0	0	−2.176
0	0	13.48	−5.51	−2.176	0
0	0	−5.51	8.719	2.406	0
0	0	−2.176	2.406	4.281	0
−5.51	−2.176	0	0	0	3.01

a (mm/MN) h(/MN)			b(/MN) d(/MNmm)		
38.4	−6.764	0	0	0	65.41
−6.764	29.23	0	0	0	8.752
0	0	101.2	58.8	18.33	0
0	0	58.8	170	−65.61	0
0	0	18.33	−65.61	279.7	0
65.41	8.752	0	0	0	458.4

A full set of stiffnesses and compliances for $\theta = -15°$ is given in Table 5.18, and $\theta = 30°$ in Table 5.19.

It can be seen that $a_{11} = a_{22} = 38.82 \, \text{mm/MN}$ at $\theta = -15°$, and we find that the maximum value of the anisotropy ratio occurs at $a_{11}/a_{22} = 1.314$. It should be noted that for the non-symmetric quasi-isotropic laminates we see that [b] and [h] are no longer symmetric even though [B] is symmetric.

5.7 RANDOM-MAT LAMINATES

We have looked at a number of examples of the angleply laminate in which plies have been aligned in several carefully chosen directions. It is quite possible to develop this concept to the extreme where the plies are notionally infinitesimally thin. These microplies can then be stacked with alignments in all possible directions in the plane of the sheet. The plies are now randomly orientated in the plane, and so we can call this a random laminate. In practice the fibres are quite short, often less than 50 mm, and sometimes only a few millimetres long. The rods cannot now pack closely together, and moreover few of the rods can ever be aligned with the direction of any particular applied force. So this is a low efficiency laminate which has the same indifferent properties in any direction in the plane.

5.7.1 Examples of random laminates

The most common polymer-based examples of the random laminate are paper and hardboard, both consisting of cellulose fibres with an appropriate glue. Because of the arrangment of the fibres, these laminates are always stronger in the plane than through it – indeed many papers and all hardboards are quite easy to delaminate. Hardboard and chipboard are difficult to chop into regular pieces because the rods have to be broken ahead of the crack tip. Felt cloth, for hats, is a random array of textile fibres held together largely by friction.

It will be noted that dry paper will not conform to a convex surface as loosely woven cloth does. Any attempt to make dry paper conform leads to creases, because a sheet of paper is too stiff in shear, whereas cloth is usually deformable. (We discussed the mechanism for this in Section 5.2.1). Wetting out the paper – the extreme is papier mâché – destroys the stiffness of the glue holding the paper fibres in place, and permits relative movement between fibres, so the wet paper is more deformable and conformable.

Other laminates from everyday experience also come to mind, such as the double thickness of cloth, perhaps with an interfacing layer, used to make shirt collars, or waistbands for skirts and trousers. This design is most effective – as is true for all engineering composites designed for stiffness – when there is a good bond between the layers.

In most modern kitchens the working surfaces are now a laminate consisting of a core of chipboard and a top surface of a melamine composition. The chipboard is itself a random composite consisting of short flakes or slivers of wood arranged randomly in the plane and bonded together with a thermosetting resin.

Notice, incidentally, that many laminates most readily show their construction when you break them by bending – whether this be a biscuit or a piece of wood.

A large part of the glass fibre reinforced plastics industry thrives on making things from a randomly-aligned arrangement of chopped glass fibres strands impregnated with a crosslinkable resin. Products range from canoes, car body panels to cladding panels for buildings. A spectacular exploitation of the properties of glass fibre composites is the hull of mine detecting vessels: these hulls can be over 50 m long and weigh over 130 tonnes. The density of the glass fibres is about 2500 kg/m^3, and the fibre diameter of the fibres is of the order of 10 microns. So if each fibre of the 65 tonnes of glass were joined end to end the straight chain would reach from the earth to the sun, and a good way back.

Other common examples from the transport industry made in a random-in-plane system include lorry cabs, and the panels for the express bus or coach. Some motorway service station buildings are clad with large one-piece panels.

The outside of the driver's cab for the British Rail High Speed Train also relies on GRP. The structure (Morton-Jones and Ellis, (1986)) is aerodynamically efficient, light in weight (it is over 230 kg lighter than the equivalent steel structure), is easily made in production runs of a few hundred, has good weather resistance, and protects the drive from missiles impacting 'corner-on' at 280 km/h. One design problem is to achieve adequate stiffness. the solution is a sandwich panel consisting of a polyurethane foam (density 32 kg/m^3) encased in GRP skins. The outer skin was made up from an outer gelcoat, then chopped strand mat random in the plane, crossplied unidirectional cloth, then some continuous filament mat having a swirled pattern, giving a laminate some 7 mm thick, with 34 mm of foam, and some 9 mm of skin of a similar construction.

Returning to the theme of pressure vessels, if we inflate a flexible sphere, the tension in the wall is the same in all directions round the circumference. In a toy balloon the long chain rubber molecules are initially randomly orientated in all possible directions, including through the thickness of the wall. The wall is typically about 0.1 mm thick. The radius is small, so it is quite difficult to start inflating the balloon. Once started, however, the bigger radius makes it easier to continue inflation. During this stage bits of the rubber molecules are stretching out and uncoiling in the plane of the balloon wall – we are gradually forming a sort of random laminate, with alignment of rods (molecules) in the plane of the balloon wall. It is quite easy to straighten out the molecules into this new alignment, but when they are as straight as they can reasonably be, the force needed to elongate them further rises dramatically. Now this happens to some bits of molecules earlier than to others, so the effect is gradual in practice. The result is that the balloon becomes harder to inflate (apart from the psychological nervousness!) and we have to blow with ever-increasing pressure even though the wall is becoming transparently thin. Before we run out of pressure, one or two molecules snap, perhaps at a weak spot – this throws more load on to the surviving molecules and some of them fail too, and rapidly. The pressure vessel fails catastrophically at almost the speed of sound. The inflater gains, and loses, tension, too.

Another inflation process occurs when we dip a ring into a detergent solution and blow bubbles. If you blow through a straw into the detergent solution, you can readily mass-produce a lot of bubbles very quickly and they are all joined up, and no longer spherical. By various means we can initiate and inflate bubbles in molten or liquid polymer systems. There is some evidence to suggest that in the thin walls of a light cellular structure, the inflation process has developed a random alignment of molecules which have become fixed during the solidification process. Such a light-weight structure forms the core of some sandwich panels which separate load-bearing facing skins; the core acts as a separator, gives a high second moment of area per unit weight, and provides adequate shear stiffness and strength.

For example, if you cut a thin strip from an expanded polystyrene ceiling tile so that there are no totally enclosed cells in the strip, and heat it on plain flour or talc to just above 100 °C, the strip shrivels up because the orientation of molecules can then relax in the heat. This occurs for exactly the same reason that we discussed for shrink wrapping of polyethylene film. What would you expect to happen of you heated a complete tile to above 100 °C?

If you slowly heat a larger piece of expanded polystyrene tile containing substantial quantities of complete cells, the gas in the sealed beads increases in pressure, and the pressure tries to expand the cells. At about 85 °C to 95 °C the polymer starts to lose its stiffness. The thinnest cells on the surface of the tile will start to blow out, and the tile bulges generally. Rapid heating gives a different effect in that only the surface will inflate to begin with.

In a closed-cell structure like the mass of detergent bubbles, most of the mass of the fluid is found near the edge of each facet of the bubble. In other words the facet is like a convex lens, thin at the centre but much thicker at the edge. Further inflation, or the effect of surface tension, or both, can cause the membrane of film in a facet to pop. In a detergent bubble the whole system then collapses. But in the right kind of plastics cellular structure, the result of this popping process is that the fluid plastic drains into what were the edges, because of surface tension, and the resulting mass is now a three-dimensional net which is widely used as a cushioning material. The struts can now be modelled roughly as rods consisting of some aligned molecules in a glue of unaligned molecules of the same kind. Under small compressive loads along their length the rods will bend, but at a higher load they will buckle. The rods in the network are pointing in all possible directions, so under an applied com-

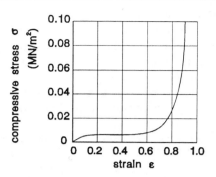

Figure 5.29 Typical stress–strain curve for cushion foam.

pressive force the ones in line with the force will buckle first. As the load is increased, progressive buckling occurs, and the cushion eventually bottoms with a pronounced increase in density compared with its initial state. Figure 5.29 shows a typical stress–strain curve for a cushion foam). An interesting feature of lightweight cushion foam is that it deforms without bulging sideways – it has a zero Poisson's ratio. For more details about the behaviour of foams, Gibson and Ashby (1988) give an exceptionally clear summary.

Let us return to toy balloons, this time the narrow cylindrical ones, and inflate one not too near to bursting. The cylindrical part becomes longer and fatter, and there is some molecular orientation round the circumference and along the length, thus forming a crossply structure. But at the spherical end the polymer molecules achieve a random-in-plane orientation like fibres in paper. The inflated balloon is stiffer than an uninflated one.

Now suppose we mould from a molten polyester plastic a sort of test tube, and let it solidify. At a special temperature we can stretch it by holding the open end and ramming a warm plug down the axis. We can then inflate the stretched tube within a bottle mould to make a cylindrical balloon shape. The result can be modelled as a crossply of molecules in the cylinder and a random array of molecules in the wall of the spherical end. Cooling the inflated bottle freezes in these arrangements and this bottle is therefore much stiffer and stronger than it would be if unstructured and of the same dimensions. This is how the clear carbonated drinks bottles are made which you see in the supermarket.

The technology and materials have been developed to the stage where large containers up to 100 litres capacity can be blown by this orientation process: the limitation currently is the manufacture of the right sort of machinery which is currently being developed, particularly by one firm in the USA in conjunction with a USA supplier of the PET. Shapes other than the easy round-ended cylinder are also being investigated: orientated polyster petrol tanks for the Ford Fiesta car have been under development.

5.7.2 Analysis and behaviour of random laminates

We shall now explore, using problems, some aspects of the analysis and stiffness of a modest range of random laminates. We shall use the tools of laminate analysis and try to represent the random laminate as a symmetric laminate based on a number of unidirectional plies.

The model of a random-in-plane composite based on a very large number of infinitesimally thin unidirectional plies orientated at all possible angles has the same maximum value of volume fraction of fibres as a unidirectional array for a single ply. For hexagonal packing, we saw in Chapter 1 that $V_{f\text{max}} \approx 90.7\%$. In practice chopped strand random mats have a much smaller V_f because the construction is more roughly made: fibres are much more widely spaced to help impregnation of the sheet and complete wetting out of the interface of the fibres), and layers of fibres overlap.

Problem 5.42: Comment on the likely success of the following attempts (not all correct) to model random-in-plane laminates based on identical thin unidirectional plies laid up at the stated angles to the reference direction:

(a) $(0°/30°/60°/90°)_T$;
(b) $(0°/30°/60°/90°)_S$;
(c) $(-90°/-60°/-30°/0°/0°/30°/60°/90°)_T$;
(d) $(-90°/-60°/-30°/0°/0°/30°/60°/90°)_S$;
(e) $(-90°/-60°/-30°/0°/30°/60°)_T$;
(f) $(-60°/-30°/0°/30°/60°/90°)_S$;
(g) $(-60°/-30°/0°/30°/60°/90°)_T$.

Problem 5.43: A laminate CRNDM7 is based on 0.1 mm plies of material PCPHMC having the elastic constants $E_1 = 208\text{GPa}$, $E_2 = 7.6\,\text{GPa}$, $G_{12} = 4.8\,\text{GPa}$, $v_{12} = 0.3$. The laminate is based on $(-60°/-30°/0°/30°/60°/90°)_S$. The ABD matrices and their inverses are given below. Does this represent a random-in-plane laminate? Calculate E_x, E_y, G_{xy}, and v_{xy}.

Stiffnesses of laminate CRNDM7

A (kN/mm) B (kN)			B (kN) D (kNmm)		
100.9	31.63	0	0	0	0
31.63	100.9	0	0	0	0
0	0	34.64	0	0	0
0	0	0	13.1	4.613	-3.073
0	0	0	4.613	9.481	-4.24
0	0	0	-3.073	-4.24	4.975

a (mm/MN) h (/MN)			b (/MN) d (/MNmm)		
10.99	−3.444	0	0	0	0
−3.444	10.99	0	0	0	0
0	0	28.87	0	0	0
0	0	0	95.02	−32.3	31.16
0	0	0	−32.3	181.4	134.7
0	0	0	31.16	134.7	335

Problem 5.44: By modelling a chopped glass strand laminate as symmetrical with a very large number of infinitesimally thin microlaminae orientated uniformly in all possible directions to a reference direction in the plane, show that the in-plane tensile modulus of this randomly aligned laminate, E_r, is approximately given by $E_r \sim 3E_1/8 + 5E_2/8$, and that the in-plane shear modulus G_r is given by $G_r \sim E_r/3$. E_1 and E_2 are the moduli of the microlamina in its principal directions. Give an estimate of the likely Poisson's ratio. What is the likely maximum volume fraction in this model composite; how and why does this compare with typical values of $V_f = 0.15$ in practice?

5.8 SOME OTHER SPECIAL LAMINATES

We have already given details of a range of examples of simple laminates. The choice of which one to use, if any of them, depends on the loading patterns in the particular application and which of the various coupling effects are thought desirable and which undesirable. In this book we cannot describe all the possible implications of choosing different orientations for plies: the interested reader will have to explore these possibilities by using commercially available computer software such as LAP. What we can do, for laminates based on unidirectional plies, is to summarize the coupling effects already described in this chapter. We can then outline some more complicated ideas which enable us to achieve special effects, full details of which are cited in the select bibliography.

5.8.1 General comments on coupling effects

The main effects we have met so far, together with the stiffness coefficients responsible for them are as follows:

Coupling effect	Coefficients
In-plane direct-shear	A_{16}, A_{26}
In-plane – bending ('stretch'-bending)	[B]

Direct extension – bending	B_{11}, B_{12}, B_{22}
Bending – twisting	D_{16}, D_{26}
Extension – twisting	B_{16}, B_{26}

Adding to the list of undesirable coupling effects restricts the choice of suitable simple laminates, and the outline answers to Problem 5.45 indicate these. Seeking prescribed different stiffnesses in different directions may rule out some or all of the remaining possibilities.

For example, to eliminate in-plane direct-shear, in-plane bending, and bending–twisting coupling effects, the most obvious choice is a symmetric crossply laminate. But Caprino and Crivelli-Visconti (1982) show that it is also possible to use a balanced antisymmetric angleply laminate, e.g. $(+\theta/-\theta/-\theta/+\theta|-\theta/+\theta/+\theta/-\theta)_T$, or $(0/+\theta/-\theta/-\theta/+\theta/0|0/-\theta/+\theta/+\theta/-\theta/0)_T$.

> **Problem 5.45:** Using the data for laminates based on unidirectional plies presented in earlier sections of this chapter, identify those lay-ups which satisfy each of the above criteria taken one-at-a-time.

> **Problem 5.46:** Satisfy yourself that in a laminate based on the same unidirectional plies the minimum number to ensure that $[B] = A_{16} = A_{26} = D_{16} = D_{26} = 0$ is eight plies.

The approach of Caprino and Crivelli-Visconti (1982) has been usefully generalized by Gunnink (1983), who worked with groups of sublaminates based on identical unidirectional plies. As an example suppose we designate a sublaminate $SL1 = (0°/+\theta/-\theta)_S$ with a thickness h_{SL1} and SL2 is some other sublaminate having a thickness h_{SL2}. Using the suffix L to denote the complete laminate having a thickness $h_L = h_{SL1} + h_{SL2}$, we can write down the interesting terms of the ABD matrix

$$(A_{16})_L = (A_{16})_{SL1} + (A_{16})_{SL2}$$
$$[B]_L = (\tfrac{1}{2})([A]_{SL2} h_{SL1} + [A]_{SL1} h_{SL2})$$
$$(D_{16})_L = (D_{16})_{SL1} + (D_{16})_{SL2} + (\tfrac{1}{4})\{(A_{16})_{SL1} h_{SL2}^2 + (A_{16})_{SL2} h_{SL1}^2\}$$

Terms in suffixes 26 have a similar form to those given for 16. If SL2 is symmetric or antisymmetric with respect to SL1, and if both sublaminates are balanced, then $[B] = 0$. In addition, when SL2 is antisymmetric with respect to SL1, then $(A_{16})_L = (A_{26})_L = (D_{16})_L = (D_{26})_L = 0$. The expressions quoted from Caprino and Crivelli-Visconti (1982) satisfy both conditions.

5.8.2 Other ways of eliminating bending–stretching coupling

As remarked earlier, and as will be discussed in greater detail in Chapter 7, hot-cured flat laminated plates will distort substantially unless bending-stretching coupling is eliminated. Symmetric laminates provide an obvious way of avoiding this type of coupling.

For sublaminates based on unidirectional plies, Kandil and Verchery (1990) offer a much more generalized approach than that discussed earlier in this chapter. They predict that $[B] = 0$ for the complete laminate when the sum of the normalized weighting factors for the bending–stretching coupling in each sublaminate is zero. From the definition of $[B]$ we therefore have the condition

$$\sum_{f=1}^{F} (h_f^2 - h_{f-1}^2)/h_{\text{SL}}^2 = 0$$

where F is the total number of plies in the sublaminate. Obviously symmetric laminates satisfy this expression. Perhaps rather suprisingly, there is a range of non-symmetric laminates which also satisfies this condition. The following examples are interesting because they provide a wider range of in-plane and bending properties than can be achieved with symmetric stacking sequences, whilst still eliminating thermal coupling.

1. Laminates based on symmetric sublaminates each of which has the same number of plies and orientations, such as $(\theta_1/\theta_2/\theta_2/\theta_1|\theta_2/\theta_1/\theta_1/\theta_2)_T$ and $(\theta_1/\theta_2/\theta_3/\theta_3/\theta_2/\theta_1|\theta_2/\theta_3/\theta_1/\theta_1/\theta_3/\theta_2)_T$. It is worth studying this in more detail. It is clear that this type of laminate must have at least eight plies. Adding a central ply (of any thickness, material and orientation) will not change $[B]$ and hence will not contribute bending-stretching coupling.

The two examples given here reduce to those given by Caprino and Crivelli-Visconti (1982) in Section 5.8.1 if the plies are assumed to be all the same thickness and material with (for the first) $\theta_2 = -\theta_1$ and (for the second) $\theta_1 = 0$, $\theta_3 = -\theta_2$.

Where $\theta_2 \neq \theta_1$, or where the plies aligned at θ_1 are made from material m_1 taken as different from m_2 at θ_2, balance is no longer possible, and hence the laminate has $[B] = 0$, but shows direct-shear and bend-twist coupling. Such a laminate has the general ABD character of an unbalanced symmetric angleply laminate (e.g. Section 5.5.2) but the potential of a far wider range of properties conferred by the use of plies made from different materials.

Although the thickness of each ply in any sublaminate must be the same, the thicknesses of ply can be different in each sublaminate: the normalization (division by the relevant sublaminate thickness) takes account of different ply thicknesses in the sublaminate.

2. A non-symmetric laminate with seven plies arranged as

$$(\theta_1/\theta_2\theta_2/\theta_3/\theta_1/\theta_1/\theta_2)_T$$

The cancellation of coefficients in $[B]$ is self-evident even though differently oriented plies may be of different materials provided the ply thickness is kept uniform.

3. Laminates consisting of n sublaminates of orientations each of n plies arranged in rotating order, such as

$$(\theta_1/\theta_2/\theta_3|\theta_2/\theta_3/\theta_1|\theta_3/\theta_1/\theta_2)_T$$

Problem 5.47: A laminate is based on two sublaminates. In the top sublaminate the arrangement is $(\theta_1/\theta_2/\theta_2/\theta_1)$ with each ply of thickness $h_L/12$. In the lower sublaminate SL2 the arrangement is $(\theta_2/\theta_1/\theta_1/\theta_2)$ with each ply of thickness $h_L/6$. Each orientation θ_i corresponds to material m_i. Show from first principles that each sublaminate has $[B] = 0$, and that for the laminate-as-a-whole $[B] = 0$.

5.8.3 Quasi-homogeneous laminates

We noted in Chapter 2 that an isotropic plate has the same value of Young's modulus E, Shear modulus G, and Poisson's ratio v, for in-plane and bending deformation. In general this does not occur in laminates based on different isotropic plies or similar unidirectional plies. For example try comparing the Poisson's ratio $v_{xy} = -a_{12}/a_{11}$ for any symmetric laminate under N_x with that calculated erroneously as $v_{xy} = -d_{12}/d_{12}$ under M_x.

A quasi-homogeneous laminate has the same normalized in-plane and bending stiffnesses matrices: $[A]/h_L = [D]/(h_L^3/12)$. If a laminate consists of two or more sublaminates each having the same number of identical plies (orientated at angles θ_i) then Kandil and Verchery (1990) show that the laminate is quasi-homogeneous when each sublaminate has the same sum of the normalized weighting factors in bending:

$$\sum_{f=1}^{F} (h_f^2 - h_{f-1}^3)/h_{SL}^3 = \text{constant}$$

Angle ply laminates such as $(\theta_1/\theta_2)_r$ and $(\theta_1/\theta_2/\theta_2/\theta_1|\theta_2/\theta_1/\theta_1/\theta_2)_T$ are quasi-homogeneous, with suffix r denoting 'repeating'. More complicated is $(\theta_1/\theta_1/\theta_2/\theta_2/\theta_3/\theta_3/\theta_4/\theta_4/\theta_4/\theta_3/\theta_2/\theta_3/\theta_4/\theta_4/\theta_4/\theta_2/\theta_3/\theta_1/\theta_1/\theta_1/\theta_2/\theta_3/\theta_2/\theta_1)_S$.

Problem 5.48: Snow that the non-symmetric laminate $(\theta_1/\theta_2/\theta_1\theta_2)_T$ is quasi-homogeneous, assuming all plies are identical but aligned at angles θ_1 or θ_2.

FURTHER READING

Caprino G. and Crivelli-Visconti, I. (1982) A note on specially orthotropic laminates, *J. Composite Materials* **16** (September) 395.

Dinwoodie, J.M. (1989) *Wood: Nature's cellular polymeric fibre composite*, The Institute of Metals, London.

Dobrin, P.B. (1978) Mechanical properties of arteries, *Physiological Reviews* **58** (April) 397–460.

French, M.J. (1988) *Invention and Evolution: design in nature and engineering*, Cambridge.

Gibson, L.J. and Ashby, H.F. (1988) *Cellular Solids: structure and properties*, Pergamon.

Gunnink, J.W. (1983) Comment on 'A note on specially orthotropic laminates', *J. Composite Materials* **17** (November) 508.

Hearle, J.W.S., Grosberg, P. and Backer, S. (1969) *Structural Mechanics of Fibres, Yarns and Fabrics,* Volume 1, Wiley.

Jones, R.M. (1975) *Mechanics of Composite Materials,* McGraw-Hill.

Kandil, N. and Verchery, G. (1990) Design of stacking sequences of laminated plates for thermoelastic effects, Proc 2nd CADCOMP, Springer Verlag.

Karaolis, N.M., Musgrove, P.J. and Jeronomidis, G. (1988) Active and passive aerodynamic power control using asymmetric fibre reinforced laminates for wind turbine blades. Proceedings of the 10th British Wind Energy Association Conference, London 22–24 March 1988, Mechanical Engineering Publications, London.

NATO (1980) Mechanics of flexible fibre assemblies, Proceedings of the NATO Advanced Study Institute, Kitini, Greece 1979. *NATO ASI Series E: Applied Sciences* **38**, Sijthoff & Noordhoff, The Netherlands.

Morton-Jones D.H. and Ellis, J.W. (1986) *Polymer Products: design, materials and processing,* Chapman & Hall.

Postle, R. Carnaby, G.A. and de Jong, S. (1988) *Mechanics of Wool Structures,* Ellis Horwood.

Stavsky, Y. and Hoff, N.J. (1969) Mechanics of Composite Structures, in A.G.H. Dietz (ed.) *Composite Engineering Laminates,* MIT Press.

Tetlow, R. (1973) Structural engineering design and applications, in M. Langley (ed), *Carbon fibres in Engineering,* McGraw-Hill.

Vincent, J.F.V. (1990) *Structural Biomaterials,* Macmillan.

Wainwright, S.A., Biggs, W.D., Currey, J.D., and Gosline, J.M. (1976) *Mechanical Design in Organisms,* Arnold, London.

Walter, J.D. (1981) Cord reinforced rubber, in S.K. Clark (ed.,) *Mechanics of Pneumatic Tires,* US Government Printing Office, Washington DC 20402.

Strength of polymer/fibre composites

6

6.1 INTRODUCTION

We have already seen that the stiffness(es) of unidirectional plies, and laminates based on them, depend greatly on the directions in which they are measured or described. The rules governing calculation of these stiffnesses from a given set of data for the individual plies are relatively straightforward, and we have discussed them and their implications in some detail in the previous four chapters.

The designer is also concerned with strength, which is much more complicated and controversial, not just for unidirectional plies, but even more so for laminates based on them. Inevitably therefore we can only introduce some of the main ideas in this book.

The analysis of strength begins with a discussion of the strength of unidirectional plies under single loads in the principal directions. It is assumed here that each ply has properties which vary linearly with stress up to failure. It continues with an introduction to the failure criteria, placing special emphasis on the Tsai – Hill failure criterion, which it is necessary to understand when a ply is loaded in more than one principal direction, and then these criteria are applied in assessment of the strength of a unidirectional ply loaded in-plane and off axis, over a range of angles. These ideas are also applied to the single ply under bending and twisting moments.

Most of the rest of this chapter applies these principles to assessment of the strength of representative simple wide crossply and angleply laminates cured and used at room temperature. Throughout we assume the reader has already mastered the principles of calculation of the stress and strain profiles in the laminate, and for angleply laminates in particular this chapter uses the convenient presentation of stress profiles in the principal directions (in each ply).

Finally some introductory remarks are made about edge effects, which assume particular importance in narrow laminates.

Chapter 7 discusses the causes, analysis and important practical consequences, of residual and internal stresses in a laminate arising from a hot cure cycle, and it is there that some remarks about the strength of hot cured laminates, and laminates cured cold but used at a different ambient temperature, are properly made.

It is not possible to cover all aspects of strength in this book, and it will be recognized that other aspects of strength are not covered here either because they are well-documented elsewhere and outside the scope of this introduction, or because they involve more advanced concepts, or because they are still the subject of vigorous research. These features include the strength of joints (Matthews, 1987), buckling under compressive stress (Agarwal and Broutman, 1990), cyclic loading and crack growth mechanics (Beaumont, 1989, interlaminar shear (Pagano, 1989), and environmental effects.

6.2 EXAMPLES OF FAMILIAR UNIDIRECTIONAL MATERIALS

Before delving into the complexities, it is worth recalling the simple concepts of strength given in Sections 1.3 and 1.4, and looking at a range of examples of everyday materials which show some of the characteristics of unidirectional materials.

We model the basic building block as a regular parallel array of stiff strong rods perfectly bonded to a relatively flexible but rather weak matrix.

If we apply a load to our unidirectional ply along the direction of the fibres, it seems reasonable to expect that the fibres take most of the load – after all that is what they are there for. This expectation is only realized when fibres are regularly spaced, well-bonded to the matrix, and when the fibres have a large ratio of length/diameter (typically greater than 40/1). And the greater the proportion of fibres present the greater the strength. If the fibres fail at a lower strain than the matrix, then as soon as the fibres reach their breaking load, the model breaks, because the matrix alone is not strong enough to take the load. If the matrix breaks at a lower strain than the fibres, then when the matrix fails, the composite may still survive in spite of the cracks because the load is taken by the fibres.

But there is another mode of failure, which can occur if the bond between the rods and the matrix is not very strong. A break may occur in some fibres which then pull out of the matrix which has also broken. This is what happens when you try to break a stick of celery by biting across the width, as shown in Figure 1.27. The fleshy part of the celery breaks quite readily because it is not very strong, but the fibres in the ridges are much stronger and pull out all too readily in an untidy fashion.

What happens if we apply loads at right angles to the fibres? The argument we applied to modulus still applies to strength, namely that there is no means of building up a load in the fibres, so almost the entire load is taken by the matrix. But the matrix is weak, so when loaded across the fibres the composite is also weak. As there is no great build-up of load in the fibres, they are unlikely to break. But there are two main modes of failure in this direction of loading. First, the matrix can break so that the failure surface is entirely fractured matrix – a clean break if the matrix is brittle (Figure 1.26). If the bond between the matrix and the rods is less than the strength of the matrix itself, it is quite likely that debonding will occur: the failure surface then consists of some broken matrix and some new fibre surfaces as shown in Figure 1.26.

An obvious example of this model of behaviour is wood: we know that wood is much stiffer and stronger in tension along the grain rather than across it, typically some thirty to fifty times for a softwood. We instinctively split logs for firewood along the grain to save effort. The designer makes sure that tensile loads are applied along the grain of wood: this is a design constraint, but there are many elegant solutions in roof support frameworks and in furniture designs in the home. We recall from Section 3.3.14 that wood is actually much more complicated than a parallel rods-in-matrix model suggests.

Grass has many of the characteristics of the Mark I model. Its damage resistance, and hence its use for lawns and playing fields, owes much to the strength of the constituent cellulose fibres. The intrinsic high strength of the fibres is not fully exploited because not all the molecules and fibres of cellulose are aligned along the leaf. It is easy to split a blade of grass along its length. But grazing animals need to break the blade across its width, and this is much more difficult. The larger grazing animals like cows break off swathes of grass by brute strength, by wrapping the tongue round and pulling. Sheep grip the grass between the lower teeth and upper palate and jerk the head backwards. Small rodents have to cut grass leaves with their teeth.

If we apply a shear load to the edge of the unidirectional sheet, then the matrix or the interface will break to give a failure similar the transverse failure described in the previous paragraph. See for example Figure 1.26.

If tensile loads are applied at an angle to the fibre direction, then the mode of failure depends on the angle, as indicated in Figure 1.35. At low angles the failure is predominantly like that when the sheet is stretched along the fibre direction. At large angles, above 50°, the character of the break is that of a transverse load. And at intermediate angles, from 5° to 50°, shear occurs. It is worth mentioning that if we try to tear a unidirectional material at some angle to the principal directions, the crack most often turns along the fibre direction.

Consider what happens when you want to peel and eat a banana. The skin is quite tough and difficult to start to break. There are two common strategies for starting to get at the fruit inside. One is to put in a starter crack round (part of) the circumference near the stalk end: this crack doesn't need to pierce the skin. On applying a bending moment via the stalk, the crack grows all the way through and then the skin splits along its length. The other way is not to put in the starter crack, but simply to apply the bending moment until splitting occurs – this is a less controllable event, and any success depends not least on the ripeness of the banana. Notice that the bending causes a split along the previously uncut banana. This is because the skin behaves like a rods-in-matrix model. It is easy to split the skin along its length because you are trying to break the matrix rather than the rods – you use the same argument when you try to split a log for firewood. But it is much more difficult and messy to tear a banana skin round the circumference – try it and see, though not at the tea table. And if you try and break the skin along its length by applying a tensile strength, you get some pull out fracture too (Figure 1.28), just as you do in celery.

When you have eaten most of the banana, the skin will be fairly well peeled back, and if the fruit is really ripe it will not be too well bonded to the inside of the skin, so if you squeeze too hard, it will extrude out and make a mess. If you completely peel a slightly underripe banana and squash the fruit in your hand, you find that the fruit itself is also a composite, and it has planes of weakness running along its length and the squashing causes the fruit to split. (Figure 1.30).

The stem of the dandelion is an interesting example of a rods-in-matrix type of structure. Casual manipulation of a freshly plucked stem reveals that, like a banana, it is much easier to split the stalk along its length than round the circumference. One way to start splitting the tube along its length is to squash it flat between finger and thumb. The split halves can then be pulled apart to obtain two separate strips (there is a technological counterpart to this pulling-apart, known as the trouser tear test). This dramatic difference in strength in the two directions results from the internal construction of the dandelion stem, which consists of sausage-like cells lined up along the length of the stem (Vincent and Jeronimidis, in press). The cells are stronger along their length than the matrix which holds them together.

Many plastics cups used in vending machines are vacuum-formed from polystyrene. The inside of the mould has the pattern required on the outside of the cup. The mould is cold and has many small diameter holes drilled in it, especially at the base corner of the cup. During manufacture the thin sheet of polystyrene is warmed until it is soft, and is then clamped over the mould from which the air is then evacuated so that the sheet is sucked into contact with the mould. The sheet has been stretched to form the cup, and the sidewalls are stretched quite considerably, resulting in some molecular alignment, most of it unidirectional in the side wall. Like the banana stalk, it is much easier to split the cup down the length than round the circumference. The state of molecular alignment is frozen-in by cooling the sheet in contact with the mould. The polystyrene just begins to start softening at about 95° to 100 °C, so you can just about release some of the alignment by pouring absolutely boiling water into the cup: the cup shortens somewhat as the stretched molecules coil up a little. If you put the cup on a tray on some flour or talc in an oven at 130° to 140 °C, the cup will soften enough for almost all the molecules to coil up to the state they were in before the forming process: the result is a disc (usually not quite flat). If you cut squares having sides about 10 mm long from different parts of a clean cup, and put them in the oven, you will be able to find out where the maximum amount of shrinkage occurs.

With these examples in mind we can now examine the behaviour of unidirectional plies and the behaviour of laminates made from them.

6.3 STRENGTH OF UNIDIRECTIONAL PLIES

6.3.1 Strength of unidirectional plies stressed in principal directions

For isotropic materials, principles for governing strength are well established. If the material yields, we use a yield criteria such as von Mises', and predict failure when a value of principal stress, or the combination of principal stresses, reaches a critical value. For a brittle linear elastic material, we resort to fracture mechanics. Even for relatively simple structures the analysis can become quite complicated. The essential point however is that failure occurs at the maximum value of (the combination of) principal stress(es), and it is on this basis that design begins. Failure in some other direction at a lower value of stress is not considered likely or important.

In fibre composites the severe directionality of properties means the failure is by no means likely in the direction of the greatest stress. It is *almost always* necessary to check that the stresses in the principal directions in any ply (and in particular the transverse and in-plane shear stresses) are not likely to be troublesome. It is quite likely that the maximum (global) stress will not cause failure in the global directions, because of the use of plies at different angles to the global stresses.

Thus for fibre composites the designer will find that the concept of principal stresses has little relevance. The concept of stresses in the principal directions is all pervasive and supremely important in design for load-bearing performance. What is needed is a criterion for failure which acknowledges the directionality of fibre composites. There are many such criteria, many of them are complicated, and we introduce only the simplest of them in this chapter.

With some considerable care and skill, the strengths of unidirectional plies can be measured both in uniaxial tension or compression and in shear. For behaviour measured in the principal directions, these represent fundamental parameters for assessing the load-bearing capabilities of such a ply. It is necessary to distinguish compressive and tensile strengths, and positive and negative shear stresses, but it is not usual to distinguish shear strengths resulting from positive and negative shear stresses applied alone. In more detail these strengths may be listed as:

- Longitudinal tensile strength, $\sigma_{1\mathrm{Tmax}}$
- Longitudinal compressive strength, $\sigma_{1\mathrm{Cmax}}$

- Transverse tensile strength, σ_{2Tmax}
- Transverse compressive strength, σ_{2Cmax}
- In-plane shear strength, τ_{12max}

In addition, for laminates it is necessary to measure the interlaminar shear strength, τ_{ILSmax}.

For design purposes it is best to measure these properties using representative samples, rather than calculate them from the properties of constituent fibres and matrix because, apart from the longitudinal tensile strength, it is unwise to rely on the methods of micromechanics.

We need to know how close to failure a given applied stress brings the material. One way of indicating this is to define a load factor F as the ratio of the strength to the applied stress so that $\sigma_{1max} = F\sigma_1$, or $\tau_{12max} = F\tau_{12}$.

Problem 6.1: A single ply having a 60% volume fraction of Kevlar fibres unidirectionally arranged in epoxy resin is 1mm thick and has the following properties:

$$E_1 = 76\,\text{GPa}, E_2 = 5.5\,\text{GPa}, G_{12} = 2.1\,\text{GPa}, v_{12} = 0.34$$
$$\sigma_{1Tmax} = 1240\,\text{MPa}, \sigma_{1Cmax} = -230\,\text{MPa}, \sigma_{2Tmax} = 30\,\text{MPa},$$
$$\sigma_{2Cmax} = -140\,\text{MPa}, \tau_{12max} = 60\,\text{MPa}$$

What are the load factors when you any one possible single stress of 20 MPa is applied in the principal directions? (Possible here means tensile, compressive, or shear, and it may be assumed that the ply will not buckle under the test conditions.)

For the record, though we shall not need the data until we examine more elaborate problems later, the AD and ad matrices for a 1mm ply are as follows:

A(kN/mm)			D(kNmm)		
76.64	1.886	0	6.387	0.1571	0
1.886	5.546	0	0.1571	0.4622	0
0	0	2.1	0	0	0.175

a(mm/MN)			d(/MNmm)		
13.16	−4.474	0	157.9	−53.68	0
−4.474	181.8	0	−53.68	2182	0
0	0	476.2	0	0	5714

6.3.2 Failure criteria for combined loading in principal directions

We also need to be able to calculate the response when a ply carries more than one stress each in the principal directions. This is a situation which arises frequently in practice. We therefore need to invoke a failure criterion, of which there are at least forty! Each criterion of failure has its advantages and disadvantages, and a full discussion lies outside the scope of this short book. We shall discuss only three, the maximum stress criterion, the maximum strain criterion and, in much greater detail, the Tsai – Hill criterion.

In discussion of the strength of unidirectional plies, there are two situations which need particular attention. One concerns combinations of stresses applied in the principal directions (Sections 6.3.2 and 6.3.3). The other concerns one or more global stresses applied off-axis which result in combinations of stresses in the principal directions: this is discussed in Sections 6.3.4 and 6.3.5.

Maximum stress failure criterion

This is one of the simplest failure criteria, though not the best, and it assumes that the composite fails when any stress in the principal directions exceeds the strength in the same direction. Bearing in mind the differences between strengths in tension and compression, there are therefore five subcriteria, each of which must be tested. Formally we may write the subcriteria for failure as:

$$\sigma_{1T} \geqslant \sigma_{1Tmax}; \ \sigma_{1C} \geqslant \sigma_{1Cmax}; \ \sigma_{2T} \geqslant \sigma_{2Tmax}; \ \sigma_{2C} \geqslant \sigma_{2Cmax}; \ \tau_{12} \geqslant \tau_{12max}$$

Failure occurs when any one of these conditions obtains.

The advantages of the maximum stress criterion are that it is simple to apply, it indicates which mode and direction of failure will occur, and standard tests will (with due care) give unidirectional strength data.

The disadvantages are that it takes no account of interactions between stresses (a 'non-interactive' criterion), and it is not useful where the stress-strain curve is non-linear (as often occurs in shear).

Maximum strain failure criterion

The maximum strain criterion is similar in approach to the maximum stress criterion. Here however the essence of this simple criterion is that the material is deemed to have failed under a combination of stresses in the principal directions when any of the strains exceeds the failure strain as measured in a test under that single stress in the principal direction. We therefore have again five subcriteria, so that failure is predicted when any of the following relationships apply:

$$\varepsilon_{1T} \geqslant \varepsilon_{1Tmax}; \ \varepsilon_{1C} \geqslant \varepsilon_{1Cmax}; \ \varepsilon_{2T} \geqslant \varepsilon_{2Tmax}; \ \varepsilon_{2C} \geqslant \varepsilon_{2Cmax}; \ \gamma_{12} \geqslant \gamma_{12max}$$

Given the applied (in-plane) stresses, we can find the strains using

$$\varepsilon_1 = (1/E_1)(\sigma_1 - v_{12}\sigma_2), \varepsilon_2 = (1/E_2)(\sigma_2 - v_{21}\sigma_1), \; \gamma_{12} = \tau_{12}/G_{12}$$

The advantages of the maximum strain criterion are that simple tests give the required experimental design data (provided due care is taken), the mode of failure (tensile, compression or shear, and a particular direction) is clearly designated, and the method gives a rather better representation where non-linear behaviour occurs (e.g. shear), compared with the maximum stress criterion.

The disadvantage is that the predictions are still not very accurate, because no account is taken of interactions between different strains (a 'non-interactive' criterion), failure strain data are usually found to have more scatter than strength data, and failure strain data are less commonly found in the literature than strength data.

Nevertheless it is worth looking at some examples of the use of the maximum strain criterion. The general approach is to apply global stresses or strains, calculate the strains present in the principal directions, and establish which ratio, R, of actual strain to corresponding failure strain gives the highest value. It is this value which is deemed to cause first-ply failure (FPF) if all the applied loads or stresses or strains are multiplied by $1/R$. It follows that the load factor F under a given set of loading conditions is given by $F = 1/R$.

Tsai – Hill failure criterion

This theory has its origins in a yield criterion for failure of anisotropic materials. Von Mises proposed that an isotropic material would yield when a certain amount of strain energy had been applied. Hill extended this theory to describe the criterion for yielding of anisotropic materials. Tsai adapted this theory to represent anisotropic materials which do not yield. The result is known as the Tsai – Hill failure criterion which states that failure will occur when the stresses in the principal directions satisfy the expression

$$(\sigma_1/\sigma_{1\max})^2 - \sigma_1\sigma_2/(\sigma_{1\max})^2 + (\sigma_2/\sigma_{2\max})^2 + (\tau_{12}/\tau_{12\max})^2 = 1$$

This Tsai-Hill criterion works quite well for a range of experimental situations and readily available materials where in-plane stresses are considered and where edge effects, stress concentrations and crack-like defects are absent.

The advantages of the Tsai–Hill failure criterion are that it includes some allowance for interaction between the different stresses present in the different principal directions (an 'interactive' criterion), and it is only one criterion, rather than five subcriteria.

More usefully, we can identify a load factor F by which all applied stresses must be equally multiplied in order that failure will just occur, so that the Tsai–Hill criterion can be rewritten as

$$(\sigma_1/\sigma_{1\max})^2 - \sigma_1\sigma_2/(\sigma_{1\max})^2 + (\sigma_2/\sigma_{2\max})^2 + (\tau_{12}/\tau_{12\max})^2 = 1/F^2$$

The reduces to the expression in Section 6.3.1 when the material carries only a uniaxial tensile stress in the longitudinal direction.

Another way of interpreting this equation is that when a ply carries loads σ_1, σ_2, τ_{12}, failure will occur when each load is multiplied by F, so that the failure loads are $F\sigma_1$, $F\sigma_2$, $F\tau_{12}$. We can now write that failure will occur when

$$(F\sigma_1/\sigma_{1max})^2 - F^2\sigma_1\sigma_2/(\sigma_{1max})^2 + (F\sigma_2/\sigma_{2max})^2 + (F\tau_{12}/\tau_{12max})^2 = 1$$

It will already be apparent that:

1. The Tsai–Hill criterion does not predict the mode of failure, only the ratio of the applied stresses which cause failure. We can however make a *rough estimate* of the failure mode by calculating the ratios σ_1/σ_{1max}, σ_2/σ_{2max}, and τ_{12}/τ_{12max}. If one of these is much larger than the others, this would suggest that the associated mode of failure would be the most likely. It must be emphasized that this is not a precise indication–this is most apparent when two (or more) stress/strength ratios have comparable values, and the physical interpretation then becomes obviously problematic.
2. To apply the Tsai–Hill criterion we need to know the stress profiles and distributions in plies and laminates in the principal directions – this is one of the main reasons for the heavy emphasis on stress profiles especially in the principal directions in earlier chapters.
3. Direct strengths have different values in different directions, so if the applied stress is tensile we use the corresponding tensile strength, and if it is compressive we use the compressive strength.
4. The second term $\sigma_1\sigma_2/(\sigma_{1max})^2$ is small and can usually be neglected in those composites where $\sigma_{1max} \gg \sigma_{2max}$, because $\sigma_2 \ll \sigma_1$ in most practical situations. There is a curious anomaly when $\sigma_2 = \sigma_1$.

Example: behaviour of Lamina 00G

Lamina 00G is a unidirectional laminate made from two 1 mm plies of EG, based on cold-cured epoxy resin and long glass fibres, for which data are given in Tables 6.1 and 6.2. Load resultants and moment resultants are applied in the principal directions of the material.

It can readily be verified that the strain responses to direct loading for a range of straightforward situations is as given in Table 6.3. The predicted strength response can be confirmed, bearing in mind that the maximum strains at failure assuming linearity, and bearing in mind the different strengths in tension and compression. With these data we can now examine two theories.

Table 6.1 Properties of ply EG

E_{11} 45.6 GPa;	E_{22} 10.73 GPa	G_{12} 5.14 GPa	ν_{12} 0.274
α_{11} 7 × 10^{-6}/K	α_{22} 3 × 10^{-5}/K		
σ_{1Tmax} 1000 MPa	σ_{2Tmax} 100 MPa	σ_{1Cmax} − 1600 MPa	σ_{2Cmax} − 270 MPa
τ_{12max} 80 MPa			

Table 6.2 Stiffnesses and compliances of Lamina 00G (total thickness 2 mm)

A(kN/mm)			D(kNmm)		
92.84	5.986	0	30.95	1.995	0
5.986	21.85	0	1.995	7.282	0
0	0	10.28	0	0	3.427

a(mm/MN)			d(/MNmm)		
10.96	−3.004	0	32.89	−9.013	0
−3.004	46.6	0	−9.013	139.8	0
0	0	97.28	0	0	291.8

Table 6.3 Failure of Lamina 00G under in-plane loading

Load (N/mm)			Strain (%)			FPF load factor	
N_1	N_2	N_{12}	ε_1	ε_2	γ_{12}	Tsai–Hill	Max strain
100			0.1096	−0.03		20	20
−100						32	31.02
	100		−0.03	0.466		2	2
	−100					5.4	5.4
	100		0.07961	0.4359		2	2.138
−100	−100					5.4	5.77
		25			0.2432	6.4	6.398
100	100	25	0.07961	0.4359	0.2432	1.909	2.138
100	100	100				1.249	1.6
100	100	138.5				1.	
−100	−100	138.5				1.13	

Maximum strain criterion of failure for Lamina 00G
The maximum strain criterion of failure is not interactive and can be tested almost by inspection using the following maximum strains which assume linear behaviour to failure from the basic data: $\varepsilon_{1tmax} = 2.192\%$; $\varepsilon_{2tmax} = 0.932\%$; $\varepsilon_{1cmax} = -3.509\%$; $\varepsilon_{2cmax} = -2.52\%$; $\gamma_{12max} = 1.556\%$. The procedure is to test the strains caused by the applied loads against the failure strain: the lowest load factor gives the direction and mode of failure.

For $N_1 = 100$ N/mm, $\varepsilon_{1tmax}/\varepsilon_1 = 2.192/0.1096 = 20$, and $\varepsilon_{2cmax}/\varepsilon_2 = -2.52/-0.03 = 84$, so clearly the lamina fails in the longitudinal direction at 20 times the applied load. But when $N_1 = -100$ N/mm, $\varepsilon_{1cmax}/\varepsilon_1 = -3.509/-0.1096 = 32$, $\varepsilon_{2tmax}/\varepsilon_2 = 0.932/0.03 = 31$, so under axial compression in the longitudinal direction the failure is in the transverse direction, and is tensile.

Combinations of stress can sometimes change the failure mode, even with a non-interactive failure criterion. For example applying $N_1 = 100$ N/mm and $N_2 = 100$ N/mm causes transverse failure; adding $N_{12} = 25$ N/mm does not change the failure mode or load factor according to the maximum strain criterion because $\varepsilon_1/\varepsilon_{2max} > \gamma_{12}/\gamma_{12max}$; but increasing the shear load to 100 N/mm causes a shear failure and a reduction in load factor. (Adding $N_{12} = 25$ N/mm does of course reduce the Tsai–Hill load factor because the interaction of stresses is now taken into account.)

Tsai–Hill criterion of failure for Lamina 00G
When the stress-strain behaviour is linear up to the point of failure, there is no difference in prediction of failure under a single stress in the principal directions from the three failure criteria examined in Section 6.3.2.

The Tsai – Hill failure criterion recognizes the contribution of some simple interactions between the various applied loads or moments (or both). This means that increasing any type of stress in the principal directions will reduce the load factor and make failure more likely. Let us consider some examples.

For a load $N_1 = 100$ N/mm, the stress on the 2 mm thick laminate is $N_1/h = 50$ MPa. We therefore calculate the load factor as $\sigma_{1tmax}/\sigma_1 = 1000/50 = 20$, and conclude that the lamina will fail in tension in the 1 direction at 20 times the applied load. If a compressive load $N_1 = -100$ N/mm were applied, the load factor would be $1600/50 = 32$, but the Tsai–Hill criterion gives no indication of the mode or direction of failure. Under biaxial load $N_1 = N_2 = 100$ N/mm, the failure load is calculated from the abbreviated Tsai – Hill form as

$$(\sigma_1/\sigma_{1max})^2 + (\sigma_2/\sigma_{2max})^2 = 1/F^2$$

i.e.

$$(50/1000)^2 + (50/100)^2 = 0.2525, \text{ so } F = 1.99$$

The full Tsai–Hill criterion, including the $-(\sigma_1\sigma_2/\sigma_{1max})^2$ gives $F = 2$.

Using the non-interactive maximum strain criterion, the load factor under the loading $N_1 = N_2 = 100$ N/mm is slightly higher at 2.14: this is expected because of the lack of interaction.

Problem 6.2: For the Kevlar/epoxy ply described in Problem 6.1, examine the load factors when all possible combinations of stress 20 MPa are applied in the principal directions, and comment on the likely mode of failure.

6.3.3 Strength of unidirectional plies in bending in principal directions

It will be apparent that for loading in the principal directions all plies fail at the same direct load, but in bending a sequence of failure events can occur because of the variation of stress and strain through the thickness. For Lamina 00G, these profiles are exemplified in the Figures 6.1 to 6.3 for selected moments, where the failure sequence assumes (incorrectly) that any failed ply still contributes its stiffness when the load exceeds the failure load for that ply. Values of curvatures and load factors arising when moments (or combinations of moments) are applied to Lamina 00G are given in Tables 6.4 and 6.5.

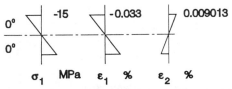

Figure 6.1 Stress and strain profiles for OOG under $M_1 = 10$ N.

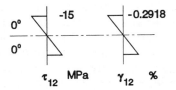

Figure 6.2 Stress and strain profiles for OOG under $M_{12} = 10$ N.

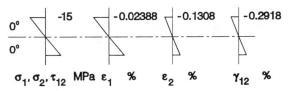

Figure 6.3 Stress and strain profiles for OOG under $M_1 = M_2 = M_{12} = 10$ N.

Table 6.4 First ply failure of Lamina 00G under bending

Moment (N)			Curvature (/m)			FPF load factor	
M_1	M_2	M_{12}	κ_1	κ_2	κ_{12}	Tsai–Hill	Max strain
10	—	—	0.3289	−0.09013	0	66.67	66.4
—	10	—	−0.09013	1.398	0	6.667	6.667
10	10	—	0.2388	1.308	0	6.667	7.13
—	—	10	0	0	2.918	5.333	5.33
10	10	10	0.2388	1.308	2.918	4.165	5.33
640	10	10	20.96	−4.37	2.918	1.018	1.674
658	10	10					1.009

When bending moments are applied, the extremes of states of stress must be examined for each ply. Thus for Lamina 00G under moments $M_1 = M_{12} = 10$ N, the bending stresses vary linearly through the thickness, being -15 MPa at the upper surface, and $+15$ MPa on the lower surface, and all are zero at the midplane. The ply surface stresses can be tested using the abbreviated

Table 6.5 Bending failure sequences for Lamina 00G

M_1	M_2	M_{12}	Place	Tsai–Hill load factor
10			Bottom	66.67
			Top	106.7
	10		Bottom	6.667
			Top	18
10	10		Bottom	6.667
			Top	18
		10	Top	5.333
			Bottom	5.333
10	10	10	Bottom	4.165
			Top	5.144
640	10	10	Bottom	1.018
			Top	1.596

Tsai–Hill form:

Top
$$(-15/-1600)^2 + (-15/-270)^2 + (15/80)^2 = 0.03833$$
$$\text{i.e. } F = 5.11$$

Bottom
$$(15/1000)^2 + (15/100)^2 + (15/80)^2$$
$$\text{i.e. } F = 4.16$$

Clearly the lower surface fails first at 4.16 times each of the applied moments; on further increasing the load, the top surface fails in compression at $5.11 \times$ each of the applied moments.

Using the maximum strain critrion, the general approach is to test each extreme value of strain against the maximum value allowed in that mode and direction.

For $M_1 = M_2 = M_{12} = 10\,\text{N}$, the maximum values of strain are clearly at the top and bottom surfaces of the lamina. Checking systematically:

Top	$\varepsilon_{1cmax}/\varepsilon_1$	$= -3.509/-0.02388$	$= 1503$
Bottom	$\varepsilon_{1tmax}/\varepsilon_1$	$= 2.192/0.02388$	$= 91.8$
Top	$\varepsilon_{2cmax}/\varepsilon_2$	$= -2.52/-0.1308$	$= 17.2$
Bottom	$\varepsilon_{2tmax}/\varepsilon_2$	$= 0.932/0.1308$	$= 7.12$
Top	$\gamma_{12max}/\gamma_{12}$	$= 1.556/0.2918$	$= 5.33$

We do not need to check the bottom surface for shear, because our version of the maximum strain criterion does not distinguish failure from positive and negative shear strain in the principal directions. On the basis of the calculations, the lamina fails under $M_1 = M_2 = M_{12} = 10\,\text{N}$ first by shear then, as the load is increased, by transverse tension, they by transverse compression, and then by axial tension. The load factor for $M_1 = M_2 = M_{12} = 10\,\text{N}$ based on the

Tsai–Hill criterion (4.16) is lower than that for the maximum strain criterion (5.33) because of the interactions allowed.

6.3.4 Strength of unidirectional plies loaded in-plane and off-axis

Although we describe properties in terms of the principal directions, we frequently apply stresses in directions (x, y) which make an angle θ to the principal directions. We can only apply a failure criterion to any ply when the stresses have been transformed into components in the principal directions.

For example, if a single stress σ_x is applied then, using $[\sigma]_{12} = [T][\sigma]_{xy}$ from Section 3.3.2, we can write

$$\sigma_1 = \sigma_x \cos^2 \theta; \quad \sigma_2 = \sigma_x \sin^2 \theta; \quad \tau_{12} = -\sigma_x \sin \theta \cos \theta$$

Clearly one global stress such as $\sigma_x(\theta)$ must produce a combination of stresses in the principal directions unless the angle θ between the (x, y) and $(1, 2)$ directions is $0°$ or $90°$.

The over-riding practical conclusion from the following detailed discussions is that in a unidirectional ply under off-axis loading, failure is most likely to occur in the transverse (2) direction or in shear along the fibre (1) direction, unless a large global stress is applied in the longitudinal (1) direction or at a small angle (θ) to the (1) direction.

Maximum stress failure criterion

Substituting into the subcriteria we predict failure will occur if any of the following inequalities hold true:

$$\sigma_x > \sigma_{1\max}/\cos^2 \theta; \quad \sigma_x > \sigma_{2\max}/\sin^2 \theta; \quad \sigma_x > \tau_{12\max}/\sin \theta \cos \theta$$

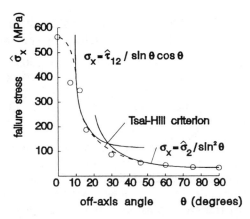

Figure 6.4 Off-axis tensile strength of a carbon fibre/epoxy unidirectional ply, $V_f \sim 0.5$. Data courtesy Sinclair & Chamis.

There are still five subcriteria, of course, because we distinguish tensile and compressive strengths.

Some experimental results for a unidirectional carbon fibre/epoxy composite are presented in Figure 6.4. These results show that the agreement between experimental results and the predictions of the maximum stress criterion are least successful in the range $20° < \theta < 45°$, not least because no account is taken of interactions between the stresses in the principal directions. We shall see later that this is unfortunate and potentially troublesome because stresses in many practical laminates are likely to occur in some plies in this range of angles.

Maximum strain failure criterion

To find the strain responses in the principal directions to an off-axis stress such as σ_x, we use the stress transformation matrix [T], so that $\varepsilon_1 = (1/E_1)(\cos^2 \theta - v_{12} \sin^2 \theta)\sigma_x$, and similarly for ε_2 and γ_{12}. We now assume a linear stress-strain curve, so that $\varepsilon_{1\max} = \sigma_{1\max}/E_1$. Thus for this test we predict failure under an off-axis in-plane stress σ_x when any one of the following occurs:

$$\sigma_x = \sigma_{1\max}/(\cos^2 \theta - v_{12}\sin^2 \theta); \ \sigma_x = \sigma_{2\max}/(\sin^2 \theta - v_{12}\cos^2 \theta); \ \sigma_x = \tau_{12\max}/\sin \theta \cos \theta$$

This means that there is assumed to be no interaction between the actual applied stresses – for example, if only one stress were applied $\sigma_1 = 0.95\sigma_{1\max}$, and then an additional $\sigma_2 = 0.95\sigma_{2\max}$ were applied, this criterion would suggest that the ply would survive with an implied load factor, whereas in practice failure would probably occur because of the interactions.

Tsai–Hill failure criterion

Substituting into the (modified) Tsai–Hill criterion (which is interactive), and re-arranging, gives the maximum allowable value of the single off-axis in-plane stress, $\sigma_x(\theta)$, as

$$\sigma_{x\max} = [\cos^4 \theta/(\sigma_{1\max})^2 + \sin^4 \theta/(\sigma_{2\max})^2 + \sin^2 \theta \cos^2 \theta/(\tau_{12\max})^2]^{-0.5}$$

For off-axis in-plane loading another advantage of the Tsai–Hill criterion now becomes evident: it gives a smooth function of failure versus θ, rather than the three separate regimes described using either the maximum stress or the maximum strain criterion.

We shall now apply the full Tsai–Hill criterion first to examine the special case of Lamina 30G loaded in different ways, and then the effect of change of angle, using Lamina θG.

Example: In-plane loading of Lamina 30G

Lamina 30G consists of two 1 mm plies of EG each laid up at $30°$ to the stress directions (x, y) (Table 6.6).

Table 6.6 Stiffnesses and compliances of Lamina 30G (total thickness 2 mm)

A(kN/mm)			D(kN)		
63.54	17.53	22.04	21.18	5.845	7.346
17.53	28.05	8.703	5.845	9.349	2.901
22.04	8.703	21.83	7.346	2.901	7.276
a(mm/MN)			d(/MNmm)		
26.19	-9.324	-22.73	78.58	-27.97	-68.18
-9.324	44.01	-8.133	-27.97	132	-24.4
-22.73	-8.133	72	-68.18	-24.4	216

The reduced stiffnesses of the single ply are:

$[Q^*(+30°)]$ GPa $\qquad\qquad$ $[Q^*(-30°)]$ GPa

31.77	8.767	11.02	$\quad$	31.77	7.767	-11.02
8.767	14.02	4.351		8.767	14.02	-4.351
11.02	4.351	10.91		-11.02	-4.351	10.91

The global stress and strain responses under stated load conditions, and the load factor according to the Tsai-Hill criterion, are shown in Table 6.7.

There seems to be no discernible pattern in Table 6.7 for the global responses to the applied in-plane loads. The curious variation in load factor between the last two entries is particularly strange, and would seem to suggest (incorrectly) that adding shear stress to a biaxial loading condition can improve the load-bearing capability, which is in direct conflict with the Tsai – Hill criterion. But of course, in order to ascertain load factors and predict failure stresses and strains, the applied stresses and strains in (x, y) *must* be transformed into the principal directions using the [T] transformation matrix: for $\theta = 30°$

$$\begin{pmatrix} \sigma_1 \text{ or } & \varepsilon_1 \\ \sigma_2 \text{ or } & \varepsilon_2 \\ \sigma_{12} \text{ or } (\gamma_{12})/2 \end{pmatrix} = \begin{pmatrix} 0.75 & 0.25 & 0.866 \\ 0.25 & 0.75 & -0.866 \\ -0.433 & 0.433 & 0.5 \end{pmatrix} \begin{pmatrix} \sigma_x \text{ or } & \varepsilon_x \\ \sigma_y \text{ or } & \varepsilon_y \\ \tau_{xy} \text{ or } (\gamma_{xy})/2 \end{pmatrix}$$

Table 6.7 Failure of Lamina 30G under in-plane loading

Load (N/mm)	σ_x (MPa)	σ_y (MPa)	τ_{xy} (MPa)	ε_x (%)	ε_y (%)	γ_{xy} (%)	T–H factor
N_x 10	5	0	0	0.02619	-0.009324	-0.02273	33.37
N_y 10	0	5	0	-0.009324	0.04410	-0.008133	21.64
N_{xy} 10	0	0	5	-0.02273	-0.008133	0.072	20.05
N_x 10 $\}$ N_y 10 $\}$	5	5	0	0.01687	0.03469	-0.03086	20
N_x 10 $\}$ N_y 10 $\}$ N_{xy} 10 $\}$	5	5	5	-0.005858	0.02655	0.04114	30.12

Table 6.8 Failure of Lamina 30G under in-plane loading

Load (N/mm)	σ_1 (MPa)	σ_2 (MPa)	τ_{12} (MPa)	ε_1 (%)	ε_2 (%)	γ_{12} (%)	$T-H$ load factor
N_x 10	3.75	1.25	-2.165	0.007437	0.009396	-0.04212	33.37
N_y 10	1.25	3.75	2.165	0.000488	0.0342	0.04212	21.64
N_{xy} 10	4.33	-4.33	2.5	0.0121	-0.04796	0.04864	28.05
$\left.\begin{array}{l} N_x\ 10 \\ N_y\ 10 \end{array}\right\}$	5	5	0	0.007961	0.04359	0	20
$\left.\begin{array}{l} N_x\ 10 \\ N_y\ 10 \\ N_{xy}\ 10 \end{array}\right\}$	9.33	0.67	2.5	0.02006	0.0006367	0.04864	30.12

For the same loading of Lamina 30G given in Table 6.7, the stresses and strains in the principal directions are given in Table 6.8. This table presents a coherent set of information. In particular we can see the effect of adding the shear load to the biaxial load: the shear load in (x, y) is resolved into $(1, 2)$, boosting the stress σ_1 at the expense of σ_2 which all but disappears.

Under biaxial loading the load factor is dominated by the σ_2 component, so almost suppressing σ_2 by the extra shear load: this throws failure on to the shear mode but at a higher load factor because the shear strength $\tau_{12\max}$ has a lower value than the compressive strength.

The reader is invited to construct a table of stress/strength ratios from the data in Table 6.8, in order to confirm the comments made in the previous paragraph.

> **Problem 6.3:** Compare the load factors you obtain using the maximum stress criterion with those already obtained using the Tsai–Hill criterion in Table 6.8.

> **Problem 6.4:** Using the Tsai–Hill criterion, drive an expression for the shear strength $\tau_{xy\max}$ of a unidirectional ply when loaded in-plane and off-axis, in terms of the angle θ between the global coordinates and the principal directions, and the strengths in the principal directions.

> **Problem 6.5:** A thin-walled tube, radius R and wall thickness h, consists of a single winding of fibre-reinforced polymer at θ to the hoop direction. How does the maximum allowable internal pressure, p, vary with θ?

Behaviour of unidirectional Lamina θK under in-plane loading

We shall now examine the predicted strength behaviour, using the Tsai–Hill criterion, for Lamina θK under in-plane loading. Lamina θK is based on a single 1 mm ply of unidirectional Kevlar aromatic polyamide fibres in an epoxy resin cured at 20 °C. The volume fraction of fibres is 0.6. The elastic constants and strengths of this ply are described in Section 6.3.1. (Problem 6.1). The in-plane extensional compliances for a 1 mm ply are given in Table 6.9, and the transformed reduced stiffnesses in Table 6.10.

Application of $N_x = 10 \, N/mm$ to Lamina θK
The responses of Lamina θK to a tensile force resultant $N_x = 10 \, N/mm$ are given in Table 6.11.

Table 6.9 Elastic in-plane compliances of Lamina θK

$\theta°$	a_{11} (mm/MN)	a_{12}	a_{22}	a_{16}	a_{26}	a_{66}
0	13.16	−4.474	181.8	0	0	476.2
15	41.47	−21.49	187.5	−101.1	+16.78	408.1
30	106.4	−55.52	190.7	−132	−14.08	272
45	165.6	−72.54	165.6	−84.33	−84.33	203.9
60	190.7	−55.52	106.4	−14.08	−132	272
75	187.5	−21.49	41.47	+16.78	−101.1	408.1
90	181.8	−4.474	13.16	0	0	476.2

Table 6.10 Transformed reduced stiffnesses of Lamina θK

$\theta°$	Q_{11}^* (GPa)	Q_{12}^*	Q_{22}^*	Q_{16}^*	Q_{26}^*	Q_{66}^*
0	76.64	1.886	5.546	0	0	2.1
15	67.5	6.262	5.933	16.47	1.307	6.476
30	45.74	15.01	10.19	22.97	7.813	15.23
45	23.59	19.39	23.59	17.77	17.77	19.6
60	10.19	15.01	45.74	7.813	22.97	15.23
75	5.933	6.262	67.5	1.307	16.47	6.476
90	5.546	1.886	76.64	0	0	2.1

Table 6.11 Strength of Lamina θK under $N_x = 10 \, N/mm$

$\theta°$	ε_x (%)	ε_y (%)	γ_{xy} (%)	σ_1 (MPa)	σ_2 (MPa)	τ_{12} (MPa)	Tsai–Hill factor
0	0.01316	−0.004474	0	10	0	0	124
15	0.04147	−0.02149	−0.1011	9.33	0.67	−2.5	20.91
30	0.1064	−0.05552	−0.132	7.5	2.5	−4.33	9.062
45	0.1656	−0.07254	−0.08433	5	5	−5	5.367
60	0.1907	−0.05552	−0.01408	2.5	7.5	−4.33	3.843
75	0.1875	−0.02149	0.01678	0.67	9.33	−2.5	3.187
90	0.1818	−0.004474	0	0	10	0	3

The strain responses in global co-ordinates follow directly from the compliances given in Table 6.9. The stresses in the principal directions are used to estimate the Tsai–Hill load factors. It is clear that under $N_x = 10$ N/mm the smooth curve of load factor versus θ decreases sharply over the range $0°$ to $30°$ and then flattens out to the minimum value at $90°$.

The mode of failure is not predicted, but inspection of the stress/strength ratios $(\sigma_1/\sigma_{1\text{max}}, \sigma_2/\sigma_{2\text{max}}, \tau_{12}/\tau_{12\text{max}})$ suggest that the following modes are likely to be dominant: $\theta \sim 0°$ longitudinal tension, $\theta = 15°$ to $30°$ shear, $\theta = 30°$ to $90°$ transverse tension. Around $30°$ it is difficult to identify a separate failure mode because the values of stress/strength ratios for transverse tension and for shear are comparable.

Problem 6.6: Calculate the Tsai–Hill load factors for a uniform in-plane compressive load of $N_x = -10$ N/mm applied to Lamina θK. Compare values for $N_x = +10$ N/mm, and comment on likely dominant modes of failure. It may be assumed that buckling failure does not occur in the intended tests.

Application of $N_{xy} = 10$ N/mm to Lamina θK
The responses of Lamina θK to a uniform in-plane shear force $N_{xy} = 10$ N/mm are given in Table 6.12.

There is a fairly weak dependence of the Tsai–Hill load factor on θ (see Problem 6.4). Using the concept of stress/strength ratios as a guide, failures are normally dominated by shear, except that at $45°$ there is no shear stress in the principal directions and failure is by transverse compression.

Problem 6.7: Investigate the effect on the shear strength $\tau_{xy\text{max}}$ of a single unidirectional EG ply tested in-plane in shear off-axis, using the Tsai–Hill failure criterion. The lamina θG has properties $E_1 = 45.6$ GPa, $E_2 = 10.73$ GPa, $v_{12} = 0.274$, $G_{12} = 5.14$ GPa, $\sigma_{1T\text{max}} = 1000$ MPa, $\sigma_{2T\text{max}} = 100$ MPa, $\sigma_{1C\text{max}} = 1600$ MPa, $\sigma_{2C\text{max}} = 270$ MPa. $\tau_{12\text{max}} = 80$ MPa.

Table 6.12 Strength of Lamina θK under $N_{xy} = 10$ N/mm

$\theta°$	ε_x (%)	ε_y (%)	γ_{xy} (%)	σ_1 (MPa)	σ_2 (MPa)	τ_{12} (MPa)	Tsai–Hill factor
0	0	0	0.4762	0	0	10	6
15	−0.1011	0.01678	0.4081	5	−5	8.66	6.72
30	−0.132	−0.01408	0.272	8.66	−8.66	5	9.592
45	−0.08433	−0.08433	0.2039	10	−10	0	13.82
60	−0.01408	−0.132	0.272	8.66	−8.66	−5	9.592
75	0.01678	−0.1011	0.4081	5	−5	−8.66	6.72
90	0	0	0.4762	0	0	−10	6

For a 1 mm ply of EG at angle θ to the principal directions, Lamina θG, the values of the compliance matrix coefficients [a] are given in Table 6.13, and for the transformed reduced stiffnesses $[Q*(\theta)]$ in Table 6.14.

6.3.5 Strength of unidirectional plies under bending off-axis

Bending of a unidirectional ply off-axis will generate three curvatures because the bending stiffness matrix [D] and its inverse [d], are fully populated. Thus in global co-ordinates we can find the curvatures induced by moment resultants per unit width, e.g.:

$$\kappa_x = d_{11}M_x + d_{12}M_y + d_{16}M_{xy}$$

from which, (carefully remembering to use $\kappa_{12}/2$ and $\kappa_{xy}/2$, Section 3.3.10), we can obtain the curvatures in the principal directions using

$$[\kappa]_{1,2} = [T][\kappa]_{x,y}$$

and hence the stresses in the principal directions, to which the failure criterion can be applied to ascertain the location (thickness co-ordinate) of failure. For a single ply, bending stresses are greatest at the upper and lower surfaces of the ply under positive pure bending moment resultants per unit width.

Table 6.13 Compliances of Lamina θG (1 mm thick EG)

$\theta°$	a_{11} (mm/MN)	a_{12}	a_{22}	a_{16}	a_{26}	a_{66}
0	21.93	−6.009	93.2	0	0	194.6
15	30.92	−10.22	92.64	−32.41	−3.222	177.7
30	52.39	−18.65	88.02	−45.45	−16.27	144
45	74.42	−22.86	74.42	−35.63	−35.63	127.1
60	88.02	−18.65	52.39	−16.27	−45.45	144
75	92.64	−10.22	30.92	−3.222	−32.41	177.7
90	93.2	−6.009	21.93	0	0	194.6

Table 6.14 Transformed reduced stiffnesses of Lamina θG

$\theta°$	Q_{11}^* (GPa)	Q_{12}^*	Q_{22}^*	Q_{16}^*	Q_{26}^*	Q_{66}^*
0	46.42	2.993	10.92	0	0	5.14
15	42.12	4.918	11.38	7.771	1.103	7.065
30	31.77	8.767	14.02	11.02	4.351	10.91
45	20.97	10.69	20.97	8.874	8.874	12.84
60	14.02	8.767	31.77	4.351	11.02	10.91
75	11.38	4.918	42.12	1.103	7.771	7.065
90	10.92	2.993	46.42	0	0	5.14

We shall first examine the special case of Lamina 30G with moments applied at 30° to the principal directions, and then discuss the effect of change of angle on load factors and modes of failure using Lamina θG.

Example: Behaviour of Lamina 30G under bending

The stiffnesses of the 2 mm ply of EG aligned at 30° to the global (x, y) co-ordinates are given in Table 6.6. Consider the strength implications from the responses to single moment resultants per unit width M_x, M_y, and M_{xy} given in Table 6.15. The curvatures are as expected.

Variations of stress and strain in the principal directions are linear through the thickness of Lamina 30G, all having the value zero at the midplane. Values at the lower ('bottom') surface are as shown in Table 6.16.

From the profiles of stress and strain in the principal directions we can now apply the Tsai–Hill failure criterion and estimate the load factor and the location where failure is first likely to occur; the results are included in Table 6.16.

It is interesting to examine stress/strength ratios to make a rough estimate of the character of failure. Making due allowance whether direct stresses are tensile or compressive we can infer the comments given in Table 6.17.

Table 6.15 Response of Lamina 30G to bending

Moment	N	κ_x/m	κ_y/m	κ_{xy}/m
M_x	10	0.7858	−0.2797	−0.6818
M_y	10	−0.2797	1.32	−0.244
M_{xy}	10	−0.6818	−0.244	2.16

Table 6.16 Response of Lamina 30G to bending (first ply failure)

Moment N	σ_1^b (MPa)	σ_2^b (MPa)	τ_{12}^b (MPa)	ε_1^b (%)	ε_2^b (%)	γ_{12}^b (%)	Tsai–Hill factor	Fail place
M_x 10	11.25	3.75	−6.495	0.02242	0.02819	−0.1264	11.12	bottom
M_y 10	3.75	11.25	6.495	0.001464	0.1026	0.1264	7.213	bottom
M_{xy} 10	12.99	−12.99	7.5	0.03629	−0.1289	0.1459	6.202	top

Note: Values at the upper surface have the opposite signs to those quoted above.

Table 6.17 First failures in Lamina 30G

Moment N	$\sigma_1/\sigma_{1\max}$	$\sigma_2/\sigma_{2\max}$	$\tau_{12}/\sigma_{12\max}$	Dominant mode
M_x 10	0.01125	0.0375	(−)0.081	shear
M_y 10	0.00375	0.1125	0.081	transv. tension/shear
M_{xy} 10	−0.0081	0.13	0.094	transv. tension/shear

Failure in bending of Lamina θG

Lamina θG is based on a 1 mm ply of EG, for which bending compliance coefficients d_{ij} as a function of θ are given in Table 6.18. The transformed reduced stiffness coefficients for EG are given in Table 6.14.

These data can be used to calculate load factors based on the Tsai–Hill criterion under one or more bending moments, and Table 6.19 presents results for $M_x = 1$ N, giving stresses in the principal directions on the top (t) surface.

It can be seen that the Tsai–Hill load factor, and inferred that the likely mode of failure (using the maximum stress criterion), varies through the range $0° < θ < 90°$. By inspection of Table 6.19, ratios of stress to strength in the principal directions confirm that first failures are mainly 0° longitudinal, 15°–45° shear, 60°–90° transverse tension, with substantial overlap of mode in the range 45°–60°.

Under combinations of moments applied off-axis to a unidirectional lamina, we find that in general the load factor, location of first ply failure, and likely dominate mode of failure all vary with θ. Fibre failure is unlikely unless θ is very small.

Problem 6.8: Tables 6.20, 6.21 and 6.22 present the stress responses of ply θG in the principal directions for the top surface together with the Tsai–Hill load factor for three loading conditions. Comment on the results and discuss your conclusions.

Problem 6.9: The Lamina θK is 1 mm thick and has the bending compliances given in Table 6.23. The transformed reduced stiffness coefficients for θK are given in Table 6.10. Under a bending moment of $M_x = 1$ N, the responses of Lamina θK are given in Table 6.24. Compare the load factors for first failure of Lamina θK with those for Lamina θG (given in Table 6.19), and comment on any differences in the likely modes of failure.

Table 6.18 Bending compliances for Lamina θG (1 mm thick EG)

$θ°$	d_{11} 1/(MNmm)	d_{12}	d_{22}	d_{16}	d_{26}	d_{66}
0	263.2	−72.11	1118	0	0	2335
15	371	−122.7	1112	−388.9	−38.67	2132
30	628.6	−223.8	1056	−545.4	−195.2	1728
45	893	−274.3	893	−427.6	−427.6	1526
60	1056	−223.8	628.6	−195.2	−545.4	1728
75	1112	−122.7	371	−38.67	−388.9	2132
90	1118	−72.11	263.2	0	0	2335

Table 6.19 Bending and failure of Lamina θG under $M_x = 10\,\text{N}$

$\theta°$	κ_x (/m)	κ_y (/m)	κ_{xy} (/m)	σ_1^t (MPa)	σ_2^t (MPa)	τ_{12}^t (MPa)	Tsai–Hill factor	Place
0	2.632	−0.7211	0	−60	0	0	16.67	bottom
15	3.71	−1.227	−3.889	−55.98	−4.019	15	5.02	bottom
30	6.286	−2.238	−5.454	−45	−15	25.98	2.781	bottom
45	8.93	−2.743	−4.276	−30	−30	30	2.082	bottom
60	10.56	−2.238	−1.952	−15	−45	25.98	1.803	bottom
75	11.12	−1.227	−0.3867	−4.019	−55.98	15	1.694	bottom
90	11.18	−0.7211	0	0	−60	0	1.667	bottom

Table 6.20 Bending and failure of Lamina θG under $M_x = 5\,\text{N} + M_{xy} = 10\,\text{N}$

$\theta°$	σ_1^t (MPa)	σ_2^t (MPa)	τ_{12}^t (MPa)	Tsai–Hill factor	Place
0	−30	0	−60	1.332	bottom
15	−57.99	27.99	−44.46	1.597	bottom
30	−74.46	44.46	−17.01	1.993	bottom
45	−75	45	15	2.014	bottom
60	−59.46	29.46	42.99	1.62	bottom
75	−32.01	2.01	59.46	1.344	bottom
90	0	−30	60	1.238	bottom

Table 6.21 Bending and failure of Lamina θG under $M_x = 2\,\text{N} + M_y = 10\,\text{N}$

$\theta°$	κ_x (/m)	κ_y (/m)	κ_{xy} (/m)	σ_1^t (MPa)	σ_2^t (MPa)	τ_{12}^t (MPa)	Tsai–Hill factor	Place
0	−0.1947	11.04	0	−12	−60	0	1.668	bottom
15	−0.4846	10.87	−1.165	−15.22	−56.78	−12	1.704	bottom
30	−0.9805	10.11	−3.043	−24	−48	−20.78	1.834	bottom
45	−0.9573	8.381	−5.131	−36	−36	−24	2.134	bottom
60	−0.1253	5.839	−5.845	−48	−24	−20.78	2.814	bottom
75	+0.9966	3.465	−3.967	−56.78	−15.22	−12	4.564	bottom
90	+1.516	2.487	0	−60	−12	0	7.607	bottom

Table 6.22 Bending and failure of Lamina θG under $M_x = 2\,\text{N} + M_y = -10\,\text{N}$

$\theta°$	κ_x (/m)	κ_y (/m)	κ_{xy} (/m)	σ_1^t (MPa)	σ_2^t (MPa)	τ_{12}^t (MPa)	Tsai–Hill factor	Place
0	1.247	−11.33	0	−12	60	0	1.665	bottom
15	1.969	−11.36	−0.391	−7.177	55.18	18	1.677	bottom
30	3.495	−11.01	0.861	6	42	31.88	1.746	top
45	4.529	−9.479	3.421	24	24	36	1.961	top
60	4.35	−6.734	5.064	42	6	31.18	2.524	top
75	3.45	−3.955	3.812	55.18	−7.177	18	4.109	bottom
90	2.958	−2.776	0	60	−12	0	7.309	bottom

Table 6.23 Bending compliances for Lamina θK

$\theta°$	d_{11} (/MNmm)	d_{12}	d_{22}	d_{16}	d_{26}	d_{66}
0	157.9	−53.68	2182	0	0	5714
15	497.7	−257.9	2250	−1213	+201.4	4897
30	1276	−666.3	2288	−1584	−169	3264
45	1987	−870.5	1987	−1012	−1012	2447
60	2288	−666.3	1276	−169	−1584	3264
75	2250	−257.9	497.7	+201.4	−1213	4897
90	2182	−53.68	157.9	0	0	5714

Table 6.24 Response of Lamina θK to $M_x = 1$ N up to first failure

$\theta°$	κ_x (/m)	κ_y (/m)	κ_{xy} (/m)	σ_1^t (MPa)	σ_2^t (MPa)	τ_{12}^t (MPa)	Tsai–Hill factor	Place
0	0.1579	−0.0537	0	−6	0	0	38.33	top
15	0.4977	−0.2579	−1.213	−5.598	−0.4019	1.5	29.07	top
30	1.276	−0.6663	−1.584	−4.5	−1.5	2.598	15.1	bottom
45	1.987	−0.8705	−1.012	−3	−3	3	8.944	bottom
60	2.288	−0.6663	−0.169	−1.5	−4.5	2.598	6.406	bottom
75	2.25	−0.2579	0.201	−0.4019	5.598	1.5	5.312	bottom
90	2.182	−0.0537	0	0	−6	0	5	bottom

6.4 INTRODUCTION TO STRENGTH OF LAMINATES

A laminate consists of a bonded stack of plies (usually taken as unidirectional in this book) in which individual plies are aligned at specific angles θ to the global reference direction x in the laminate. Given the ABD and abhd matrices for the laminate as a whole, the global stress and strain profiles in the fth ply can be calculated from the laminate midplane strains and curvatures $[\varepsilon°, \kappa]_L$ and the ply transformed reduced stiffnesses $[Q^*]_f$. The stresses in the principal directions can then be calculated using the $[T]$ matrix, and the combination (or combinations) of these stresses at any thickness co-ordinate, z, can be tested to establish the load factor by which all plies must be scaled up to achieve failure, using the chosen failure criterion. This set of calculations is repeated for each ply, to find in which ply the lowest load factor obtains and in which, under the given global applied load or moment ratios, first ply failure is predicted.

Even though one ply has failed, the laminate may still be capable of sustaining additional load. There continues to be debate on the best way to calculate behaviour after the first ply has failed. One might argue that one should set the stiffness of the failed ply to zero, recalculate the ABD matrix and work out the new stress and strain profiles, then reapply the failure criterion. This argument is not entirely plausible under in-plane loading, because even though a ply may be cracked, it does retain stiffness, especially in other principal directions and (with good bonding at the interface between fibres and matrix) this stiffness can transmit loads between plies. Setting the stiffnesses of the failed ply to zero has perhaps even less justification when bending loads are applied, because of the significant contribution of the thickness co-ordinates of any ply to bending stiffness. In practice, however, many software packages tend to retain the original ABD matrix and simply increase the load beyond first ply failure. This gives a first approximation to what is likely to happen, and is the approach followed here.

6.5 CROSSPLY LAMINATES

The strength properties of a crossply laminate depend not only on the properties of each of the plies in the stack and its relation to the directions of the applied stresses but also on the proportions of plies in the laminate and the bonding of the plies. The general discussion which follows relates only to symmetric crossply laminates stressed along the directions of the rods. We shall reach the amazing conclusion that it is possible to grow a crack in such a crossply in the same direction as the applied force.

Consider as an example a crossply laminate consisting of four identical plies each of which is much stronger along the direction of the rods than across it. The outside plies are lined up in the x-direction, and the inside plies are in the y-direction. Let us also suppose that the matrix is rather brittle.

It is helpful to look first at the behaviour of a faulty laminate in which there is no bonding between the plies. On firmly gripping the edges of the stack and applying a tensile force in the x direction, all plies will stretch uniformly until the plies with rods in the y direction reach their failure strain which is very small. At some point of local weakness the middle plies then break, and thereafter take no load whatsoever. It is, however, still possible to transfer load between grips because the outer plies have yet to reach their failure strain. In this system there is a concealed crack within the stack perpendicular to the direction of the applied load.

If we take a laminate with a similar arrangement of plies, but this time all properly bonded together, what now happens? At first the behaviour is the same as before. At a point of local weakness the middle plies break once. But there is still a good bond between the sets of plies, so load is still transferred through the bonds into the transverse plies: as the load is increased, more cracks develop in the transverse plies. You will recall here that the major lateral contraction ratio is much bigger than the minor one. The effect of the bond is that the transverse plies do not want to contract along the rods direction as much as the outside plies want to perpendicular to the direction of the applied load. If the interply bond is weak, it will break, and the outside plies will carry the load as before. But if the bond is a good one, the Poisson's ratio effect may induce cracks in the outer plies in the direction of the applied force. The inner plies are already cracked, so the effect is to produce a very porous structure – not much use if one is trying to make a tank or pipe to contain a fluid or gas. Further increase in load will develop an ever more porous structure, but there will continue to be transfer of load until the rods in the outer plies break.

6.5.1 Symmetric crossply laminate 0/90SG

Laminate 0/90SG consists of four identical plies of EG each 0.5 mm thick and laid up in the sequence $(0/90)_s$. The stiffness and compliance matrices are as follows:

	A (kN/mm)			D (kNmm)	
57.34	5.986	0	27.99	1.995	0
5.986	57.34	0	1.995	10.24	0
0	0	10.28	0	0	3.427

	a (mm/MN)			d (MNmm)	
17.63	− 1.84	0	36.23	−7.06	0
− 1.84	17.63	0	− 7.06	99.03	0
0	0	97.28	0	0	291.8

The stress distributions for several in-plane loading situations are given in Figure 6.5, and the midplane strains and failure sequences are given in Table 6.25. Failure locations are designated by ply number counting from the top down. As expected the behaviour of this laminate under a single direct load resultant is the same in either principal direction.

The stress distributions in global co-ordinates in this symmetric crossply

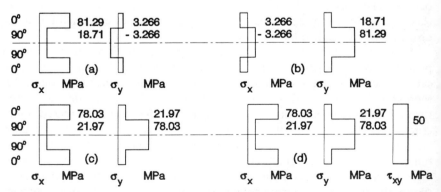

Figure 6.5 Stress profiles for symmetric laminate 0/90SG: (a) $N_x = 100$ N/mm; (b) $N_y = 100$ N/mm; (c) $N_x = N_y = 100$ N/mm; (d) $N_x = N_y = N_{xy} = 100$ N/mm.

Table 6.25

Load (N/mm)	ε_x (%)	ε_y (%)	γ_{xy} (%)	Tsai–Hill factor	Failure place
N_x 100	0.1763	−0.0184	0	5.343	2, 3
				11.62	1, 4
N_y 100	−0.0184	0.1763	0	5.343	1, 4
				11.62	2, 3
N_x 100 N_y 100	0.1579	0.1579	0	4.358	all plies
N_{xy} 100	0	0	0.9728	1.6	all plies
N_x 100 N_y 100 N_{xy} 100	0.1579	0.1579	0.9728	1.502	all plies

laminate are as expected. Under $N_x = 100 \, \text{N/mm}$, most of the load is carried in the outer plies which have the higher modulus. The top and bottom plies 1 and 4 have a major Poisson's ratio of $v_{12} = 0.274$, and under the resulting global strain $\varepsilon_x = 0.1763\%$ they would (if not bonded to the middle layers) wish to contract to $\varepsilon_{y1} = \varepsilon_{y4} = -0.274 \times 0.1763\% = -0.04831\%$. The central layers 2 and 3 have a minor Poisson's ratio $v_{21} = v_{12}E_2/E_1 = 0.274 \times 10.73/45.6 = 0.06447\%$ and (if not bonded to the outer plies) would wish to contract to $\varepsilon_{y2} = \varepsilon_{y3} = -0.06447 \times 0.1763 = -0.01137\%$. For all plies to have the same strain plies 2 and 3 need to be compressed more and the outer plies 1 and 4 less; so the transverse stress is tensile in plies 1 and 4, and compressive in plies 2 and 3, and shown in Figure 6.6. The net force in the transverse direction is zero, as there is no applied load in the y direction.

Under bending moments, the stress distributions in global co-ordinates for the symmetric crossply laminate 0/90SG are shown in Figure 6.7, and the midplane curvatures and some failure sequences are given in Table 6.26. Location of failure is denoted by the ply number counting from the top down and by the ply surface (top or bottom).

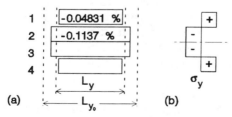

(a)   (b)

Figure 6.6 Response of symmetric crossply arrangement to load resultant N_x: (a) unbonded plies; (b) bonded plies.

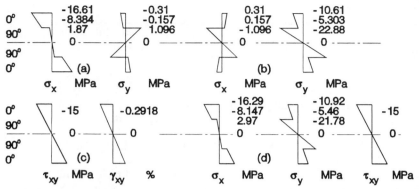

Figure 6.7 Stress profiles in global directions for symmetric laminate 0/90SG: (a) $M_x = 10$ N; (b) $M_y = 10$ N; (c) $M_x = M_y = 10$ N; (d) $M_x = M_y = M_{xy} = 10$ N.

Table 6.26

Moment (N)	κ_x (/m)	κ_y (/m)	κ_{xy} (/m)	Tsai–Hill factor	Place	Angle °
M_x 10	0.3623	−0.0706	0	53.14	3b	90
				59.72	4b	0
				96.65	1t	0
M_{xy} 10	0	0	2.918	5.333	1t, 4b	0
				10.67	remainder	
M_x 10 ⎫				4.605	4b	0
M_y 10 ⎬	0.2917	0.9197	2.918	5.211	1t	0
M_{xy} 10 ⎭				9.209	4t	0

The bending strain profiles caused by $M_x = 10$ N are linear and continuous through the thickness of this symmetric crossply laminate 0/90SG. The bending stress σ_x reflects the high modulus of plies 1 and 4 and the smaller modulus of plies 2 and 3. If we consider the interface between plies 3 and 4, both plies have the same strain ε_x at the interface. If at the same strain ε_x but not bonded, ply 4 would wish to contract more than ply 3. To create the same strain, ply 4 is stretched more in the y direction at the bonded interface, and ply 3 is compressed in the y direction at the bonded interface. At the midplane, the bending strains ε_x and ε_y are zero, so $\sigma_y = 0$. The net transverse force is zero as there is no external force in the y direction.

Under $M_x = 10$ N, the first failure is dominated by the large transverse compressive stress σ_y at the lower surface of ply 3, and the next failure occurs at a higher load dominated by the large tensile longitudinal stress at the bottom of ply 4. It is interesting to note that, in comparison with the behaviour of an isotropic plate, first ply failure in 0/90SG under M_x does *not* occur at the bottom surface 4B.

Problem 6.10: Stress profiles given so far in this chapter refer to global co-ordinates. It is extremely helpful (especially in bending) in assessments of strength and likely modes of failures to see at-a-glance the stresses in each ply in their principal directions. Replot stress profiles in ply principal directions from the data for 0/90SG under $N_x = 100 \, \text{N/mm}$ and under $M_y = 10 \, \text{N}$. Hence calculate the first three failure events for 0/90SG under $M_y = 10 \, \text{N}$, using the Tsai–Hill criterion.

6.5.2 Non-symmetric crossply laminate 0/90NSG

This non-symmetric crossply laminate 0/90NSG is made from two identical plies of EG each 1 mm thick in the sequence $(0/90)_T$. The stiffness and compliance matrices for 0/90NSG are given in Table 6.27.

The stress distributions for two in-plane loading situations are given in Figure 6.8, and for two bending situations in Figure 6.9. The midplane strains and curvatures with Tsai-Hill load factors are given in Table 6.28.

Table 6.27 Stiffnesses and compliances of crossply laminate 0/90NSG

A (kN/mm)			B (kN)		
	B (kN)			D (kNmm)	
57.34	5.986	0	−17.75	0	0
5.986	57.34	0	0	17.75	0
0	0	10.28	0	0	0
−17.75	0	0	19.11	1.995	0
0	0	0	1.995	19.11	0
0	0	0	0	0	3.427
a (mm/MN)			*b* (/MN)		
	h (/MN)			*d* (/MNmm)	
24.85	−2.594	0	23.08	0	0
−2.594	24.85	0	0	−23.08	0
0	0	97.28	0	0	0
23.08	0	0	74.56	−7.783	0
0	−23.08	0	−7.783	74.56	0
0	0	0	0	0	291.8

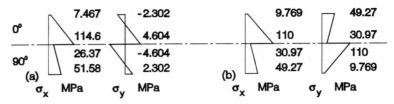

Figure 6.8 Stress profiles in global co-ordinates for nonsymmetric laminate 0/90NSG under in-plane loads: (a) $N_x = 100 \, \text{N/mm}$; (b) $N_x = N_y = 100 \, \text{N/mm}$.

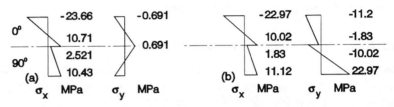

Figure 6.9 Stress profiles in principal directions for nonsymmetric laminate 0/90NSG under bending moments: (a) $M_x = 10 \, \text{N}$; (b) $M_x = M_y = 10 \, \text{N}$.

6.6 ANGLEPLY LAMINATES

The strength behaviour of angleply laminates depends on the stacking sequence, ply properties and the angle $(+\theta/-\theta)$ to the reference x direction. The following introductory remarks relate to symmetrical angleply laminates.

The main task is to calculate the stresses in the principal directions in each ply, and then test the Tsai–Hill criterion to find which ply fails first because it has the lowest load factor by which all stresses are scaled up to just cause failure.

The behaviour for extreme values of θ (0° and 90°) have already been discussed, because these represent unidirectional laminates outlined in Sections 6.3.2 and 6.3.3. For intermediate values of θ in the range $0° < \theta < 90°$ we can develop a similar approach to that given in Section 6.5 for crossply laminates.

Consider a symmetric angleply laminate based on four identical plies $(\theta/-\theta)_s$, in which there is perfect bonding between the plies. On firmly gripping the edges of the stack and applying a tensile force in the x direction, all plies will stretch uniformly until the combination of stresses in the principal directions in either the $+\theta$ or $-\theta$ plies reaches the predicted failure criterion according to Tsai–Hill. Because of the bonding between the plies, no shear is developed in global co-ordinates in each direction, although shear stress *is* developed in the principal directions. The sign of this shear stress will be

Table 6.28

Load/moment (N/mm) or (N)	ε_x (%)	ε_y (%)	γ_{xy} (%)	κ_x (/m)	κ_y (/m)	κ_{xy} (/m)	Tsai–Hill factor	Place
$N_x\,100$	0.2485	−0.02594	0	2.308	0	0	1.939 3.791	2B 2T
$N_x\,100$ $N_y\,100$	0.2226	0.2226	0	2.308	−2.308	0	2.031 3.092	1T, 2B 2T, 1B
$N_{xy}\,100$	0	0	0.9728	0	0	0	1.6	All
$N_x\,100$ $N_y\,100$ $N_{xy}\,100$	0.2226	0.2226	0.9728	2.308	−2.308	0	1.257 1.421	1T, 2B 1B, 2T
$M_x\,10$	0.02308	0	0	0.7456	−0.0778	0	9.583 39.71	2B 2T
$M_x\,10$ $M_y\,10$	0.02308	−0.02308	0	0.6677	0.6677	0	8.953 23.55	2B 1T
$M_{xy}\,10$	0	0	0	0	0	2.918	5.333	1T, 2B
$M_x\,10$ $M_y\,10$ $M_{xy}\,10$	0.02308	−0.02308	0	0.6677	0.6677	2.918	4.582 5.202	2B 1T

positive or negative depending on whether θ is positive or negative and on the numerical value of θ.

We therefore expect the failure will occur under uniform axial tension σ_x at the same load in all plies, because the Tsai–Hill criterion we use does not differentiate between failures under positive and negative shear stresses. In practice the first failure event is therefore most likely to be at some point of local weakness: a single crack occurs in one ply along the fibre direction (assuming that θ is not so small that longitudinal failure occurs). Most of this cracked ply can still transfer load from end to end of the specimen, but the share of the load taken before failure by the material in the failed ply in the region of the crack must now be redistributed between the other plies. A small increase in applied load will cause other plies to crack, so that each ply will develop multiple cracks, to form a porous net-like mesh of nodes.

Although the structure loses some stiffness, it fails safe from a load-bearing point of view. Load transfer is still possible, initially by a shear mechanism over the surviving area of bond between plies at the nodes. As the load is further increased, the area of the nodes decreases and eventually becomes too small to transfer a load either by in-plane shear between plies or, because the mesh is now starting to behave like lazy tongs, by rotational shear about the axis of the centre of the node which is perpendicular to the plane of the laminate.

First, we shall look at aspects of the strength of a specific symmetrical angleply laminate ($\theta = \pm\, 30°$), then at the effect of change of angle. Finally, we shall discuss the behaviour of a non-symmetric angleply laminate.

In the context of stiffness, the stress responses in global co-ordinates were discussed in general, and with many examples, in Sections 5.3, 5.5 and 5.6. Because we are concerned here with strength, we present all the following data for stresses in principal directions. The reader is reminded that in stress profiles in principal directions the real directions of stresses changes by step at each ply interface where the angle of orientation to the x axis changes. We can often see from the three profiles at-a-glance the likely failure events before doing the strength calculations.

6.6.1 Symmetric angleply laminate 30/-30SG

Laminate 30/−30SG consists of four identical plies of EG each 0.5 mm thick and laid up in the sequence $(30°/-30°)_s$. Transformed reduced stiffness coefficients $[Q^*(30°)]$ for EG are given in Table 6.14. The stiffness and compliance matrices for laminate 30/−30SG are as follows:

A (kN/mm)			D (kNmm)		
63.54	17.53	0	21.18	5.845	5.51
17.53	28.05	0	5.845	9.349	2.176
0	0	21.83	5.51	2.176	7.276

	a (mm/MN)			d (/MNmm)	
19.02	−11.89	0	66.74	−32.21	−40.91
−11.89	43.09	0	−32.21	130.5	−14.64
0	0	45.81	−40.91	−14.64	172.8

The stresses in principal directions and failure sequences (for Tsai–Hill) are given in Table 6.29. The stress distributions for several direct loading situations denoted by the asterisk (*) are given in Figure 6.10.

The data in Table 6.29 remind us that in order to apply the Tsai–Hill criterion we must transform stresses in the reference (x, y) co-ordinates to the principal directions in each layer.

For an applied load $N_x = 10$ N/mm ($\sigma_x = 5$ MPa), the stiffness response is qualitatively similar to that for laminate RRX30S. The direct load N_x in an unbonded stack would cause negative shear τ_{xy} in the $+30°$ plies and positive shear in the $-30°$ plies. To bond the plies together in an unsheared state, a positive shear has to be applied to the $+30°$ plies and a negative shear to the $-30°$ plies. On releasing these applied shears after bonding, there will be induced a negative shear stress and strain in the $+30°$ plies and positive shear stress and strain in the $-30°$ plies. This stress τ_{xy} is transformed into values of τ_{12} (of the same sign) aligned in the principal directions.

We can now apply the Tsai–Hill criterion to each layer:

$$[(5.117/1000)^2 - (5.117 \times -0.1168)/(1000)^2 + (-0.1168/-270)^2$$
$$+ (-1.376/80)^2)]^{-0.5} = 55.6$$

Table 6.29

Load (N/mm)	σ_1 (MPa)	σ_2 (MPa)	τ_{12} (MPa)	Tsai–Hill factor	Place
N_x 10	5.117	−0.1168	*	55.66	all
N_y 10	1.739	3.261	*	22.38	all
N_{xy} 10	*	*	1.177	44.8	$(-30°)$
				54.4	$(+30°)$

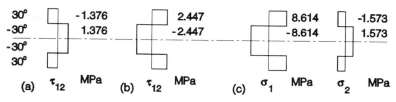

Figure 6.10 Stress profiles in principal directions for symmetric laminate 30/−30SG: (a) $N_x = 10$ N/mm; (b) $N_y = 10$ N/mm; (c) $N_{xy} = 10$ N/mm.

It is evident by inspection that shear would dominate the failure if the loads were scaled up by the load factor. The change in sign of the shear stress τ_{12} between the $+30°$ and $-30°$ plies has no effect on the magnitude of the load factor.

For an applied load $N_y = 10\,N/mm$, the compressive and shear stresses in the principal directions combine to cause failure, the lower load factor (compared with N_x) arising from the higher values of (both) stresses.

For $N_{xy} = 10\,N/mm$, there are no direct strains ($\varepsilon_x = \varepsilon_y = 0$), but there are direct stresses σ_x and σ_y caused by the off-axis loading response. Because the corresponding direct stresses in the principal directions (σ_1 and σ_2) have different signs (and different strengths in tension and compression), the load factors are different in the $+30°$ and $-30°$ plies.

The midplane curvatures and failure sequences caused by several bending moment resultants are shown in Table 6.30, and the stress responses in the ply principal directions are shown in detail in Figure 6.11.

Under a bending moment $M_x = 10\,N$, the three curvatures are as expected, and the three strain responses in the x, y directions are all linear through the thickness and zero at the midplane. The bending stresses in global co-ordinates are not continuous through the laminate because of the change in sign of the angle of the plies. A discussion of the detailed reasons for the stress distributions was given in Section 5.5.1 for laminate RRX30S.

Referring to the stress distributions σ_1, σ_2 and τ_{12} in Figure 6.11(a), and bearing in mind the associated stress/strength ratios, the first failures are clearly dominated by the shear stresses at the lower surface 4B and at the top surface 1T – other stresses contribute little. Once these two surfaces have failed, the applied moment can be almost doubled before the inside surfaces of the same $+30°$ plies fail, again dominated by shear. Even on the surviving surfaces 3B and 2T shear stresses will dominate as the moment is further increased.

Table 6.30

Moment (N)	κ_x (/m)	κ_y (/m)	κ_{xy} (/m)	Tsai–Hill factor	Place	Angle degrees
M_x 10	0.6674	-0.3221	-0.4091	14	4B	30
				14.52	1T	30
				27	4T	30
M_x 10 $\Big\}$ M_y 10	0.3453	0.983	-0.5554	7.96	4B	30
				15.93	4T	30
				18.86	3B	-30
M_{xy} 10	-0.4091	-0.1464	1.728	8.55	1T	30
				12.4	4B	30
				17.1	1B	30

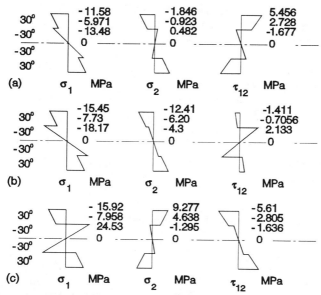

Figure 6.11 Stress profiles in principal directions for symmetric laminate 30/− 30SG: (a) $M_x = 10$ N; (b) $M_x = M_y = 10$ N; (c) $M_{xy} = 10$ N.

Under bending moments $M_x = M_y = 10$ N, the laminate deforms with positive synclastic curvature and negative twisting curvature. The first ply failure will occur at the bottom surface 4B where the dominant stress is σ_2, and then at an increased moment at 4T, again because of σ_2. The next ply surface to fail will be 3B, and here there is some contribution from shear stress, although σ_2 is still dominant. On slightly increasing the moment further, the top ply fails at 1T, dominated firmly by the transverse compressive stress σ_2.

Under the twisting moment $M_{xy} = 10$ N, the laminate 30/− 30SG deforms with negative synclastic curvature superposed on positive twisting curvature. First failure occurs at the upper surface 1T with comparable contributions from the tensile transverse stress ($\sigma_2/\sigma_{2max} \sim 0.09$) and the shear stress ($\tau_{12}/\tau_{12max} \sim 0.07$). On further increasing the moment, the lower surface 4B fails mainly because of shear stress: the now compressive transverse stress is only at a much smaller proportion (0.034) of its compressive strength. This is a good example where the sign of the direct stress is important.

Problem 6.11: A torque tube is to be made from the same prepreg EG having a (room temperature) cured ply thickness of 0.125 mm, for which data are given in Table 6.1. Which of the following arrangements gives the highest torque to first ply failure: (a) four plies laid up as ($+45°/-45°$)$_S$; (b) eight plies arranged (0°/90°/$+45°$/$-45°$)$_S$?

The transformed reduced stiffness coefficients $[Q(45°)]$ for EG are given in Table 6.14.

Laminate (a), $(45°/-45°)_s$, has the following properties:

A (kN/mm)			D (kNmm)		
10.49	5.346	0	0.2185	0.1114	0.06933
5.346	10.49	0	0.1114	0.2185	0.06933
0	0	6.42	0.06933	0.06933	0.1337

a (mm/MN)			d (/MNmm)		
128.9	−65.7	0	6608	−2731	−2010
−65.7	128.9	0	−2731	6608	−2010
0	0	155.8	−2010	−2010	9561

Laminate (b), $(0°/90°/45°/-45°)_s$, has the following properties:

A (kN/mm)			D (kNmm)		
24.82	6.843	0	2.725	0.3296	0.06933
6.843	24.82	0	0.3296	1.893	0.06933
0	0	8.99	0.06933	0.06933	0.5085

a (mm/MN)			d (/MNmm)		
43.6	−12.02	0	375.8	−63.87	−42.52
−12.02	43.6	0	−63.87	541.7	−65.15
0	0	111.2	−42.52	−65.15	1981

6.6.2 A practical example of an angleply laminate

On a lighter note, there is an interesting opportunity to observe pressure vessels in action, albeit not too scientifically. The British sausage wears one of two skins. The traditional skin comes from the intestine of the pig, and behaves as if it were a helically filament-wound laminate to contain the internal pressure it must withstand: molecular orientation in an angleply arrangement. The modern and much cheaper skin is extruded so that any molecular orientation is along the length of the tube, having the character of a unidirectional ply or laminate.

How do you tell which type of skin your sausages are wearing? The apparatus consists of one frying pan, with lid, heated to the desired test temperature. Place one sausage in the pan, replace lid, and cook in the usual way. The sausage with the cheap skin will split along its length as the meat expands because of the pressure generated during heating. The executive sausage resists the pressure and does not split – it relieves the pressure eventually by extruding meat out at each end. Either process can be a bit explosive – hence the need for the lid. For the same reason the cook normally sabotages this experiment by pricking the sausages many times with a fork. Our test relies on unpunctured pressure vessels.

6.6.3 Strength of symmetrical laminate (θ/-θ)SG as function of angle

The purpose of this section is to review how the failure load factors vary with angle, θ, to the reference direction, x, and to comment, with less precision, on the likely dominant mode of failure. The general four-ply symmetrical angleply laminate $(\theta/-\theta)_s$ is based on 0.5 mm thick plies of cold-cured EG having the properties given in Table 6.1. The coefficients of extensional compliance [a], and the bending compliance [d], are presented in Tables 6.31 and 6.32 for angles in the range $0° \leqslant \theta \leqslant 90°$.

Application of $N_x = 10$ N/mm to (θ/$-\theta$)SG

The response of laminate $(\theta/-\theta)$SG to the force resultant $N_x = 10$ N/mm is presented in Table 6.33.

Table 6.31 Extensional compliance coefficients of laminate $(\theta/-\theta)$SG

$\theta°$	a_{11} (mm/MN)	a_{12}	a_{22}	a_{16}	a_{26}	a_{66}
0	10.96	−3.004	46.6	0	0	97.28
15	12.5	−5.405	46.29	0	0	70.77
30	19.02	−11.89	43.09	0	0	45.81
45	32.21	−16.42	32.21	0	0	38.94
60	43.09	−11.89	19.02	0	0	45.81
75	46.29	−5.405	12.15	0	0	70.77
90	46.6	−3.004	10.96	0	0	97.28

Table 6.32 Bending compliance coefficients for laminate $(\theta/-\theta)$SG

$\theta°$	d_{11} (/MNmm)	d_{12}	d_{22}	d_{16}	d_{26}	d_{66}
0	32.89	−9.013	139.8	0	0	291.8
15	41.99	−15.77	138.9	−32.8	−3.261	239.8
30	66.74	−32.21	130.5	−40.91	−14.64	172.8
45	103.2	−42.67	103.2	−31.4	−31.4	149.4
60	130.5	−32.21	66.74	−14.64	−40.91	172.8
75	138.9	−15.77	41.99	−3.261	−32.8	239.8
90	139.8	−9.013	32.89	0	0	291.8

Table 6.33 Response of lamina $(\theta/-\theta)$SG to $N_x = 10$ N/mm

θ (°)	ε_x (%)	ε_y (%)	γ_{xy} (%)	σ_1^t (MPa)	σ_2^t (MPa)	τ_{12}^t (MPa^a)	Tsai–Hill factor	Place
0	0.01096	−0.003004	0	5	0	0	200	all
15	0.0125	−0.005405	0	5.121	−0.121	−0.4602	128.9	all
30	0.01902	−0.01189	0	5.117	−0.1168	−1.376	55.66	all
45	0.03221	−0.01642	0	3.901	1.099	−2.5	30.04	all
60	0.04309	−0.01189	0	1.739	3.261	−2.447	22.38	all
75	0.04629	−0.005405	0	0.3803	4.62	−1.329	20.38	all
90	0.0466	−0.003004	0	0	5	0	20	all

Note: ^{a} Shear stress has opposite sign in plies at $-\theta$.

Problem 6.12: By comparing the stress ratios $\sigma_i/\sigma_{i\max}$ deduce the most likely mode(s) of failure of laminates $(\theta/-\theta)$SG under N_x.

Problem 6.13: Confirm by sample calculation that the Tsai–Hill load factors for laminates $(\theta/-\theta)$SG loaded in in-plane compression $N_x = -10$ N/mm have the following values for first ply failure, and compare them with those already presented for in-plane tension $N_x = +10$ N/mm. It may be assumed that buckling does not occur.

Application of $N_y = 10$ N/mm to $(\theta/-\theta)$SG

The response of laminate $(\theta/-\theta)$SG to the in-plane force resultant $N_y = 10$ N/mm is given in Table 6.34.

Problem 6.14: Explore how the Tsai–Hill load factor for a pipe under internal pressure varies with winding angle for a four-ply symmetric laminate $(\theta/-\theta)$SG based on EG. Suggest the dominant likely modes of failure, and comment on interesting features of the stresses and strains in the system.

Table 6.34 Response of lamina $(\theta/-\theta)$SG to $N_y = 10$ N/mm

θ (°)	ε_x (%)	ε_y (%)	γ_{xy} (%)	σ_1^t (MPa)	σ_2^t (MPa)	τ_{12}^t (MPa)	Tsai–Hill factor	Place
0	−0.003004	0.0466	0	0	5	0	20	all
15	−0.005405	0.04629	0	0.3803	4.62	1.329	20.38	all
30	−0.01189	0.04309	0	1.739	3.261	2.447	22.38	all
45	−0.01642	0.03221	0	3.901	1.099	2.5	30.04	all
60	−0.01189	0.01902	0	5.117	−0.1168	1.376	55.66	all
75	−0.005405	0.0125	0	5.121	−0.121	0.4602	128.9	all
90	−0.003004	0.01096	0	5	0	0	200	all

Table 6.35 Response of laminate $(\theta/-\theta)$SG to $N_{xy} = 10$ N/mm

θ (°)	ε_x (%)	ε_y (%)	γ_{xy} (%)	σ_1^t (MPa)	σ_2^t (MPa)	τ_{12}^t (MPa)	Tsai–Hill factor	Place
0	0	0	0.09728	0	0	5	16	all
15	0	0	0.07077	7.684	1.403	3.15	23.74	2, 3
							24.64	1, 4
30	0	0	0.04581	8.614	−1.573	1.177	44.8	2, 3
							54.37	1, 4
45	0	0	0.03894	8.456	−1.544	0	60.69	2, 3
							92.35	1, 4
60	0	0	0.04581	8.614	−1.573	−1.177	44.8	2, 3
							54.37	1, 4
75	0	0	0.07077	7.684	−1.403	−3.15	23.74	2, 3
							24.64	1, 4
90	0	0	0.09728	0	0	−5	16	all

Note: Stresses quoted above are for the top $(+\theta)$ ply.

Application of $N_{xy} = 10$ N/mm to $(\theta/-\theta)SG$

The response of laminate $(\theta/-\theta)$SG to an in-plane shear force resultant $N_{xy} = 10$ N/mm is presented in Table 6.35. Note that stresses σ_1 and σ_2 have opposite signs for negative θ.

Problem 6.15: (a) Why does a different load factor arise in the inner plies 2 and 3 compared with the outer plies 1 and 4 for Lamina $(\theta/-\theta)$SG? (b) What will be the effect of applying a negative shear load $N_{xy} = -10$ N/mm to $(\theta/-\theta)$SG rather than a positive one?

Problem 6.16: Using the Tsai–Hill criterion, calculate the load factors in laminate $(45/-45)$SG (total thickness 2 mm) under the loading $N_x = 10$ N/mm and $N_{xy} = 10$ N/mm.

Problem 6.17: Which mode of failure dominates Lamina $(\theta/-\theta)$SG under an in-plane shear force resultant?

Application of $M_x = 10$ N to laminate $(\theta/-\theta)SG$

The response of laminate $(\theta/-\theta)$SG to an applied moment resultant $M_x = 10$ N is given in Table 6.36.

It is interesting to note that although the first failure is always predicted to occur on the lower surface of this laminate under a positive moment $M_x = 10$ N, the subsequent failure locations vary somewhat as θ increases.

Problem 6.18: What are the most likely dominant failure mechanisms of first failure for the events predicted in Table 6.36?

Table 6.36 Response of laminate $(\theta/-\theta)$SG to $M_x = 10$ N

θ (°)	κ_x (/m)	κ_y (/m)	κ_{xy} (/m)	σ_1^t (MPa)	σ_2^t (MPa)	τ_{12}^t (MPa)	Tsai–Hill factor	Place
0	0.3289	−0.09013	0	−15	0	0	66.67	4b
							106.7	1t
							133.3	3b, 4t
							213.3	2t, 1b
15	0.4199	−0.1577	−0.328	−13.78	−0.4915	2.944	25.3	4b
							26.45	1t
							50.61	4t
							52.9	1b
30	0.6674	−0.3221	−0.4091	−11.58	−1.846	5.456	14	4b
							14.52	1t
							27.99	4t
							29.04	1b
45	1.032	−0.4267	−0.314	−8.147	−5.46	7.5	9.21	4b
							10.42	1t
							18.42	4t
							19.92	3b
60	1.305	−0.3221	−0.1464	−3.869	−10.57	6.867	7.35	4b
							10.61	1t
							14.7	4t
							14.82	3b
75	1.389	−0.1577	−0.03261	−0.984	−13.94	3.83	6.785	4b
							13.57	3b, 4t
							14.21	1t
90	1.398	−0.09013	0	0	−15	0	6.667	4b
							13.33	4t, 3b
							18	1t
							36	2t, 1b

Note: Stresses are profiles which vary through the thickness; the top surface stress in global co-ordinates is not necessarily the maximum value in the principal direction.

Application of $M_{xy} = 10$ N to laminate $(\theta/-\theta)$SG

The response of laminate $(\theta/-\theta)$SG to an applied twisting moment resultant per unit width, $M_{xy} = 10$ N, is presented in Table 6.37.

Note that as expected, the failure modes have their greatest values for $\theta = 45°$, and the values of stress in the range $0 < \theta < 45°$ are reflected for $45° > \theta > 90°$, apart from the sign of the shear stress.

Problem 6.19: What are the most likely dominant mechanisms of first failure for the events predicted in Table 6.37?

Table 6.37 Response of lamina $(\theta/-\theta)$SG to $M_{xy} = 10$ N

θ (°)	κ_x (/m)	κ_y (/m)	κ_{xy} (/m)	σ_1^t (MPa)	σ_2^t (MPa)	τ_{12}^t (MPa)	Tsai–Hill factor	Place
0	0	0	2.918	0	0	−15	5.33 10.67	1t, 4b 2t, 4t, 1b, 3b
15	−0.328	−0.03261	2.398	−11.57	6.428	−11.43	6.401 6.874 12.8	1t 4b 1b
30	−0.4091	−1.464	1.728	−15.92	9.277	−5.61	8.55 12.4 17.1 24.81	1t 4b 1b 4t
45	−0.314	−0.314	1.494	−16.92	10.29	0	9.634 19.27 22.86 39.11	1t 1b 4b 3b
60	−1.464	−0.4091	1.728	−15.92	9.277	5.61	8.55 12.4 17.1 24.81	1t 4b 1b 4t
75	−0.0326	−0.328	2.398	−11.57	6.248	11.43	6.401 6.874 12.8 13.72	1t 4b 1b 4t
90	0	0	2.918	0	0	15	5.333 10.67	1t, 4b 2t, 4t, 1b, 3b

Note: Stresses are profiles which vary through the thickness of the laminte; the top surface stress in global co-ordinates is not necessarily the maximum value in the principal direction.

The special angleply laminate $(0/\theta/-\theta)_s$ made from 0.5 mm plies of EG is of interest in demonstrating the possibility of first ply failures occurring within the laminate rather than at the outer surface (where the global bending stress is a maximum). The failure predictions for a bending moment of $M_x = 10\,N$ are as follows:

Ply angle $\theta°$	0	15	30	45	60	75	90
Tsai–Hill factor	150	132	96	81	77	76	76
First failure place	6b	6b	5b	5b	5b	5b	5b

For $\theta = 0$ to $15°$ the failure is on the bottom surface of the laminate as expected. Above $15°$ the fifth ply fails at its lowest surface because it is weaker under the prevailing stresses in its principal directions than the sixth ply is under the prevailing stresses in its principal directions.

6.6.4 Non-symmetric angleply laminate 30/−30NSG

Laminate $30/-30$NSG is made from two identical plies of EG each 1 mm thick. The top ply is at $30°$ to the reference direction, and the bottom ply at $-30°$. The stiffness and compliance matrices are given in Table 6.38.

The stress distributions for several direct loading situations are given in Figure 6.12, and the midplane strains and curvatures (and the failure sequences predicted by Tsai–Hill) are given in Table 6.39.

Under $N_x = 10\,N/mm$, laminate $30/-30$NSG extends, and contracts trans-

Table 6.38 Stiffnesses and compliances of laminate 30/−30NSG

A (kN/mm) B (kN)				B (kN) D (kNmm)		
63.54	17.53	0	0	0	−11.02
17.53	28.05	0	0	0	−4.351
0	0	21.83	−11.02	−4.351	0
0	0	−11.02	21.18	5.845	0
0	0	−4.351	5.845	9.349	0
−11.02	−4.351	0	0	0	7.276

a (mm/MN) h (/MN)				b (/MN) d (/MNmm)		
23.73	−10.21	0	0	0	29.83
−10.21	43.69	0	0	0	10.67
0	0	63	29.83	10.67	0
0	0	29.83	71.18	−30.62	0
0	0	10.67	−30.62	131.1	0
29.83	10.67	0	0	0	189

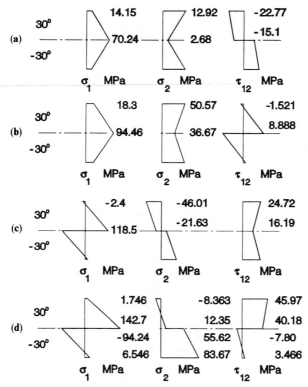

Figure 6.12 Stress profiles in principal directions for 30/−30NSG: (a) $N_x =$ 100 N/mm; (b) $N_x = N_y = 100$ N/mm; (c) $N_{xy} = 100$ N/mm; (d) $N_x = N_{xy} = 100$ N/mm.

versely, but because the laminate is non-symmetric it also takes up a positive twisting curvature, as shown in Figure 6.12(a). The shear stress dominates first failure as 1T and 2B, and the second pair of failures at 2T and 1B at the increased load, even though the longitudinal stress is $\sigma_1 = 70.24$ MPa at the midplane.

Under equal biaxial loading $N_x = N_y = 100$ N/mm, the non-symmetric laminate shows positive synclastic curvature with superposed twisting curvature, as shown in Figure 6.12(b). The failures are dominated by the large transverse tensile stresses σ_2.

Under in-plane shear loading $N_{xy} = 100$ N/mm (Figure 6.12(c)), the non-symmetric laminate shears and develops positive synclastic curvature. First failure at 2B is caused mainly by the massive transverse tensile stress σ_2. The second failure is at 1T, where the large compressive transverse stress σ_2/σ_{2max} is assisted by the large shear stress (τ_{12}/τ_{12max}).

Under the combination of in-plane transverse and shear loads $N_y =$

Table 6.39 Responses of laminate $30/-30$NSG to in-plane loads

Load (N)	ε_x (%)	ε_y (%)	γ_{xy} (%)	κ_x (/m)	κ_y (/m)	κ_{xy} (/m)	Tsai–Hill factor	Place	Angle (°)
N_x 100	0.2373	−0.1021	—	—	—	2.983	3.1	1T	30
							3.1	2B	−30
							4.9	2T	−30
							4.9	1B	30
N_x 100 N_y 100	0.1353	0.3349	—	—	—	4.05	1.9	1T	30
							1.9	2B	−30
N_{xy} 100	—	—	0.63	2.983	1.067	—	1.8	2B	−30
							2.8	1T	30
N_y 100 N_{xy} 100	−0.102	0.4369	0.63	2.983	1.067	1.067	1.1	2B	−30
							1.7	1T	30
							1.7+	2T	−30
							1.8	1B	30

$N_{xy} = 100$ N/mm (Figure 6.12(d)), the panel extends (ε_y), contracts (ε_x), shears, develops positive synclastic curvature and positive twist. First failure at 2B is almost exclusively caused by the high transverse tensile stress σ_1. The second failure at 1T is dominated by the shear stress τ_{12}. Complete failure of this laminate is then predicted after only a small further increase in load.

6.7 QUASI-ISOTROPIC LAMINATES

6.7.1 Symmetric quasi-isotropic laminate $60/0/-60$SG

The symmetric quasi-isotropic laminate $60/0/-60$SG is made from six identical plies of EG each 0.333 mm thick and laid up in the sequence $(60/0/-60)_s$. The stiffness and compliance matrices are given in Table 6.40.

Table 6.40 Stiffnesses and compliances of laminate $60/0/-60$SG

A (kN/mm)			D (kNmm)		
49.59	13.67	0	14.9	4.832	1.928
13.67	49.59	0	4.832	17.52	4.883
0	0	17.96	1.928	4.883	6.259
a (mm/MN)			d (/MNmm)		
21.82	−6.016	0	74.06	−17.97	−8.796
−6.016	21.82	0	−17.97	77.27	−54.74
0	0	55.68	−8.796	−54.24	205.2

Table 6.41 Responses of laminate $60/0/-60$SG to in-plane loads

Load (N/mm)	ε_x (%)	ε_y (%)	γ_{xy} (%)	Tsai–Hill factor	Angle degrees
N_x 100	0.2182	−0.06016	—	4.4	60, −60
				10	0
N_x 100 } N_y 100 }	0.1581	0.1581	—	4.2	All
N_{xy} 100	—	—	0.5568	2.7	0
				3.7	−60
				4.6	+60

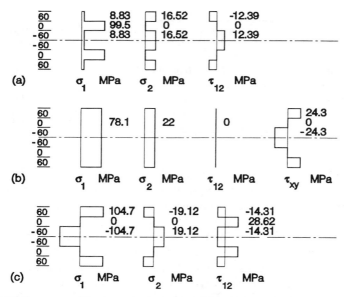

Figure 6.13 Stress profiles in principal directions for symmetric laminate $60/0/-60$SG: (a) $N_x = 10$ N/mm; (b) $N_x = N_y = 100$ N/mm; (c) $N_{xy} = 100$ N/mm.

The stress distributions under several different direct loads are given in Figure 6.13, and the midplane strains and failure sequences are given in Table 6.41.

Under direct load $N_x = 100$ N/mm, the 60° and −60° plies fail due to almost equal contributions from shear ($\tau_{12}/\tau_{12\max}$) and from transverse tensile stress ($\sigma_2/\sigma_{2\max}$). the change of sign of shear stress has no effect on the Tsai–Hill criterion. The failure of the 0° plies at the much increased load is solely caused by the longitudinal tensile stress σ_1.

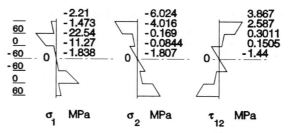

	-2.21		-6.024		3.867
60	-1.473		-4.016		2.587
0	-22.54		-0.169		0.3011
-60	-11.27		-0.0844		0.1505
-60	-1.838		-1.807		-1.44
60					
0					
60					

σ_1 MPa σ_2 MPa τ_{12} MPa

Figure 6.14 Stress profiles in principal directions of $60/0/-60SG$ under $M_x = 10\,N$.

Table 6.42

Moment (N)	κ_x (/m)	κ_y (/m)	κ_{xy} (/m)	Tsai–Hill factor	Place	Angle degrees
M_x 10	0.7406	−0.1797	−0.08796	12	6B	60
				18	1T	60
				19	6T	60
				28	1B	60
				39	4B	−60

Under equal biaxial loading $N_x = N_y = 100\,N/mm$, all plies carry the same direct tensile stresses σ_1 and σ_2, and no shear stress, so all plies fail at the same load factor.

Under the in-plane shear load $N_{xy} = 100\,N/mm$, the first failure occurs in the $0°$ plies due to the shear τ_{12}. Increasing the load, the $-60°$ plies fail next, mainly because of the combination of shear and transverse stress σ_2. Finally the $60°$ plies fail, and here the tensile stress σ_1 has rather more effect than the transverse compressive stress. This is another good example of the importance of sign of the direct stresses when using the Tsai–Hill model.

Applying a bending moment $M_x = 10\,N$ gives the stress distributions in Figure 6.14, and the curvatures and some early failures using the Tsai–Hill criterion in Table 6.42.

Applying a bending moment of $M_x = 10\,N/mm$ to $60/0/-60SG$ gives anticlastic curvature, with a small amount of negative twisting curvature. Failure occurs first at the lower surface 6B of the laminate due mainly to a combination of σ_2 and τ_{12}. At a slightly larger moment, failure occurs at 1T, where σ_2 is negative. The reader is invited to ascertain subsequent causes of failure as the moment is increased.

6.7.2 Non-symmetric quasi-isotropic laminate 60/0/ − 60NS

Laminate 60/0/ − 60NS is a non-symmetric quasi-isotropic laminate made from three identical plies of EG each 0.667 mm thick laid up in the sequence $(60/0/ - 60)_T$. The stiffness and compliance matrices are given in Table 6.43.

The stress distributions under two different in-plane direct loads are given in Figure 6.15, and the midplane strains and curvatures (together with Tsai–Hill failure sequences) are given in Table 6.44.

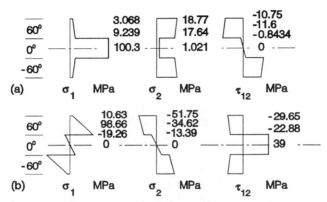

Figure 6.15 Stress profiles in principal directions for laminate 60/0/ − 60NSG: (a) $N_x = 100 \, \text{N/mm}$; (b) $N_{xy} = 100 \, \text{N/mm}$.

Table 6.43 Stiffnesses and compliances of laminate 60/0/ − 60NS

A (kN/mm) B (kN)				B (kN) D (kNmm)	
49.65	13.69	0	0	0	− 3.868
13.69	49.65	0	0	0	− 9.796
0	0	17.98	− 3.868	− 9.796	0
0	0	− 3.868	10.15	5.703	0
0	0	− 9.796	5.703	20.67	0
− 3.868	− 9.796	0	0	0	7.135

a (mm/MN) h (/MN)				b (/MN) d (/MNmm)	
21.92	− 5.073	0	0	0	4.923
− 5.073	28.8	0	0	0	36.79
0	0	75.88	10.31	33.12	0
0	0	10.31	118	− 27.67	0
0	0	33.12	− 27.67	71.71	0
4.923	36.79	0	0	0	193.3

Table 6.44 Responses of 60/0/−60NSG to in-plane loads

Load (N)	ε_x (%)	ε_y (%)	γ_{xy} (%)	κ_x (/m)	κ_y (/m)	κ_{xy} (/m)	Tsai–Hill factor	Place	Angle (°)
N_x 100	0.2192	−0.05073	—	—	—	0.4923	4.3	1,3	60/−60
							9.8	2	0
N_x 100	—	—	0.7588	1.031	3.312	—	1.5	3B	−60
							1.9	2B	0
							2.0	2T	0
							2.2	3T	−60
							2.4	1T	60
							3.0	1B	60

Under $N_x = 100 \, \text{N/mm}$ the laminate extends and contracts, and there is positive twisting curvature as the result of the lack of symmetry. First failure occurs in the $+60°$ and $-60°$ plies from the combination of transverse tensile stress σ_2 and the shear stress τ_{12}. On increasing the load the $0°$ ply fails because of the large longitudinal stress σ_1.

Under an in-plane shear load $N_{xy} = 100 \, \text{N/mm}$, the laminate shears, and there is superposed a positive synclastic curvature, κ_x and κ_y, because of the non-symmetric stacking sequence. First failure at 3B is caused by the combination of substantial values of transverse tensile stress σ_2 and shear stress τ_{12}. On increasing the shear load, failure at 2B and 2T in the $0°$ ply is swamped by the high shear stress τ_{12}.

6.8 EDGE EFFECTS IN LAMINATES

In Chapter 1 we gave a very brief outline of the strength of wide composite laminates, and preceding parts of this chapter are based on the same concept. Because of the stress distributions within individual plies when external (or thermal) loads are applied, narrow laminates can be prone to delamination at their edges, which can reduce their strength considerably. It is sometimes possible to reduce these edge effects somewhat by choice of stacking sequence.

The edge effects described here relate to a width roughly equal to the thickness of the laminate. It is generally reckoned that, in the absence of holes through the thickness, a plate is 'wide' if its width is greater than about ten times the plate thickness. Calculation of the values of stresses involves difficult analysis outside the scope of this book. For details see Pagano (1989). What we shall do here is introduce a simplified analysis which will clearly indicate the important features in qualitative terms, from which some important conclusions can be drawn.

The essence of the problem is that when external loads are applied to a laminate in which adjacent plies are orientated at different angles (or are made from different materials), there are, as we have seen in Chapter 5, different in-plane stresses in different plies in any given direction. In classical lamination theory we assume that these stresses are uniform across the laminate. But at the edges of the laminate there may be no applied force in some directions, and in these directions the stress at the edge of the laminate must therefore be zero. To compensate the local decay in stress (over a width of about the thickness of the laminate), shear stresses are set up, and these give rise to interply shearing at the edges of angleply laminates, and stresses through the thickness of crossply laminates at the edges causing interply splitting.

6.8.1 Edge stresses in symmetric crossply laminate

Consider a symmetric crossply laminate made from four identical unidirectional plies arranged $(0°/90°)_s$ and strained in the x direction (Figure 6.16). The main cause of interlaminar stresses is the mismatch of Poisson's ratios between the 0° and 90° plies, which sets up tensile transverse stresses σ_y in the 0° plies and compressive transverse stresses in the 90° plies, as shown in Figure 6.17 (the reasoning for this was explained in Section 4.4.1). These stresses cannot exist at the free edges because there is no externally applied load in the y direction. Hence in the region of the edge there is a transverse stress gradient from the 'wideplate' value of σ_y down to zero at the free edge.

If we take a deliberately simplified view, we can construct the following helpful explanation.

In Figure 6.18 the region AB'C'D on the top 0° ply from the centre line to the start of the edge region takes the full transverse tensile stress, σ_y. In the region B'BCC' which is about one laminate thickness wide, we find σ_y reduces to zero

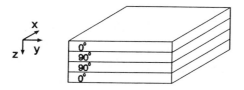

Figure 6.16 The laminate 0/90SG.

Figure 6.17 Transverse stress in wide 0/90SG laminate.

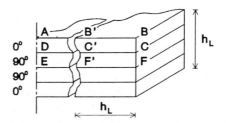

Figure 6.18 Element of 0/90SG showing edge region.

at the stress-free surface B'B. Load is therefore transferred from the 0° ply to the 90° ply by the presence of a shear stress (Figure 6.19 (a)) – actually a build-up of shear stress from zero at C' to the maximum at the free edge C; we shall take an average value, τ_{yz}. Taking moments about the interface CC' for the 0° ply, the uniform stress σ_y exerts an internal anticlockwise moment $M_{fi} = \sigma_y h_f \cdot h_f/2$ (Figure 6.19 (b)). To maintain equilibrium of moments we must therefore apply a clockwise couple M_f at the interface acting, for convenience here about a point $h_f/2$ from the free edge and represented by a uniform through thickness stress σ_z. By the nature of the distribution, vertical equilibrium is achieved, as there is no net force in the z direction acting on the top of the 0° ply. In particular the edge stress σ_z is tensile, which is troublesome. In the 90° ply immediately under the 0° ply there is developed by the same reasoning an anticlockwise couple M_f which also leads to tensile stress σ_z acting at the edge.

The picture which emerges from this crude analysis of stresses at the edge of this four-ply $(0°/90°)_s$ laminate is that the central two plies at their edges are compressed together, whereas the outer 0° plies carry tensile stresses at their edges which act to pull them apart from the 90° plies. As the bond between plies contains no reinforcing fibres, the interface has at best the strength of the matrix, and may be much less if the edge contains imperfections caused by sawing, routing, or drilling of holes, let alone any imperfection caused by careless lamination. The conclusion is that even at modest longitudinal stress σ_x, there is likely to be a risk of progressive longitudinal delamination, as sketched in Figure 6.20. Narrow laminates are more prone to this delamina-

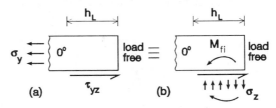

Figure 6.19 Free body diagram of free-edge region of 0° ply.

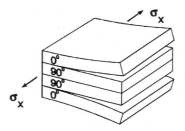

Figure 6.20 Delamination of edge of $(0°/90°)_s$ under direct load N_x.

tion than wide plates, hence wide plates are preferred for the measurement of tensile stress, with the width at least sixteen times the laminate thickness.

Problem 6.20: How would you expect the behaviour of a narrow four-ply $(90°/0°)_S$ crossply laminate to behave under longitudinal load, compared with a $(0°/90°)_S$ layup?

6.8.2 Edge stresses in symmetric angleply laminate

In the simplest wide symmetric angleply laminate $(\theta/-\theta)_S$ we saw in Section 5.5.1 that under an in-plane force resultant N_x, there were developed direct stresses σ_x and shear stresses τ_{xy}, profiles of which are shown in Figure 5.15. Compared with the crossply laminate, there are no transverse stresses to worry about at the free edge, but in the angleply laminate it is the behaviour of the shear stresses which concern us.

When a force resultant N_x is applied, the sides of the laminate are load free. The wide specimen is under shear stress τ_{xy}, and to ensure this stress runs out to zero at the free edge, we appeal to the development of a localized shear stress τ_{xz} acting on the underside of the $+\theta$ ply over a region about h_L wide (Figure 6.21). Note that we cannot apply this stress τ_{xz} at the top surface of this ply because this surface is also load free.

Similarly an equal but opposite local shear stress τ_{xz} acts on the top of the second $-\theta$ ply. The mismatch in sign of τ_{xz} follows the mismatch of signs of τ_{xy}

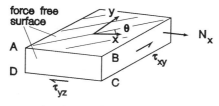

Figure 6.21 Shear stress τ_{xz} developed locally on lower side of top $+\theta$ ply to ensure ABCD is force-free.

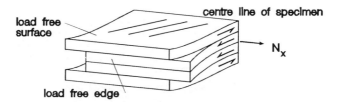

Figure 6.22 Debonding at edge of $(\theta/-\theta)_s$ under direct load N_x.

which originate in the opposite signs of Q^*_{16} and Q^*_{26} in the $+\theta$ and $-\theta$ plies. The large the value of Q^*_{16} and Q^*_{26}, the greater the values of τ_{xy} and τ_{xz}, and hence the greater the risk of debonding at the free edge as the two plies try to develop 'interply shear' (Figure 6.22). For many composites the peak in Q^*_{16} is reached at about 30°.

FURTHER READING

Agarwal, B.D. and Broutman, L.J. (1990) *Analysis and Performance of Fibre Composites*, Wiley.

Beaumont, P.W.R. (1989a) Damage and fracture of fibre composites, in L.N. Phillips (ed) *Design with Advanced Composite Materials*, Design Council London and Springer Verlag.

Beaumont, P.W.R. (1989b) The failure of fibre composites: an overview, *J. Strain Analysis* **24** (4), 189–205.

Datoo, M.H. (1991) *Mechanics of Fibrous Composites*, Elsevier.

Matthews, F.L. (1987) *Joining Fibre-Reinforced Plastics*, Elsevier.

Matthews, F.L. (1990) *The Failure of Reinforced Plastics*, Mechanical Engineering Publications, London.

Pagano, N.J. (ed) (1989) *Interlaminar Response of Composite Materials*, Elsevier.

Tsai, (1988) *Composites Design* (4th edn) Think Composites.

Vincent, J.F.A. and Jeronimidis, G. (in press) The mechanical design of fossil plants, in Rayner (ed.) *Biomechanics in Evolution*.

Effect of change of temperature 　7

7.1 INTRODUCTION

How will a ply or a laminate behave when it undergoes a change of temperature? Will clearances in the plane, or curvatures causing out-of-plane displacements, be acceptable? Will the product warp? Will it warp too much? What modes of deformation might we expect to see in a given ply or laminate? How does laminate construction affect the deformation response to a change of uniform temperature or a temperature gradient through the thickness of the ply or laminate? For example, which of the following midplane strain and curvature responses in Figure 7.1 is likely from which initially-flat single ply or laminate constructions under a change of uniform temperature? What types of stress will be set up in the plies and what effect will this have on the strength of the laminate, bearing in mind the simultaneous application of mechanical loads?

Analysis of the effect of change of temperature in composite materials is important and it is disappointing that it is usually dismissed briefly and often rather inadequately in many standard texts on both solid body mechanics and composite mechanics.

This chapter examines principles for the stiffness and (albeit very briefly) strength behaviour of plies and laminates which undergo a temperature change. The discussion provides a clear statement about the resulting midplane strains and curvatures, and the associated internal stress and strain profiles through the thickness. For simplicity it is assumed here that the stiffness (and strength) properties of the basic ply material are unaffected by change of temperature, and that the thermal expansion coefficients are constant with temperature. (It is not difficult to develop more elaborate models which take into account the effect of temperature on properties and non-linear thermal expansion coefficients, but the principles remain the same.) A brief remark is also made about the effect of change of moisture content on the change of dimensions of plies and laminates.

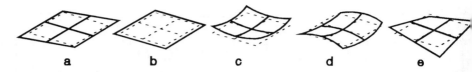

Figure 7.1 Effect of uniform increase in temperature on initially flat plates: (a) off-axis unidirectional ply; (b) symmetric crossply laminate; (c) non-symmetric laminates of isotropic plies; (d) non-symmetric crossply laminate; (e) non-symmetric angleply laminate.

First, we shall discuss the effect of temperature change using the language and concepts of solid body mechanics given in Chapter 2. We shall then translate these concepts into thermal force and moment resultants, to provide a clear link with the concepts of ply and laminate analysis given in Chapters 3 to 5. The next step is to provide representative examples of the application of these resultants in describing the behaviour of plies and laminates as the result of temperature change. Each main subsection of this chapter ends with a small library of representative patterns of behaviour.

There is a large number of numerical examples and illustrations presented towards the end of this chapter. These explore a wide range of interesting effects and challenges to the designer. The theoretical analysis can be applied to each example, and several such workings are presented. It is strongly recommended that the reader check by hand and step-by-step at least one example for which a full outline is not given – this might be either a crossply or an angleply laminate – in order to demonstrate a thorough grasp of the principles involved, and to gain from the confidence of having done so. Thereafter the use of laminate software is recommended.

7.2 ELEMENTARY PRINCIPLES FOR TEMPERATURE CHANGE

The essential starting point is that when a single isotropic or unidirectional sheet is free to change its dimensions, to bend or to twist freely, under an imposed change of temperature, then there are developed no stresses in that body. This is a fundamental principle of physics.

We shall therefore start this chapter with a brief discussion of free expansion resulting from temperature change in an isotropic body and in a unidirectional ply. This free expansion is important because it is necessary to take account of it to prevent mismatch of dimensions of products in service and changes in clearance between adjacent parts.

We shall then look at examples in conventional solid body mechanics where a body is restrained from free expansion as the result of temperature changes. The restraint means that internal stresses arise in the body when temperature changes, even though there are no externally applied forces. This is important in isotropic plates and tubes, but it is particularly important in composite materials, either at the single ply level or in laminates in service. Curing stresses in composite materials arise from restraints on free shrinkage when cooling from curing temperature down to ambient temperature and are often accompanied by shrinkage of the matrix on to the fibre (which promotes good adhesion). The curing stresses can be substantial, which has a considerable (adverse) effect on the strength of the material, especially in the transverse direction.

7.2.1 Free expansion

Change of temperature can cause changes in dimensions or bending in an isotropic material, which can cause articles to malfunction by obstructing clearances or by distortion.

The effect of a change from a uniform temperature T_0 to a different uniform temperature T_1 is to cause a change in all linear dimensions. Denoting $T_1 - T_0 = \Delta T$, the thermal strain in an isotropic material is given by $\varepsilon^T = \alpha \Delta T$. The thermal strain is tensile if $T_1 > T_0$, and compressive if $T_1 < T_0$.

If the upper surface of a thin isotropic sheet initially at T_0 is held at a steady temperature T_1 and the lower surface held at T_2, and the sheet is totally free to expand, then the sheet will expand in its midplane, and develop synclastic curvature.

Problem 7.1: Show that the fractional increase in volume of a block of isotropic material is approximately $3\alpha\Delta T$.

Problem 7.2: A sheet 3 mm thick made from a unidirectional sheet of carbon fibre reinforced epoxy has $\alpha_1 = 6 \times 10^{-6}/\text{K}$ and $\alpha_2 = 3.5 \times 10^{-5}/\text{K}$. A circular hole 40 mm diameter is made at 15 °C. What are the leading dimensions of the hole at 60 °C?

Problem 7.3: A isotropic thin rectangular sheet of material is 3 mm thick and is used as the vertical separator between two stirred inert unpressurized fluids x and y of the same density maintained at 15 °C(x) and 35 °C(y). The sheet is held in place by flexible gaskets which permit free expansion of the sheet at all four edges. The expansion coefficient of the sheet material is α/K. The sheet is flat and has the dimensions 500 mm × 250 mm at 20 °C. Estimate the dimensions of the sheet in service, assuming that the fluids do not cause the sheet to swell.

7.2.2 Effect of restraint on temperature change

A change of temperature alone does not induce shear strain, and does not induce any stress in an isotropic material free to expand or contract. But if expansion is prevented, then a direct stress is induced. This is particularly important in laminates, and is a major concept on which much of this chapter depends.

The induced stress is a contribution to the load-bearing capability of the material. Using a suitable failure criterion we can then establish the load factor needed to cause failure – either occasioned by adding more mechanical load or by further changing the temperature. We should recall that the simple notion of failure stress being independent of temperature is rarely true, but we assume this for simplicity and convenience within this chapter. Allowance for the temperature dependence of strength can be readily made where data are available.

Homogeneous materials

We distinguish two types of homogeneous material: the isotropic material, and the unidirectional ply treated as homogeneous in the scale of interest but having orthotropic properties.

In a long slender isotropic rod the total strain is the sum of the mechanical strain and the thermal strain, given by $\varepsilon = \sigma/E + \alpha\Delta T$. If the rod is prevented from expanding, then the mechanical strain is zero, and the stress set up in the rod is given by $\sigma = -E\alpha\Delta T$, the negative sign confirming a compressive strain.

In a thin isotropic plate, we can write the total strain in the form:

$$\begin{pmatrix} \varepsilon_1 \\ \varepsilon_2 \\ \gamma_{12} \end{pmatrix} = \begin{pmatrix} 1/E & -v/E & 0 \\ -v/E & 1/E & 0 \\ 0 & 0 & 1/G \end{pmatrix} \begin{pmatrix} \sigma_1 \\ \sigma_2 \\ \tau_{12} \end{pmatrix} + \begin{pmatrix} \alpha\Delta T \\ \alpha\Delta T \\ 0 \end{pmatrix}$$

where for the purposes of this discussion it is assumed that α, E, v, and G are independent of temperature. Thus where the sheet is totally prevented from expanding in the 1 and 2 directions we have $(\varepsilon_1 =)0 = \sigma_1/E - v\sigma_2/E + \alpha\Delta T$, and $(\varepsilon_2 =)0 = -v\sigma_1/E + \sigma_2/E + \alpha\Delta T$, and the thermally induced stresses are given by $\sigma_1 = \sigma_2 = -E\alpha\Delta T/(1-v)$.

The influence of mechanical stress and thermal strain on the total strain behaviour of a thin unidirectional sheet may be calculated from

$$\begin{pmatrix} \varepsilon_1 \\ \varepsilon_2 \\ \gamma_{12} \end{pmatrix} = \begin{pmatrix} S_{11} & S_{12} & 0 \\ S_{12} & S_{22} & 0 \\ 0 & 0 & S_{66} \end{pmatrix} \begin{pmatrix} \sigma_1 \\ \sigma_2 \\ \tau_{12} \end{pmatrix} + \begin{pmatrix} \alpha_1\Delta T \\ \alpha_2\Delta T \\ 0 \end{pmatrix}$$

i.e.

$$\begin{pmatrix} \varepsilon_1 \\ \varepsilon_2 \\ \gamma_{12} \end{pmatrix} = \begin{pmatrix} 1/E_1 & -v_{12}/E_1 & 0 \\ -v_{12}/E_1 & 1/E_2 & 0 \\ 0 & 0 & 1/G_{12} \end{pmatrix} \begin{pmatrix} \sigma_1 \\ \sigma_2 \\ \tau_{12} \end{pmatrix} + \begin{pmatrix} \alpha_1\Delta T \\ \alpha_2\Delta T \\ 0 \end{pmatrix}$$

or in the inverted form

$$\begin{pmatrix} \sigma_1 \\ \sigma_2 \\ \tau_{12} \end{pmatrix} = \begin{pmatrix} Q_{11} & Q_{12} & 0 \\ Q_{12} & Q_{22} & 0 \\ 0 & 0 & Q_{66} \end{pmatrix} \begin{pmatrix} \varepsilon_1 - \alpha_1\Delta T \\ \varepsilon_2 - \alpha_2\Delta T \\ \gamma_{12} \end{pmatrix}$$

Again we see that the change of temperature alone does not induce shear strain.

Dissimilar materials bonded together

We shall discuss later a more formal treatment of laminates but it is helpful here to summarize the effect of heating up two dissimilar materials in three different circumstances of restraint. The first involves only axial restraint; the second involves restraint in the plane of a symmetric laminate, and third deals with the behaviour of a nonsymmetric bimetallic strip where bending occurs. It will be apparent that all three problems can be solved by consideration of equilibrium, geometrical compatibility and the relevant stress-strain relationships. This approach was used in Chapter 2.

Axially constrained unbonded thin-walled tubes

Consider two concentric tubes made from materials a and b fastened together at the ends at temperature $T°$. What are the axial stresses in each tube when the temperature is changed by ΔT? You may assume that there is a small (concentric) gap between the two tubes (Figure 7.2).

By equilibrium, there is no external axial force, so $F_a + F_b = 0$, i.e. $\sigma_a A_a + \sigma_b A_b = 0$.

If the ends are bonded together then both tubes change length to the same extent, i.e. $\varepsilon_a = \varepsilon_b$.

The stress-strain relationships for each material are

$$\varepsilon_a = \sigma_a/E_a + \alpha_a \Delta T \quad \text{and} \quad \varepsilon_b = \sigma_b/E_b + \alpha_b \Delta T$$

Combining these equations gives

$$\sigma_a = +[E_a E_b A_b(\alpha_b - \alpha_a)\Delta T]/(E_a A_a + E_b A_b)$$
$$\sigma_b = -[E_a E_b A_a(\alpha_b - \alpha_a)\Delta T]/(E_a A_a + E_b A_b)$$

We do not need to consider the effect of circumferential changes because there is no bond between the tubes across which load transfer can occur.

Symmetrical bonded laminates

Consider a three layer symmetric laminate consisting of two skins of material a firmly bonded at temperature $T°$ to a core sheet of material b (Figure 7.3). The

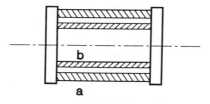

Figure 7.2 Two concentric tubes fastened at each end.

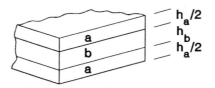

Figure 7.3 Three-ply symmetric laminate.

total thickness of each material is h_a and h_b, and both materials are isotropic. What stresses are set up when the laminate is heated by ΔT and free to expand?

Although the laminate as a whole is free to expand, the individual plies are not free, because they are bonded to and restrained by other material in the plane. The plies are free to expand through the thickness, and so there are no stresses developed in the z direction.

There are no external forces so in the plane we have $F_{xa} + F_{xb} = 0$, i.e. $\sigma_{xa}h_a + \sigma_{xb}h_b = 0$, and similarly $F_{ya} + F_{yb} = 0$.

The strains in each ply are the same: $\varepsilon_{xa} = \varepsilon_{xb}$, and $\varepsilon_{ya} = \varepsilon_{yb}$.

Because the sheet is under biaxial stress, the stress/strain relationships relating to biaxial loading must be used here. Expansion occurs in x, y, and z directions, but there is no mutual restraint in the thickness direction. Thus we have

$$\varepsilon_{xa} = \sigma_{xa}/E_a - \nu_a\sigma_{ya}/E_a + \alpha_a\Delta T$$

$$\varepsilon_{xb} = \sigma_{xb}/E_b - \nu_b\sigma_{yb}/E_b + \alpha_b\Delta T$$

Combining these equations, and recognizing that for the special case of isotropic plies $\sigma_{xa} = \sigma_{ya}$, leads to equations such as

$$\sigma_{xa} = + [(\alpha_b - \alpha_a)\Delta T]/[(1 - \nu_a)E_a + (1 - \nu_b)h_a/E_bh_b]$$

and

$$\sigma_{xb} = - \sigma_{xa}h_a/h_b$$

For the special case where $h_a = h_b$ we find

$$\sigma_{xa} = [E_aE_b(\alpha_b - \alpha_a)\Delta T]/[E_a(1 - \nu_b) + E_b(1 - \nu_a)]$$

Non-symmetrical bonded long thin narrow strip

Consider a two layer strip of isotropic materials a and b, each having the same thickness h_1 and firmly bonded at $T°$. (Figure 7.4). This arrangement is used as the bimetallic strip in thermostats, where a change of temperature induces curvature, such that the displacement operates a switch. How will this laminated strip behave when heated by ΔT?

The strip as a whole is free to expand and there are no external forces involved, but each strip is constrained by the bond at the interface. The physical picture is that the restrained expansion induces bending. The bending

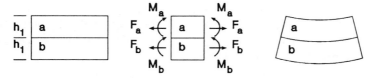

Figure 7.4 Non-symmetric two-ply laminate.

can be analysed if the correct forces and moments are ascribed in separate free body diagrams for each strip, as shown in Figure 7.4.

$$F_a + F_b = 0 \tag{7.1}$$

These forces are accompanied by internal moments M_a and M_b, and by equilibrium of moments, taking sagging moments as positive, we therefore have

$$M_a + M_b - F_a h_1/2 + F_b h_1/2 = 0$$

i.e.

$$M_a + M_b = F_a h_1 \tag{7.2}$$

Geometrical compatibility requires that the direct strains ε_a^i and ε_b^i at the interface ($z_a^i = + h_1/2$ and $z_b^i = - h_1/2$) be the same:

$$\varepsilon_a^i = \varepsilon_b^i \tag{7.3}$$

The stress strain relationships are as used before:

$$\varepsilon_a = \sigma_a/E_a + \alpha_a \Delta T \quad \text{and} \quad \varepsilon_b = \sigma_b/E_b + \alpha_b \Delta T \tag{7.4}$$

Considering the bending of each separate strip about its centroid, the stresses in each strip may be calculated as:

$$\sigma_a = F_a/A_a + M_a z_a/I_a \quad \text{and} \quad \sigma_b = F_b/A_b + M_b z_b/I_b \tag{7.5}$$

Using compatibility the strains at the interface now become:

$$F_a/E_a A_a + M_a h_1/2E_a I_a + \alpha_a \Delta T = F_b/E_b A_b + M_b(- h_1/2)/E_b I_b + \alpha_b \Delta T \tag{7.6}$$

If we assume that the radius of curvature of the neutral axis of each separate strip is the same (and in practice the differences in radius are negligibly small for moment calculations but *not* for bending strain calculations), we know from the engineer's theory of bending that

$$1/R_a = M_a/E_a I_a = M_b/E_b I_b = 1/R_b = 1/R = \kappa \tag{7.7}$$

and hence from Equation (7.2) we find

$$M_a + M_b = M_a(1 + E_b I_b/E_a I_a) = F_a h_1 \tag{7.8}$$

For the special case where each strip has the same thickness, then $A_a = A_b = bh_1$, and $I_a = I_b = bh_1^3/12$.

The procedure is now straightforward, but the algebra now becomes rather unwieldy. We can proceed by solving Equations (7.6) and (7.8) for F_a, which enables us to calculate F_b, M_a, M_b, and thus the curvature $\kappa = 1/R$ can be found in terms of the ratio $\phi = E_b/E_a$:

$$\kappa = \{12\phi(\alpha_a - \alpha_b)\Delta T\}/\{h_1[(1 + \phi^2) + 12\phi]$$

The forces and moments are thus

$$F\alpha = \kappa b h_1^2 E_a(1 + \phi)/12 \quad \text{and} \quad F_b = -F_a$$
$$M_a = E_a I_a \kappa \quad \text{and} \quad M_b = E_b I_b \kappa$$

We can now express the stress profiles using Equation (7.5) as

$$\sigma_a(z) = (\kappa h_1 E_a/12)((1 + \phi) + 12z_a/h_1)$$
$$\sigma_b(z) = -(\kappa h_1 E_a/12)((1 + \phi) + 12\phi z_b/h_1)$$

where z_a or z_b are measured from the centroid of the component strip a or b. The stresses at the interface $z_a^i = +h_1/2$ and $z_b^i = -h_1/2$ are:

$$\sigma_a^i = (\kappa h_1 E_a/12)((1 + \phi) + 6)$$
$$\sigma_b^i = (\kappa h_1 E_a/12)((1 + \phi) + 6\phi)$$

Provided that $\alpha_a > \alpha_b$, the curvature is sagging in the co-ordinates shown, so that σ_a^i is tensile and σ_b^i is compressive. Substituting into Equation (7.6) confirms that the strains at the interface are identical, as required.

The discussions presented in this section can of course be developed much further, but the algebra becomes even more cumbersome and clouds the simple principles involved. In the next section we shall discuss how these principles involving restraint can be absorbed into the analysis of laminates.

Problem 7.4: A bimetallic strip consists of an aluminium strip and a steel strip, each 0.75 mm thick and 10 mm wide, firmly bonded together along their long face at 20 °C. Calculate the curvature and stress profiles when the strip has a uniform temperature of 60 °C. The elastic constants are $E_S = 208$ GPa, $E_A = 70$ GPa, $\alpha_S = 1.1 \times 10^{-5}/K$ and $\alpha_A = 2.3 \times 10^{-5}/K$.

Problem 7.5: The bimetallic strip of Problem 7.4 is now replaced by one in which the aluminium component is 1 mm thick and the steel in 0.5 mm thick. What difference does this make to the quantities previously calculated?

7.3 PRINCIPLES FOR EFFECT OF CHANGE OF TEMPERATURE IN A PLY

We shall restrict outselves here to two special cases of temperature change. The first is where there is a change of uniform temperature, so that the complete plate changes its temperature by ΔT. The second is the practical situation where a temperature gradient is imposed through the thickness of the sheet.

We shall see later that, for laminates only, we need to discuss a third case where the laminate is made by curing (crosslinking) the resin hot, so that cooling to ambient (service) temperature results in curing or residual stresses. We shall be concerned to determine midplane strains and curvatures for the laminate as a whole, and the stress and strain profiles set up in each ply (these profiles being especially important when considering strength).

7.3.1 Thermal force and moment resultants in a single ply

In examining the behaviour of a single ply under forces caused by a change of temperature (and later the more practically important behaviour of laminates) it is extremely convenient to use the concept of a notional thermal force resultant, N^T, and a notional thermal moment resultant M^T. These fictitious quantities enable us to use easily all the formal laminate analysis presented in Chapters 4, 5 and 6.

Ply thermal force resultants per unit width

The concept of the notional thermal force resultant per unit width in a ply which undergoes a change of temperature can be developed as follows. The free thermal strain induced in an elemental layer of the ply f of thickness dz is

$$[\varepsilon(z)^T]_f = [\alpha]_f \Delta T$$

and hence the notional thermal stress associated with this is

$$[\sigma(z)_f = [Q]_f [\alpha]_f \Delta T$$

The notional thermal force resultant over the complete ply is obtained by integrating over the ply thickness in the same way as was discussed in Section 3.1.3:

$$[N^T]_f = \int_{z=-h/2}^{+h/2} [Q]_f [\alpha]_f \Delta T \, dz$$

For a uniform temperature change in a ply of thickness h we find that

$$[N]_f^T = [Q]_f [\alpha]_f \Delta T \cdot h$$

We use this notional thermal force resultant to calculate the midplane strains in the ply as

$$[\varepsilon^\circ]_f = [a]_f [N^T]_f$$

This is an unnecessarily long argument where only a temperature difference is applied to a single ply, but we shall see that this approach is most useful when dealing with laminates. Where external loads (or restraints) are also applied to the single ply, e.g. $[N^a]_f$, then the midplane strains are calculated as

$$[\varepsilon^\circ]_f = [a]_f[N^a]_f + [a]_f[N^T]_f$$

and hence the thermal stresses $[\sigma^T]_f$ can be found as the difference between the total strains and the free expansion strains in the ply:

$$[\sigma^T]_f = [Q]_f[\varepsilon]^e = [Q]_f([\varepsilon^\circ]_f - [Q]_f[a]_f[N^T]_f)$$

where $[\varepsilon]^e$ may be termed the *effective stress-inducing strains*. It is apparent that when no external load is applied, then strains are free to develop in a single ply, but there are induced no thermal stresses, which corresponds to practical experience.

Ply thermal moment resultants per unit width

We can proceed similarly to define a notional thermal moment resultant $[M]_f^T$. The steps are as follows. Given the stresses $[\sigma]_f^T$, we can write down their moments $[\sigma]_f^T z\, dz$, and integrate over the thickness of the ply to obtain the ply moment resultant (per unit width):

$$[M]_f^T = \int_{z=-h/2}^{+h/2} [\sigma]_f^T z\, dz = \int_{z=-h/2}^{+h/2} [Q]_f[\alpha]_f \Delta T z\, dz$$

Clearly $[M]^T$ is zero for a single ply or a symmetric laminate under a change of uniform temperature through the thickness.

We now calculate the curvatures caused by a change of temperature as:

$$[\kappa]_f = [d]_f[M]_f^T$$

7.3.2 Isotropic ply

Change of uniform temperature

If no external mechanical load is applied, only thermal strains are present in a single sheet. For a single sheet the stresses in it can only result from the applied loads and not from the change in temperature. The stresses are based on the overall strains less the thermal strains caused by the change of uniform temperature alone. We therefore have

$$[\sigma] = [Q][\varepsilon^\circ] - [Q][\alpha]\Delta T$$

The last two terms are consistent with observation and experiment: if no external load is applied then $[\varepsilon^\circ] = [a][N]^T$ and $[\sigma] = 0$; and if a load is applied without any change of uniform temperature, then $[\varepsilon^\circ] = [a][N]^a$, and $[\sigma] = [Q][\varepsilon^\circ]$.

For example, consider the single ply NR2, which is 2 mm thick, and for which data are given in Section 3.1.5. For $\Delta T = 30\,K$ and $\alpha\ (= \alpha_1 = \alpha_2) = 2.2 \times 10^{-4}/K$, we have $\varepsilon_1^T = \varepsilon_2^T = 6.6 \times 10^{-3} = 0.66\%$.

More formally, and using stiffness data from Table 3.1, we find $N_1^T = Q_{11}\alpha\Delta Th + Q_{12}\alpha\Delta Th = (Q_{11} + Q_{12})\alpha\Delta Th = (1.316 + 0.6448) \times 10^6 \times 2.2 \times 10^{-4} \times 30 \times 2 \times 10^{-3} = 25.88$ N/m $= 0.02588$ N/mm. Hence, using data for the inverse stiffness matrix from Table 3.2, $\varepsilon_1^T = [a][N]^T = a_{11}N_1^T + a_{12}N_2^T = (5 \times 10^2 - 2.45 \times 10^2) \times 10^{-3} \times 0.02588 = 6.6 \times 10^{-3}$.

All stresses are zero because there are no applied loads. More formally, the only midplane strains developed are those caused by the thermal expansion, so $\varepsilon^\circ = \alpha\Delta T$.

Temperature gradient across the ply thickness

A simple approach

Where a temperature gradient is applied between the top and bottom surface of a single horizontal ply initially at T°, so that surface temperatures are T^t and T^b, the midplane thermal strain $\varepsilon^{\circ T}$ and the midplane thermal curvature κ^T are defined by $\varepsilon^{\circ T} = (\varepsilon^t + \varepsilon^b)/2$ and $\kappa^T = (\varepsilon^b - \varepsilon^t)/h$.

In the special circumstance of a single ply it is then possible to follow the simple approach and calculate $[M]^T = [D][\kappa]^T$ so that $M_1^T = D_{11}\kappa_1^T + D_{12}\kappa_2^T$ and $M_2^T = D_{12}\kappa_1^T + D_{22}\kappa_2^T$.

To suppress any chosen curvature, we have to establish what value of externally applied moment $M^a = -M^T$ would be needed to apply an additional negative value of that curvature. Thus to obtain a curvature $\kappa_1^a = -\kappa_1^T$, we might seek $\kappa_1^a = -\kappa_1^T = d_{11}M_1^a$, i.e. $M_1^a = -\kappa_1^T/d_{11}$, and at the same time we would find the positive curvature κ_2^a associated with the negative value of M_1^a acting alone: $\kappa_2^a = d_{12}M_1^a$ (recalling that d_{12} is negative). The result of the temperature change and the application of M_1^a is that $\kappa_1 = \kappa_1^T + \kappa_1^a = 0$ as required, and $\kappa_2 = \kappa_2^T + \kappa_2^a$, which is greater than ΔT acting alone.

Formal approach

But working from curvatures via directly calculated surface thermal strains is acceptable only as long as there is one ply: this procedure cannot give a result where laminates of different plies (or plies of the same material in different orientations) are used. A more general method will now be introduced.

Assuming that the temperature profile $T(z)$ is linear through the thickness, we may write $\Delta T(z) = a + bz$, where $a = (\Delta T^b + \Delta T^t)/2$ is the average temperature change in the sheet, and $b = (\Delta T^b - \Delta T^t)/h = (T^b - T^t)/h$ is the temperature gradient.

For a single ply, integrating over the thickness, we therefore find the thermal force resultant

$$[N]^T = [Q][\alpha]\int_{-h/2}^{+h/2}(a + bz)\mathrm{d}z = [Q][\alpha]ah = [Q][\alpha](\Delta T^b + \Delta T^t)h/2$$

Similarly the thermal moment resultant can be derived for a ply where the temperature varies linearly through the thickness:

$$[M]^T = [Q][\alpha] \int_{-h/2}^{+h/2} (a + bz)z\,dz = [Q][\alpha]bh^3/12 = [Q][\alpha](T^b - T^t)h^2/12$$

It is clear that for an isotropic ply which sustains a temperature difference between surfaces, the in-plane force and moment resultants, and the midplane strains and curvatures, have equal values in the two (principal) co-ordinate directions in the sheet, e.g. $M_1^T = M_2^T$, and the sheet displays synclastic curvature.

Problem 7.6: A flat 5 mm thick sheet of natural rubber NR5 is initially at 20 °C. Calculate the stresses and deformations in the sheet when the surfaces are held at 2 °C and 12 °C. Data for this material are given in Table 3.1, and $\alpha = 2.2 \times 10^{-4}/K$.

The [A, D] and [a, d] are as follows for the 5 mm ply of material NR5:

[A] kN/mm			[D] kNmm		
0.00658	0.003224	0	0.01371	0.06717	0
0.003224	0.00658	0	0.06717	0.01371	0
0	0	0.01665	0	0	0.03469

[a] mm/MN			[d]/MNmm		
200000	−98000	0	96000	−47040	0
−98000	200000	0	−47040	96000	0
0	0	600600	0	0	288300

Problem 7.7: The flat 5 mm thick sheet of natural rubber NR5 is initially at 20 °C. The surfaces are held at $T^t = 12$ °C and $T^b = 2$ °C, and the sheet carries in-plane force resultants $N_x^a = 0.05$ N/mm and $N_y^a = 0.025$ N/mm Calculate the stress and strain profiles.

Problem 7.8: A natural rubber tube is 5 mm thick with a mean diameter 50 mm. Under an internal pressure of 2 kPa the passage of cold fluid induces rubber surface temperatures of 2 °C and 12 °C. Calculate the stress and strain profiles in the rubber.

7.3.3 Unidirectional ply aligned in principal directions

The effect of change of temperature on the behaviour of a unidirectional ply is only slightly more complicated than that of an isotropic sheet. Because of the reinforcement, there must be different values of expansion coefficient in the principal directions, α_1 and α_2. Similarly we use the reduced stiffness matrix where $Q_{11} \neq Q_{22}$. It follows therefore that the free expansion after a change of uniform temperature is different in the two different directions and that, after a temperature gradient has been established through the thickness, the mid-plane strains and curvatures will also be different in the two principal directions.

For a unidirectional ply, we have to distinguish the expansion coefficients α_1 and α_2 in the two principal directions. In polymer fibre composites the thermal expansion coefficient of fibres is usually much less than that of the polymer matrix. The difference in coefficients is smaller for crosslinkable resins, but larger for partially crystallizable thermoplastics and lightly crosslinkable rubber compounds. The value of α_1 is usually low and dominated by the properties of the fibres, whereas α_2 is dominated by the matrix.

Based on micromechanics and energy considerations, Schapery (1968) suggests the following basis for calculating the thermal expansion coefficients for a unidirectional ply based on the properties of the fibres and the matrix:

$$\alpha_1 = (\alpha_f E_f V_f + \alpha_m E_m V_m)/(E_f V_f + E_m V_m)$$
$$\alpha_2 = (1 + v_f)\alpha_f V_f + (1 + v_m)\alpha_m V_m - \alpha_1(v_f V_f + v_m V_m)$$

We can examine here the behaviour of unidirectional materials under a change of uniform temperature where the co-ordinates relate to the principal directions of the material. We shall examine in the next section the thermal equivalent of the off-axis in-plane loading problem.

Problem 7.9: What is the fractional increase in volume of a unidirectional material having coefficients of thermal expansion α_1 and α_2 under a uniform temperature change ΔT?

Problem 7.10: A pultruded rod is made from unidirectional fibres in a crosslinked resin, with the following properties at $20\,°C$: $E_1 = 45.6\,GPa$, $E_2 = 10.73\,GPa$, $v_{12} = 0.274$, $\alpha_1 = 7 \times 10^{-6}/K$, $\alpha_2 = 3 \times 10^{-5}/K$. If the rod is $10\,mm$ diameter and $2\,m$ long at $20\,°C$, what are its dimensions at $65\,°C$?

Problem 7.11: For the pultruded rod in Problem 7.10, what is the axial stress in the rod if it is rigidly clamped at each end at $20\,°C$ and then cooled down to $-25\,°C$?

Problem 7.12: What are the in-plane stresses induced when a sheet of unidirectional material changes its uniform temperature by ΔT and the sheet is prevented from freely expanding?

7.3.4 Unidirectional ply aligned off-axis

In this sheet the fibres are aligned in the plane at some angle θ to a chosen reference direction, x, and we assume that the sides of the rectangular sheet are parallel to the (x, y) directions.

We have already noted in Section 3.3.2 that the strains in the $(1, 2)$ directions are related to the strains in (x, y) by

$$\begin{pmatrix} \varepsilon_x \\ \varepsilon_y \\ \gamma_{xy}/2 \end{pmatrix} = [T]^{-1} \begin{pmatrix} \varepsilon_1 \\ \varepsilon_2 \\ \gamma_{12}/2 \end{pmatrix}$$

It therefore follows that thermal strains will behave similarly, i.e.

$$\begin{pmatrix} \varepsilon_1^T \\ \varepsilon_y^T \\ \gamma_{xy}^T/2 \end{pmatrix} = \begin{pmatrix} \alpha_x \\ \alpha_y \\ \alpha_{xy}/2 \end{pmatrix} \Delta T = [T]^{-1} \begin{pmatrix} \alpha_1 \\ \alpha_2 \\ 0 \end{pmatrix} \Delta T$$

where α_{xy} is the coefficient of apparent thermal shear. It is easy to see what happens if you cut a square unidirectional sheet with its principal axes at 45° to the sides of the sheet; on heating, the fibres expand less than the matrix, and hence diagonals lengthen unequally and there is a change of angle given by $\gamma_{xy}^T = \alpha_{xy}\Delta T$, which is shown in Figure 7.5(a), and in a more stylized form in Figure 7.5(b).

If a temperature gradient is set up across the thickness of the plate, $\Delta T(z) = a + bz$, then the resulting changes in dimensions can be calculated by including the temperature profile inside the integration as before to obtain the thermal force and moment resultants. We find $[N]^T = [Q^*][\alpha]ah$, and $[M]^T = [Q^*][\alpha](T^b - T^t)h^2/12$. The algebra now is straightforward but cumbersome, leading to such calculations as

$$N_x^T = (Q_{11}^*\alpha_x + Q_{12}^*\alpha_y + Q_{16}^*\alpha_{xy})(\Delta T^t + \Delta T^b)h/2$$

We can then calculate the midplane strains and curvatures due to the temperature gradient using

$$\begin{pmatrix} \varepsilon_x^T \\ \varepsilon_y^T \\ \gamma_{xy}^T \end{pmatrix} = \begin{pmatrix} a_{11} & a_{12} & a_{16} \\ a_{12} & a_{22} & a_{26} \\ a_{16} & a_{26} & a_{66} \end{pmatrix} \begin{pmatrix} N_x^T \\ N_y^T \\ N_{xy}^T \end{pmatrix} + \begin{pmatrix} b_{11} & b_{12} & b_{16} \\ b_{12} & b_{22} & b_{26} \\ b_{16} & b_{26} & b_{66} \end{pmatrix} \begin{pmatrix} M_x^T \\ M_y^T \\ M_{xy}^T \end{pmatrix}$$

$$\begin{pmatrix} \kappa_x^T \\ \kappa_y^T \\ \kappa_{xy}^T \end{pmatrix} = \begin{pmatrix} h_{11} & h_{12} & h_{16} \\ h_{12} & h_{22} & h_{26} \\ h_{16} & h_{26} & h_{66} \end{pmatrix} \begin{pmatrix} N_x^T \\ N_y^T \\ N_{xy}^T \end{pmatrix} + \begin{pmatrix} d_{11} & d_{12} & d_{16} \\ d_{12} & d_{22} & d_{26} \\ d_{16} & d_{26} & d_{66} \end{pmatrix} \begin{pmatrix} M_x^T \\ M_y^T \\ M_{xy}^T \end{pmatrix}$$

For example, where the ply NE30 ($\alpha_1 = 10^{-5}/K$, $\alpha_2 = 2 \times 10^{-4}/K$) has been

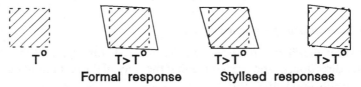

Formal response   Stylised responses

Figure 7.5 Response of off-axis ply to an increase in temperature.

cured at room temperature, $20\,°C$, and the surface temperatures are set at $T^t = 20\,°C$ and $T^b = 50\,°C$, we obtain the following expected strains and curvatures:

$$\varepsilon_x^t = 0 \quad \varepsilon_x^o = \quad 0.08625\% \quad \varepsilon_x^b = \quad 0.1725\% \quad \kappa_x = \quad 0.8625/m$$

$$\varepsilon_y^t = 0 \quad \varepsilon_y^o = \quad 0.2288\% \quad \varepsilon_y^b = \quad 0.4575\% \quad \kappa_y = \quad 2.288/m$$

$$\gamma_{xy}^t = 0 \quad \gamma_{xy}^o = -0.2468\% \quad \gamma_{xy}^b = -0.4936\% \quad \kappa_{xy} = -2.468/m$$

Problem 7.13: The unidirectional sheet NE30 has thermal expansion coefficients $\alpha_1 = 10^{-5}/K$ and $\alpha_2 = 2 \times 10^{-4}/K$. At $10\,°C$ the rectangular sheet has dimensions $L_x = 350\,mm$, $L_y = 230\,mm$, $x = 2\,mm$. What are the leading dimensions of the sheet after it has been heated until it reaches a uniform temperature of $45\,°C$?

Problem 7.14: What forces are required to prevent the expansions and shear occurring under the change of uniform temperature in Problem 7.13?

Problem 7.15: A tube is made from a single ply of NE30, and at $20\,°C$ it is $2\,mm$ thick, $1\,m$ long and has a mean diameter of $30\,mm$. The reinforcing fibres are at $30°$ to the hoop direction. Calculate the change in leading dimensions and sketch the interesting stress and strain profiles when:

(a) the tube temperature is increased by a uniform $+30\,K$;
(b) the tube surface temperatures are $-10\,°C$ (internal), and $+13\,°C$ (external);
(c) the tube is under an internal pressure of $40\,kPa$ with surface temperatures of $-10\,°C$ and $+13\,°C$.

Elastic constants and coefficients of the [ABD] and [abhd] matrices are given in Section 3.3.12. The expansion coefficients for NE30 are $\alpha_1 = 10^{-5}/K$ and $\alpha_2 = 2 \times 10^{-4}/K$.

7.4 LAMINATES OF ISOTROPIC PLIES UNDER CHANGE OF TEMPERATURE

For a laminate which undergoes a change of uniform temperature ΔT the calculation of thermal force resultants $[N]_L^T$ and thermal moment resultants $[M]_L^T$ involves the summation of all the respective resultants for each ply. Letting the fth ply be the general case, the thermal force resultants per unit width for a laminate of F plies are then:

$$[N]_L^T = (\Delta T) \sum_{f=1}^{F} [Q]_f [\alpha]_f (h_f - h_{f-1})$$

and the thermal moment resultants become:

$$[M]_L^T = (\Delta T/2) \sum_{f=1}^{F} [Q]_f [\alpha]_f (h_f^2 - h_{f-1}^2)$$

It is clear that $[N]_L^T$ is finite for any non-zero ΔT for any laminate. $[M]_L^T$ is always zero for a symmetric laminate, and hence there is no warping or bending curvature consequent solely upon a change of uniform temperature; this is a major virtue of symmetry.

We can now calculate the thermal midplane strains and curvatures in the laminate as

$$[\varepsilon^\circ]_L^T = [a][N]_L^T + [b][M]_L^T$$
$$[\kappa]_L^T = [h][N]_L^T + [d][M]_L^T$$

The stress-including strains in the ply f are the laminate midplane strains less the free thermal strains of the ply alone:

$$[\varepsilon(z)]_f^e = [\varepsilon^\circ]_L^T + z[\kappa]_L^T - [\varepsilon]_f^T$$

and hence we calculate the thermal stresses in the fth ply as

$$[\sigma(z)]_f^T = [Q]_f[\varepsilon(z)]_f^e$$

For laminates which are subject to a temperature gradient through the thickness we should calculate the more general forms of the thermal resultants as:

$$[N]_L^T = \sum_{f=1}^{F} \int_{z=-h_f}^{h_f} [Q]_f[\alpha]_f \Delta T \, dz$$

$$[M]_L^T = \sum_{f=1}^{F} \int_{z=-h_f}^{h_f} [Q]_f[\alpha]_f \Delta T z \, dz$$

These can be evaluated for the special case of a uniform temperature profile using the formal approach of Section 7.3.2.

7.4.1 Symmetric laminate SAL4S

We discuss here the behaviour of laminates based on isotropic plies in order to establish the principles of laminate behaviour under temperature change without the complications of directional effects in the plane. Sections 7.5 and 7.6 concentrate on the behaviour of crossply and angleply laminates subject to temperature change.

Change of uniform temperature

For a four-ply $(S/AL)_S$ laminate (two steel plies each h_s thick and two aluminium plies each of thickness h_A), with relevant data given in Section 4.3, we calculate terms such as

$$N_1^T = \Delta T\{(Q_{11S}\alpha_{1S} + Q_{12S}\alpha_{2S})\cdot 2h_S + (Q_{11A}\alpha_{1A} + Q_{12A}\alpha_{2A})\cdot 2h_A\}$$

Where $h_S = h_A = 0.25$ mm, $\alpha_S = 1.1 \times 10^{-5}/K$, $\alpha_A = 2.3 \times 10^{-5}/K$, $\Delta T = 30$ K and we use values of [Q] given in Tables 4.1 and 4.2, and the values of the abhd matrix in Table 4.3, we find $N_1^T = N_2^T = 83.709$ N/mm, and hence $\varepsilon_1^\circ = a_{11}N_1^T + a_{12}N_2^T = 4.2558 \times 10^{-2}\%$.

The effective stress-inducing strains causing stresses in the plies are now found from:

$$\varepsilon_{1S}^{oe} = \varepsilon_1^\circ - \alpha_S \Delta T = 4.2558 \times 10^{-4} - 1.1 \times 10^{-5} \times 30 = 9.558 \times 10^{-5}$$
$$\varepsilon_{1A}^{oe} = 4.2558 \times 10^{-4} - 2.3 \times 10^{-5} \times 30 = -2.6442 \times 10^{-4}$$

and hence the thermal stresses in the plies are

$$\sigma_{1S}^T = (Q_{11S} + Q_{12S})\varepsilon_{1S}^{oe} = (225.7 + 63.19) \times 10^9 \times 9.558 \times 10^{-5} = 27.61 \text{ MPa}$$
$$\sigma_{1A}^T = (78.55 + 25.92) \times 10^9 \times -2.6442 \times 10^{-4} = -27.62 \text{ MPa}$$

In this example, σ_1^T is the only stress, so $\sigma_1^T = \sigma_1^{tot}$.

Problem 7.16: Calculate the thermal expansion coefficients of the symmetric laminate SAL4S.

Change of uniform temperature plus applied load

Consider now the same laminate SAL4S subjected to both a change of uniform temperature $\Delta T = 30$ K and also an applied force resultant $N_1^a = 20$ N/mm. We have already seen that ΔT alone induces $\varepsilon_1^T = \varepsilon_2^T = 4.2558 \times 10^{-4}$. In addition there are the strains caused by the applied force resultant N_1^a, namely

$$\varepsilon_1^a = a_{11}N_1^a = 7.19 \times 10^{-6} \times 20 = 1.438 \times 10^{-4}$$
$$\varepsilon_2^a = a_{12}N_1^a = -2.106 \times 10^{-6} \times 20 = -4.212 \times 10^{-5}$$

The total (midplane) strains are therefore

$$\varepsilon_1^\circ = 4.2558 \times 10^{-4} + 1.438 \times 10^{-4} = 5.6938 \times 10^{-4}$$
$$\varepsilon_2^\circ = 4.2558 \times 10^{-4} - 4.212 \times 10^{-5} = 3.8346 \times 10^{-4}$$

The strains responsible for internal stress within the laminate are given by terms such as $\varepsilon_1^\circ - \alpha_f \Delta T$, i.e.

$$\varepsilon_{1S} = 5.6938 \times 10^{-4} - 3.3 \times 10^{-4} = 2.3938 \times 10^{-4}$$
$$\varepsilon_{2S} = 3.8346 \times 10^{-4} - 3.3 \times 10^{-4} = 5.346 \times 10^{-5}$$

and hence the (uniform) stresses within the steel plies are given by terms such as $Q_{11}\varepsilon_1^\circ + Q_{12}\varepsilon_2^\circ$, i.e.

$$\sigma_{1S} = 225.7 \times 10^9 \times 2.3938 \times 10^{-4} + 63.19 \times 10^9 \times 5.346 \times 10^{-5} = 57.406 \text{ MPa}$$
$$\sigma_{2S} = 63.19 \times 10^9 \times 2.3938 \times 10^{-4} + 225.7 \times 10^9 \times 5.346 \times 10^{-5} = 27.192 \text{ MPa}.$$

Similarly for the aluminium:

$$\varepsilon_{1A} = 5.6938 \times 10^{-4} - 6.9 \times 10^{-4} = -1.2062 \times 10^{-4}$$

$$\varepsilon_{2A} = 3.8346 \times 10^{-4} - 6.9 \times 10^{-4} = -3.0654 \times 10^{-4}$$

$$\sigma_{1A} = 78.55 \times 10^9 \times -1.2062 \times 10^{-4} + 25.92 \times 10^9 \times -3.0654 \times 10^{-4}$$

$$= -17.419 \,\text{MPa}$$

$$\sigma_{2A} = 25.92 \times 10^9 \times -1.2062 \times 10^{-4} + 78.55 \times 10^9 \times -3.0654 \times 10^{-4}$$

$$= -27.205 \,\text{MPa}$$

Curing or residual stresses in laminate of isotropic plies

We have already seen in Chapter 4 that under load the laminate responds as an entity. Each individual ply is not able to deform as it would on its own under its share of the applied load or moment resultant. This explains why stress distributions arise in the laminate.

A feature of many fibre/polymer composite laminates is that they are processed or crosslinked (cured) hot, then allowed to cool down to ambient temperature. Because the different plies want to shrink differently from the laminate-as-a-whole, it follows that residual or curing stresses are then developed in the laminate. These residual stresses can be large, and it is particularly important that they are carefully taken into account when designing to avoid mechanical failure. In extreme circumstances a product can tear itself apart on cooling down from the cure temperature, with disastrous results. This is a major problem for those high performance resins which need to be cured or processed at very high temperatures and which then have to be used in low temperature environments. Poorly made laminates cured at ordinary high temperatures are also at risk.

In this section we shall look at artificial curing stresses in laminates made from isotropic plies, so that the phenomena of curing stresses can be clearly appreciated. In later sections we shall study the effect of real laminates made from unidirectional plies with the associated real curing effects. Our assumptions in the following simplified analysis are that the elastic constants and the thermal expansion coefficients are independent of temperature and independent of time under thermal or mechanical load.

The basic approach to curing stresses is exactly the same as that for calculating stresses caused by a change of uniform temperature, namely to work out the residual thermal force and moment resultants, N^c and M^c, caused by the change from cure temperature to ambient, ΔT^c (usually negative). The associated midplane strains and curvatures in the laminate can then be calculated using

$$\begin{pmatrix} \varepsilon^{oc} \\ \kappa^c \end{pmatrix} = \begin{pmatrix} a & b \\ h & d \end{pmatrix} \begin{pmatrix} N^c \\ M^c \end{pmatrix}$$

Note that for a symmetric laminate $[M]^c = 0$ and $[B] = 0 = [b] = [h]$. We can now calculate the 'effective stress-inducing strains' ε^e which will cause residual stresses in the fth ply:

$$\begin{pmatrix} \varepsilon_1^e \\ \varepsilon_2^e \\ \gamma_{12}^e \end{pmatrix}_f = \begin{pmatrix} \varepsilon_1^{oc} + z\kappa_1^c \\ \varepsilon_2^{oc} + z\kappa_2^c \\ \gamma_{12}^{oc} + z\kappa_{12}^c \end{pmatrix}_L - \begin{pmatrix} \alpha\Delta T \\ \alpha\Delta T \\ 0 \end{pmatrix}_f$$

and hence the cure (residual) stresses within any given ply, using:

$$\begin{pmatrix} \sigma_1^c \\ \sigma_2^c \\ \tau_{12}^c \end{pmatrix} = \begin{pmatrix} Q_{11} & Q_{12} & 0 \\ Q_{12} & Q_{22} & 0 \\ 0 & 0 & Q_{66} \end{pmatrix}_f \begin{pmatrix} \varepsilon_1^e \\ \varepsilon_2^e \\ \gamma_{12}^e \end{pmatrix}_f$$

It will be apparent that the magnitude of the residual stresses depend on the value of ambient temperature which may vary, as well as the cure temperature (which is a set constant). The greatest values of residual stress occur at the lowest ambient temperature.

Less apparent from many texts mentioning curing or residual stresses, (though documented in some research papers), the curing temperature is not entirely unambiguous. For polymer/fibre composites based on crosslinkable resins such as polyester or epoxy, the cure temperature is very close to the temperature at which the resin may be deemed to have locked solid on to the fibres. Where stacks of isotropic plies are bonded with a crosslinking adhesive, the same comment applies. But for thermoplastics matrices, where structures are made hot or in the melt phase, it will be the crystallization or glass transition temperature (whichever is the greatest) which will rather approximately describe the temperature at which the matrix will lock on to the fibres, and it is this temperature which we shall loosely describe as the 'cure' temperature in later discussion.

One way of identifying the effective cure temperature of a hot-cured, or hot-assembled, system is to make a hot non-symmetric laminate within a flat mould. On cooling down, the laminate will bend. If it slowly warmed up, it will become flatter as the internal thermal stresses decrease. When it is flat, this indicates a stress-free ('cure') temperature.

We should also mention residual deformations in hot-cured laminates after they have cooled to ambient temperature.

In a single ply (which can be either isotropic or unidirectional), on cooling the ply shrinks in the plane. If a laminate is symmetric, it will only shrink in-plane, and by an amount dictated by the overall laminate thermal 'expansion' coefficient(s). If a laminate is non-symmetric, then the thermal force and moment resultants induce both in-plane shrinkages and also out-of-plane curvatures.

Examples of SAL4S under temperature change

There now follows a small library of representative examples, for which the following nomenclature is used. There are four key temperatures: Room temperature T° is taken as 20 °C; ambient conditions are described by the assumed surface temperature of the top T^t and the bottom T^b of a horizontal sheet; cure temperature T^c is taken as T° unless otherwise stated.

The laminate SAL4S consists of four plies each 0.25 mm thick, two outer plies of mild steel sandwiching two inner plies of aluminium. The properties of each ply and of the laminate are given in Tables 4.1 to 4.3.

To demonstrate that the principles are really understood, it is strongly recommended that the reader confirm step-by-step from first principles the results of *one* of the following examples for a symmetric, and for a non-symmetric, laminate. Experience suggests that the insight gained by so doing is well worth the effort.

1. SAL4S: $T^\circ = T^t = T^b = T^c = 20\,°C,\ N_1 = 20\,N/mm$

This is a repeat of what has appeared in Section 4.3, but it provides now for an introduction to the fuller nomenclature and is a reference for later studies involving change of temperature. The laminate is assembled at the room temperature of 20 °C. The midplane strains are $\varepsilon_x^\circ = 0.01438\%$, $\varepsilon_y^\circ = -0.004212\%$. Figure 7.6 confirms that there is no net applied force in the y direction, so the mechanical stresses in the steel and aluminium cancel each other. There are no changes of temperature, so thermal and curing stresses and strains are zero.

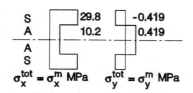

Figure 7.6 Response of SAL4S to all $T = 20\,°C$, $N_1 = 20\,N/mm$.

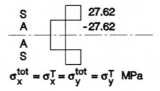

Figure 7.7 Response of SAL4S to $T^\circ = T^c = 20\,°C$, $T^t = T^b = 50\,°C$.

2. SAL4S: $T^o = 20\,°C$; $T^t = T^b = 50\,°C$; $T^c = 20\,°C$

The midplane strains are $\varepsilon^o_x = \varepsilon^o_y = 0.04256\%$. The raising of the temperature to a uniform 50 °C provides no net force on the laminate and hence total and thermal stresses in each direction cancel each other (Figure 7.7); even with a modest increase in uniform temperature the thermal stresses are substantial. The plies are isotropic so there is no shear strain or stress, and the direct stresses and strains are the same in all directions in the plane. The symmetric laminate under uniform temperature remains flat.

The thermal stress profiles arise because of the unequal thermal expansion coefficients of the steel and of the aluminium. If the plies were unbonded, the steel would expand less than the aluminium. To achieve the same thermal strain in each unbonded ply, a tensile force has to be applied to the steel and a compressive force to the aluminium, and Figure 7.7 confirms this.

3. SAL4S: $T^o = 20\,°C$; $T^t = T^b = 50\,°C$; $T^c = 20\,°C$; $N_1 = 20\,N/mm$

The midplane strains are now $\varepsilon^o_x = 0.05694\%$ and $\varepsilon^o_y = 0.03835\%$. The thermal responses are the same as in Example 2 and the mechanical responses the same as in Example 1. The total response is the sum of the two separate responses. Both thermal and mechanical stresses in the y direction are balanced because there is no net force in this direction.

4. SAL4S: $T^o = 20\,°C$; $T^t = 35\,°C$; $T^b = 10\,°C$; $T^c = 20\,°C$

The laminate midplane strains are $\varepsilon^o_x = \varepsilon^o_y = 0.003547\%$, and $\kappa_x = \kappa_y = -0.2897/m$. The temperature gradient across the sheet gives a midplane temperature of 2.5 °C above 20 °C, and hence there is a thermal and total bending strain profile offset at the midplane (Figure 7.8), and the stress profiles reflect the change of modulus between plies.

5. SAL4S: $T^o = T^t = T^b = 20\,°C$; $T^c = 140\,°C$

This laminate has been assembled hot at 140 °C to simulate a cure. The analysis (see Problem 7.17 below) shows that the laminate suffers substantial residual stress: the steel plies take $\sigma^c_s = -110.5\,MPa$ and the aluminium plies $\sigma^c_A = +110.5\,MPa$. There is no net applied force, so the direct stresses balance each other. The midplane strains induced by cooling and curing to 20 °C are $\varepsilon^c_x = \varepsilon^c_y = -0.001702$.

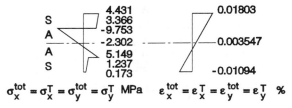

Figure 7.8 Response of SAL4S to $T^o = T^c = 20\,°C$, $T^t = 35$, $T^b = 10\,°C$.

Problem 7.17: Sketch the curing stress profile in SAL4S in the x direction where the laminate has been bonded at 140 °C and cooled to 20 °C. Confirm your result by calculation from first principles.

6. SAL4S: $T^o = 20\,°C;\ T^t = T^b = 50\,°C;\ T^c = 140\,°C$

With a temperature datum of 20 °C the midplane strains are solely due to the temperature rise of 30 K, giving $\varepsilon_x^\circ = \varepsilon_y^\circ = 0.04256\%$. The curing stress distribution is the same as that for Example 5 and the service temperature the same as for Example 3, so by addition the total stress profiles are as shown in Figure 7.9.

The amount of shinkage (ε_o^s) from cure temperature to 20 °C is found by treating this as a contraction problem, e.g.:

$$N_1^s = \{(Q_{11s} + Q_{12s})\alpha_s + (Q_{11A} + Q_{12A})\alpha_A\} \cdot 4h_s\Delta T$$

and hence $\varepsilon_1^{os} = [a]N_1^s$.

7. SAL4S: $T^o = 20\,°C;\ T^t = 40\,°C;\ T^b = 5\,°C;\ T^c = 140\,°C$

Taking the temperature datum as 20 °C the response to the temperature gradient is $\varepsilon_x^\circ = \varepsilon_y^\circ = 0.003547\%$, and $\kappa_x = \kappa_y = -0.4056/m$, indicating negative synclastic curvature.

The cure stress profile is given in Problem 7.17. The thermal stress profile is the addition of a thermal bending stress to a uniform stress as shown in Figure 7.10; adding the thermal and curing stresses gives the total stress profile in Figure 7.10. The total and thermal strain profiles are identical and as expected.

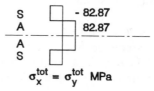

Figure 7.9 Response of SAL4S to $T^\circ = 20\,°C,\ T^c = 140\,°C,\ T^t = T^b = 50\,°C$.

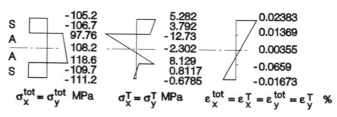

Figure 7.10 Response of SAL4S to $T^\circ = 20\,°C,\ T^c = 140\,°C,\ T^t = 40\,°C,\ T^b = 5\,°C$.

7.4.2 Non-symmetric laminate of isotropic plies

Change of uniform temperature

The calculation of internal stresses caused in a non-symmetric laminate made from isotropic plies is based on the same principles as that for a symmetric laminate discussed in Section 7.4.1. It is now necessary to calculate both the thermal force resultant $[N]^T$ and the thermal moment resultant $[M]^T$, and noting that $[B] \neq 0$, the full set of calculations, using $[ABD]$ or $[abhd]$, must be done. It will be apparent that synclastic curvature will result from a change of uniform temperature, as well as the normal midplane strains.

If we consider the two-ply laminate SAL2NS, consisting of $h_s = 0.5$ mm and $h_A = 0.5$ mm, for which stiffness data are given in Table 4.6, we can now predict the result of a change of uniform temperature of $\Delta T = 30$ K.

$$N_1^T = \Delta T\{(Q_{11s}\alpha_s + Q_{12s}\alpha_s)h_s + (Q_{11A}\alpha_A + Q_{12A}\alpha_A)(h_A)\}$$
$$= 83.709 \text{ N/mm} = N_2^T$$

Denoting the top surface coordinate h_{os} and the lower h_{2A}, we have

$$M_1^T = (\Delta T/2)\{(Q_{11s}\alpha_s + Q_{12s}\alpha_s)(0 - (-h_{os})^2) + (Q_{11A}\alpha_A + Q_{12A}\alpha_A)(h_{2A}^2 - 0)\}$$
$$= -2.9062 \text{ N} = M_2^T$$
$$\varepsilon_1^\circ = a_{11}N_1^T + a_{12}N_2^T + b_{11}M_1^T + b_{12}M_2^T = 0.048475\% = \varepsilon_2^\circ$$
$$\kappa_1 = h_{11}N_1^T + h_{12}N_2^T + d_{11}M_1^T + d_{12}M_2^T = 0.5045/\text{m} = \kappa_2$$

Thus under a change of uniform temperature this non-symmetric laminate develops synclastic curvature.

Linear temperature gradient through thickness

For a two-ply laminate of total thickness h, with a continuous linear profile $T(z) = a + bz$, we obtain:

$$N_1^T = [Q]_s[\alpha]_s \int_{z=-h/2}^{0} (a + bz)dz + [Q]_A[\alpha]_A \int_{z=0}^{+h/2} (a + bz)dz$$

$$= [Q]_s[\alpha]_s(+ah/2 - bh^2/8) + [Q]_A[\alpha]_A(ah/2 + bh^2/8) = 38.948 \text{ N/mm} = N_2^T$$

$$M_1^T = \sum_{f=1}^{F} [Q]_f[\alpha]_f \int_{z=-h/2}^{+h/2} T(z)z\,dz = \sum_{f=1}^{F} [Q]_f[\alpha]_f \int_{z=-h/2}^{+h/2} (az + bz^2)dz)$$

$$= [Q]_s[\alpha]_s(-ah^2/8 + bh^3/24) + [Q]_A[\alpha]_A(ah^2/8 + bh^3/24) = 5.5862 \text{ N} = M_2^T$$

We can now calculate the midplane strains and curvatures using:

$$\varepsilon_1^\circ = a_{11}N_1^T + a_{12}N_2^T + b_{11}M_1^T + b_{12}M_2^T$$
$$\kappa_1 = h_{11}N_1^T + h_{12}N_2^T + d_{11}M_1^T + d_{12}M_2^T$$

Residual stresses

For a non-symmetric laminate, the curing stresses can be calculated following the scheme outlined for the symmetric laminate in Section 7.4. Here it is important to realize that under a change of uniform temperature there will be a residual moment resultant $[M]^c$ in a non-symmetric laminate. As $[B] \neq 0$, the calculation of equivalent strains and the stress-generating 'mechanical strains' will involve the use of the fully populated $[abhd]$ matrix. It follows therefore that a non-symmetric laminate made from isotropic plies cured hot will show synclastic curvature (but not twisting curvature) at a lower (or higher) temperature.

This section has discussed in some detail how to calculate the response to change of temperature of symmetric and non-symmetric laminates based on isotropic plies. The worked examples provide added confidence in applying the basic theory to different situations.

Examples of regular non-symmetric laminate SAL2NS

The laminate SAL2NS is based on two plies each 0.5 mm thick, the top being mild steel and the bottom aluminium, for which stiffnesses are given in Table 4.6.

Because of the lack of symmetry, $[B] \neq 0$, and hence almost any thermal load will induce bending, and stress and strain profiles confirm this.

1. SAL2NS: $T^o = 20\,°C;\ T^t = T^b = 50\,°C;\ T^c = 20\,°C$
Under a change of uniform temperature by $+ 30\,K$, the midplane strains become $\varepsilon_x^o = \varepsilon_y^o = 0.04847\%$ and the curvatures $\kappa_x = \kappa_y = 0.5045/m$ are positive, because the aluminium is the lower ply and has a higher expansion coefficient than the upper steel ply. The stress and strain profiles are given in Figure 7.11.

The slender bimetallic strip works on the same principle (but only has one usable dimension).

2. SAL2NS: $T^o = 20\,°C;\ T^t = T^b = 50\,°C;\ T^c = 20\,°C;\ N_2 = 15\,N/mm$
Under this temperature gradient of 30 K across the thickness of the laminate,

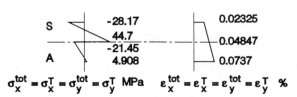

$$\sigma_x^{tot} = \sigma_x^T = \sigma_y^{tot} = \sigma_y^T \text{ MPa} \qquad \varepsilon_x^{tot} = \varepsilon_x^T = \varepsilon_y^{tot} = \varepsilon_y^T \%$$

Figure 7.11 Response of SAL2NS to $T^o = T^c = 20\,°C,\ T^t = T^b = 50\,°C$.

the midplane strains are $\varepsilon_x^° = 0.04437\%$, $\varepsilon_y^° = 6.172\%$, and the curvatures $\kappa_x = 0.4359/m$ and $\kappa_y = 0.7015/m$.

The thermal stress profiles are the same as for Example 1. The mechanical stress and strain profiles have similar profiles to those discussed for Figure 4.12 (noting that this figure relates to the application of N_1). The total strain profile is linear. The total stress profile is shown in Figure 7.12.

Problem 7.18: Two identical non-symmetrical laminates SAL2NS both have the same surface temperatures of $10\,°C$ and $35\,°C$; for one the top is at $10\,°C$, for the other $35\,°C$. In (a) $\varepsilon_x^° = \varepsilon_y^° = 0.0005362\%$ and $\kappa_x = \kappa_y = -0.3619/m$; in (b) $\varepsilon_x^° = \varepsilon_y^° = 0.007543\%$ and $\kappa_x = \kappa_y = 0.446/m$. Stress and strain profiles are given in Figures 7.13 and 7.14. Which thermal loading conditions apply to which responses? Sketch the shapes of the deformed laminates. Explain the differences in stress and strain profiles for the two thermal loading conditions.

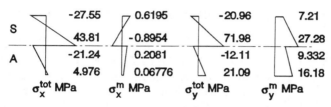

Figure 7.12 Response of SAL2NS to $T^° = T^c = 20\,°C$, $T^t = T^b = 50\,°C$, $N_2 = 15\,N/mm$.

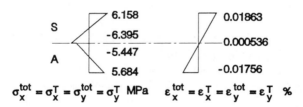

Figure 7.13 Response of SAL2NS to $T^° = T^c = 20\,°C$. Surface temperatures $10°$ and $35\,°C$.

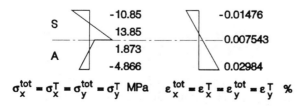

Figure 7.14 Response of SAL2NS to $T^° = T^c = 20\,°C$. Surface temperatures $10°$ and $35\,°C$.

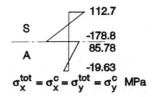

$$\sigma_x^{tot} = \sigma_x^c = \sigma_y^{tot} = \sigma_y^c \ \text{MPa}$$

Figure 7.15 Response of SAL2NS to $T^\circ = T^t = T^b = 20\,°C$, $T^c = 140\,°C$.

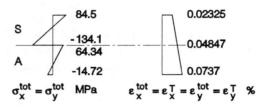

$$\sigma_x^{tot} = \sigma_y^{tot} \ \text{MPa} \qquad \varepsilon_x^{tot} = \varepsilon_x^T = \varepsilon_y^{tot} = \varepsilon_y^T \ \%$$

Figure 7.16 Response of SAL2NS to $T^\circ = 20\,°C$, $T^t = T^b = 50\,°C$, $T^c = 140\,°C$.

3. SAL2NS: $T^\circ = 20\,°C$; $T^t = T^b = 20\,°C$; $T^c = 140\,°C$
In this example the laminate is assembled at 140 °C. Because the temperature datum is 20 °C, all calculated strains and curvatures are zero. The stress profile in Figure 7.15 clearly shows the bending caused by the mismatch of free contraction strains in a non-symmetric laminate. The actual shrinkages and curvatures solely from a datum of 140 °C are $\varepsilon_x^\circ = \varepsilon_y^\circ = -0.1939\%$ and $\kappa_x = \kappa_y = -2.018/\text{m}$.

4. SAL2NS: $T^\circ = 20\,°C$; $T^t = T^b = 50\,°C$; $T^c = 140\,°C$
Cooling down from assembly at 140 °C to 20 °C and then increasing the ambient temperature to 50 °C gives $\varepsilon_x^\circ = \varepsilon_y^\circ = 0.04847\%$ and $\kappa_x = \kappa_y = 0.5045/\text{m}$. The thermal strains and stresses have already been given in Figure 7.11. The curing stress profile was given in Figure 7.15. By superposition the net result for the total stress profile (Figure 7.16) is the expected one that all values are reduced. In effect, this is the result of a smaller interval between cure and ambient compared with Example 3.

7.5 EFFECT OF CHANGE OF TEMPERATURE ON STIFFNESS OF CROSSPLY LAMINATES

The behaviour of crossply laminates is broadly similar to that of laminates based on isotropic plies. The main differences are that the properties in the two principal directions may be different in the laminate, and are certainly different in each ply. A change in uniform temperature will introduce shrinkage in the plane and, if the laminate is non-symmetric, anticlastic curvatures; the shrinkages and curvatures are usually different in the principal directions, and there is no thermal shear or twisting. Care therefore must be taken to use the correct values of stiffnesses and thermal expansion coefficients relating to the particular orientation of the ply under consideration. The procedures for calculating stress and strain profiles, and residual stresses, are identical to those for laminates based on isotropic plies, as is indicated in the full outline to the following problem, and in the examples below.

Problem 7.19: A carbon fibre reinforced epoxy unidirectional prepreg is 0.125 mm thick and has the following properties: $E_1 = 138$ GPa, $E_2 = 9$ GPa, $G_{12} = 7.1$ GPa, $v_{12} = 0.3$, $\alpha_1 = -3 \times 10^{-7}$/K, $\alpha_2 = 28.1 \times 10^{-6}$/K. The in-plane stiffness and compliance coefficients are:

[A] kN/mm			[a] mm/MN		
39.23	2.037	0	25.53	−0.7255	0
2.037	71.67	0	−0.7255	13.97	0
0	0	5.325	0	0	187.8

It is to be made into a flat crossply laminate with the stacking sequence $(0°/90°/90°)_s$ and cured at 140 °C. Calculate the stress profiles in the principal directions in the laminate at 20 °C.

7.5.1 Regular balanced symmetrical crossply laminate C + 4S

For symmetric crossply laminates, the approach to calculation of residual stress or change of temperature is similar to that for symmetric laminates based on isotropic plies. The relevant thermal force and moment resultants are calculated for the appropriate circumstances. The main differences are that $Q_{11} \neq Q_{22}$ (unless there is the same number of plies in the 0° and 90° directions) and hence midplane strains (and any curvatures) are different in the two principal directions. By virtue of symmetry, there are no curvatures caused by change of uniform temperature or by curing at uniform temperature.

Examples of regular balanced symmetrical crossply laminate C + 4S

The following examples are based on laminate C + 4S consisting of four identical plies of carbon fibre in epoxy resin, each 0.125 mm thick arranged as $(0°/90°)_s$, for which data are given in Table 5.1 and 5.2. The thermal expansion coefficients of the unidirectional ply are $\alpha_1 = 6 \times 10^{-6}$/K and $\alpha_2 = 3.5 \times 10^{-5}$/K.

1. C + 4S: $T° = 20 °C$; $T^t = T^b = 50 °C$; $T^c = 20 °C$
The midplane strains are uniform in both directions: $\varepsilon_1° = \varepsilon_2° = 0.02269\%$, and this value is both the total and the thermal strain throughout the laminate. The

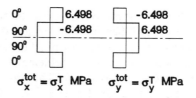

Figure 7.17 Response of C + 4S to $T° = T^c = 20°C$, $T^t = T^b = 50°C$.

total and thermal stress profiles in the laminate are shown in Figure 7.17, being equal and opposite in the two directions. The net force in each principal direction is zero as there are no applied force (resultants). In the x direction a free 0° ply would expand by $\alpha_1 \Delta T$ and a free 90° ply would expand by $\alpha_2 \Delta T$. $\alpha_1 \ll \alpha_2$, so to give the same expansion in both plies a tensile force would be applied to the 0° ply and a compressive force to the 90° ply. The stress profile for σ_x^T confirms this.

2. $C + 4S$: $T° = 20°C$; $T^t = T^b = 50°C$; $T^c = 20°C$; $N_1 = 10\,N/mm$
The midplane total strains are $\varepsilon_x° = 0.0439\%$ and $\varepsilon_y° = 0.02215\%$. The thermal stresses are the same as in Example 1, so these are added to the mechanical stresses to build up the total stress profile in Figure 7.18. The uniform mechanical strains are $\varepsilon_x^m = 0.02121\%$ and $\varepsilon_y^m = -0.0005414\%$.

3. $C + 4S$: $T° = 20°C$; $T^t = 25°C$; $T^b = 50°C$; $T^c = 20°C$
The midplane temperature and the temperature profile are responsible for $\varepsilon_x° = \varepsilon_y° = 0.01324\%$, $\kappa_x = 0.323/m$ and $\kappa_y = 0.6569/m$.

Problem 7.20: Calculate the thermal expansion coefficients for the laminate C + 4S.

Problem 7.21: Why are the midplane strains the same, but the curvatures quite different and of the same sign, in Example 3? Why are the total stress and total strain profiles in the x and y directions (Figure 7.19) so different?

Problem 7.22: What moment resultant is needed to just suppress the curvature κ_2? Sketch the shape of the sheet and compare it with that for Example 3.

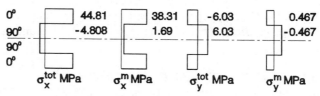

Figure 7.18 Response of C + 4S to $T° = T^c = 20°C$, $T^t = T^b = 50°C$; $N_1 = 10\,N/mm$.

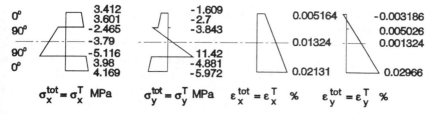

0°	90°	90°	0°

3.412
3.601
-2.465
-3.79
-5.116
3.98
4.169

$\sigma_x^{tot} = \sigma_x^T$ MPa

-1.609
-2.7
-3.843

11.42
-4.881
-5.972

$\sigma_y^{tot} = \sigma_y^T$ MPa

0.005164

0.01324

0.02131

$\varepsilon_x^{tot} = \varepsilon_x^T$ %

-0.003186
0.005026
0.001324

0.02966

$\varepsilon_y^{tot} = \varepsilon_y^T$ %

Figure 7.19 Response of C + 4S to $T° = T^c = 20\,°C$, $T^t = 25\,°C$, $T^b = 50\,°C$.

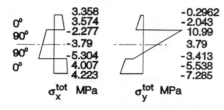

3.358
3.574
-2.277
-3.79
-5.304
4.007
4.223

σ_x^{tot} MPa

-0.2962
-2.043
10.99
3.79
-3.413
-5.538
-7.285

σ_y^{tot} MPa

Figure 7.20 Response of C + 4S to $T° = T^c = 20\,°C$, $T^t = 25\,°C$, $T^b = 50\,°C$, $\kappa_2 = 0$.

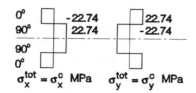

-22.74
22.74

$\sigma_x^{tot} = \sigma_x^c$ MPa

22.74
-22.74

$\sigma_y^{tot} = \sigma_y^c$ MPa

Figure 7.21 Response of C + 4S to $T° = T^t = T^b = 20\,°C$, $T^c = 125\,°C$.

Problem 7.23: Explain the shape of the stress profile $\sigma_y^{tot}(z)$ for C + 4S in Figure 7.20.

4. C + 4S: $T° = T^t = T^b = 20\,°C$; $T^c = 125\,°C$
The curing stress profiles are shown in Figure 7.21, and confirm that there is no net applied force. Bearing in mind that the transverse strength of the ply is about 80 to 100 MPa, it will be appreciated that the load-bearing capacity is much reduced by curing at 125 °C rather than at 20 °C.

It is important to ensure that the co-ordinates match the plies, when calculating laminate behaviour. Here we choose the reference direction x to correspond with the 1 direction in the 0° plies and to the 2 direction in the 90° plies. Thus $Q_{11}(90°) = Q_{22}(0°)$. With this in mind we can calculate the curing force resultants in the laminate:

$$N_x^c = \Delta T\{(Q_{11}(0°)\alpha_1 + Q_{12}(0°)\alpha_2)h/2 + (Q_{11}(90°)\alpha_2 + Q_{12}(90°)\alpha_1)h/2\}$$

and using the data for [Q] for C + 4S in Table 5.1, we find $N_x^c = -38.433$ N/mm ($= N_y^c$ by commonsense).

The laminate strains are $[\varepsilon^c] = [a][N^c]$, where values of $[a]$ for C + 4S are in Table 5.2 : $\varepsilon_x^c = a_{11}N_x^c + a_{12}N_y^c = -7.944 \times 10^{-4}$. For the midplane, which is oriented at 90° to the reference direction we have $\varepsilon_x^e = \varepsilon_x^c - \alpha_2 \Delta T = 2.8806 \times 10^{-3}$, and $\varepsilon_y^e = \varepsilon_y^c - \alpha_1 \Delta T = -1.644 \times 10^{-4}$. Hence $\sigma_x^c(90°) = Q_{11}(90°) \varepsilon_x^e(90°) + Q_{12}(90°)\varepsilon_x^e(90°) = (23.14 - 0.396) \times 10^6 = 22.744$ MPa.

Problem 7.24: A thin-walled tube is made from the crossply laminate C + 4S cured at 125 °C and operating with a bore surface temperature of 15 °C and an external surface temperature of + 40 °C. No internal or external pressures are applied to the tube. Calculate the strain and stress profiles in the tube, assuming the ply principal directions and aligned with the hoop and axial directions.

7.5.2 Balanced non-symmetric crossply laminate C + 4NS

Because $[B] \neq 0$, a change of uniform temperature or curing at elevated temperature will induce curvature, anticlastic because the terms B_{11} and B_{22} have opposite signs. For the special case where there are equal numbers of plies in each direction, $B_{11} = -B_{22}$ and the curvatures have equal values but opposite signs.

We ought to mention here that although the theory does predict anticlastic curvature, there are circumstances where it is not always possible to actually achieve it: under certain (usually large) ratios of plate width or length to thickness, out-of-plane displacements are very large (tens or even hundreds of time the plate thickness) so the assumptions of small displacements are invalid, and the plate can adopt *either* one curvature or the other. It is as if the single curvature produces such a large second moment of area that the other curvature is suppressed even though no moment is applied. On flattening such a thin plate by hand, the curvature flips to the other mode. The plate is stable in either position: the instability when flat is a consequence of too much strain energy, but a detailed discussion of this complex but fascinating phenomenon lies outside the scope of this book. Further detail is given in papers by Hyer (1981, 1982).

Examples of balanced non-symmetric crossply laminate C + 4NS

The following examples are based on laminate C + 4NS, consisting of four plies of 0.125 mm carbon fibre/epoxy prepreg arranged in the sequence (0°/0°/90°/90°), stiffnesses of which are given in Table 5.4.

1. C + 4NS: $T^o = 20$ °C; $T^t = T^b = 50$ °C; $T^c = 20$ °C
The response of this non-symmetric crossply laminate to this increase of uniform temperature is $\varepsilon_x^o = \varepsilon_y^o = 0.03451\%$, $\kappa_x = 1.14$/m and $\kappa_y = -1.14$/m,

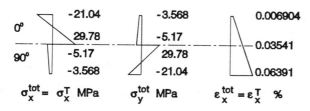

Figure 7.22 Response of C + 4NS to $T^\circ = T^c = 20\,^\circ\text{C}$, $T^t = T^b = 50\,^\circ\text{C}$.

which induces anticlastic curvature. Thermal and total strain profiles reflect the bending contribution superposed on an average strain, with corresponding profiles for total and thermal stress (Figure 7.22). Note the substantial value of σ_x for the 0° ply at the interface, even though the midplane strain is only 0.03451%.

2. $C + 4NS$: $T^\circ = 20\,^\circ C$; $T^t = T^b = 50\,^\circ C$; $T^c = 20\,^\circ C$; $N_2 = 10\,N/mm$
The response to the applied load N_2 is additional to the change of temperature described in Example 1, giving $\varepsilon_x^\circ = 0.03395\%$, $\varepsilon_y^\circ = 0.09244\%$, $\kappa_x = 1.14/\text{m}$ and $\kappa_y = -4.271/\text{m}$, thus giving the total and mechanical stress profiles in Figure 7.23. The mechanical strain profile is uniform. The mechanical strain ε_x^m is uniform under the applied force resultant N_2, and hence the curvature κ_x is the same as in Example 1.

3. $C + 4NS$: $T^\circ = 20\,^\circ C$; $T^t = T^b = 50\,^\circ C$; $T^c = 20\,^\circ C$; $\kappa_1 = \kappa_2 = 0$
The moments needed to suppress the curvatures have the expected equal values but opposite signs, namely $M_1 = -0.4061\,\text{N}$ and $M_2 = +0.4061\,\text{N}$, giving total midplane strains $\varepsilon_x^\circ = \varepsilon_y^\circ = 0.02269\%$. In the absence of curvatures, the uniform temperature rise together with the moments lead to uniform total stresses of $|6.498|$ MPa in each ply in each direction, with a net force of zero in both 1 and 2 directions.

Problem 7.25: Sketch the total stress profile $\sigma_x^{tot}(z)$, and the mechanical stress profile $\sigma_x^m(z)$, in C + 4S for the temperature conditions in Example 3.

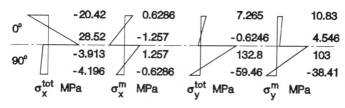

Figure 7.23 Response of C + 4NS to $T^\circ = T^c = 20\,^\circ\text{C}$, $T^t = T^b = 50\,^\circ\text{C}$, $N_2 = 10\,\text{N/mm}$.

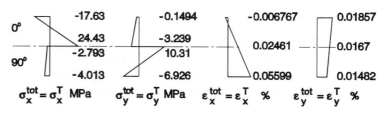

Figure 7.24 Response of C + 4NS to $T^o = T^c = 20\,°C$, $T^t = 25\,°C$, $T^b = 50\,°C$.

4. $C + 4NS$: $T^o = 20\,°C$; $T^t = 25\,°C$; $T^b = 50\,°C$; $T^c = 20\,°C$

The temperature gradient induces $\varepsilon_x^o = 0.02461\%$, $\varepsilon_y^o = 0.0167\%$, $\kappa_x = 1.255/m$ and $\kappa_y = -0.07496/m$. The total and thermal strain profiles are linear, with the corresponding stress profiles given in Figure 7.24.

7.6 EFFECT OF CHANGE OF TEMPERATURE ON STIFFNESS OF ANGLEPLY LAMINATES

The behaviour of angleply laminates follows the principles discussed above, but now the thermal force and moment resultants use the transformed stiffness matrix $[Q^*]$ and the transformed thermal expansion coefficient $[\alpha^*]$ for each ply. In particular, the opportunity for thermal shear may occur in addition to direct strains caused by change of uniform temperature, and twisting will occur in non-symmetric laminates as well as bending curvatures. As most high-performance laminates are cured hot and used at lower temperatures, it is especially a good principle to use symmetric laminates so that hot-cured flat plates will stay flat on cooling, unless there is an over-riding reason for using non-symmetric laminates.

7.6.1 Balanced symmetric angleply laminate 30/−30SG

The thermal behaviour of balanced symmetric angleply laminates inevitably has some of the character of the unidirectional ply loaded in-plane off-axis. The $[Q^*]$ matrix is now fully populated, and this must be taken into account when calculating thermal force and moment resultants.

Examples of balanced symmetric angleply laminate 30/−30SG

The following examples are based on laminate 30/−30SG, made from four 0.25 mm plies of EG (unidirectional E glass in epoxy resin) laid in the sequence $(30°/−30°)_s$, for which ply properties and ply stiffnesses $[Q^*(30°)]$ and $[Q^*(−30°)]$ are given in Table 7.1, and the ABD matrix and its inverse in Table 7.2.

Table 7.1 Properties of EG glass/epoxy prepreg

	[Q(0°)] GPa		
$E_1 = 45.6\,\text{GPa}$	46.42	2.993	0
$E_2 = 10.73\,\text{GPa}$	2.993	10.92	0
$G_{12} = 5.14\,\text{GPa}$	0	0	5.14
$\nu_{12} = 0.274$			
$\alpha_1 = 7 \times 10^{-6}/\text{K}$			
$\alpha_2 = 3 \times 10^{-5}/\text{K}$			

$\sigma_{1T\text{max}}1000\,\text{MPa}\ \sigma_{2T\text{max}}100\,\text{MPa}$
$\sigma_{1C\text{max}}1600\,\text{MPa}\ \sigma_{2C\text{max}}270\,\text{MPa}\ \tau_{12\text{max}}80\,\text{MPa}$

[Q*(+30°)]			[Q*(−30°)]		
31.77	8.767	11.02	31.77	8.767	−11.02
8.767	14.02	4.351	8.767	14.02	−4.351
11.02	4.351	10.91	−11.02	−4.351	10.91

Table 7.2 Stiffnesses and compliances of balanced regular symmetric angleply laminate $(30°/-30°)_s$ based on 0.25 mm plies of EG

A (kN/mm) B (kN)			B (kN) D (kNmm)		
31.77	8.767	0	0	0	0
8.767	140.2	0	0	0	0
0	0	10.91	0	0	0
0	0	0	2.648	0.7306	0.6887
0	0	0	0.7306	1.169	0.272
0	0	0	0.6887	0.272	0.9095

a (mm/MN) h (/MN)			b (/MN) d (/MNmm)		
38.04	−23.78	0	0	0	0
−23.78	86.18	0	0	0	0
0	0	9.62	0	0	0
0	0	0	533.9	−257.7	−327.2
0	0	0	−257.7	1044	−117.1
0	0	0	−327.2	−117.1	1382

1. 30/−30SG: $T^o = 20\,°C$; $T^t = T^b = 20\,°C$; $T^c = 20\,°C$; $N_y = 10\,N/mm$
These loading conditions serve to revise the behaviour of a balanced symmetric angleply laminate under mechanical loading alone. The midplane strains are $\varepsilon_x^o = -0.02378\%$ and $\varepsilon_y^o = +0.08618\%$, giving uniform direct stresses $\sigma_x = 0$, $\sigma_y = 10\,\text{MPa}$, $\sigma_1 = 3.478\,\text{MPa}$ and $\sigma_2 = 6.522\,\text{MPa}$, with uniform direct strains $\varepsilon_1 = 0.003709\%$ and $\varepsilon_2 = 0.05869\%$. The shear stresses and strains in

individual plies arise from $Q^*_{16}(-30°) = -Q^*_{16}(+30°)$, and the profiles in Figure 7.25 are therefore as expected and give no net in-plane shear force resultants.

2. 30/−30SG: $T^o = 20\,°C$; $T^t = T^b = 50\,°C$; $T^c = 20\,°C$
The increase in uniform temperature by 30 K induces midplane strains $\varepsilon^o_x = 0.01939\%$ and $\varepsilon^o_y = 0.066\%$, and uniform $\varepsilon_1 = 0.03104\%$ and $\varepsilon_2 = 0.05435\%$. Stresses $\sigma_x = \sigma_y = 0$, but $\sigma_1 = 3.594\,\text{MPa}$ and $\sigma_2 = -3.594\,\text{MPa}$. The shear stresses and strains are shown in Figure 7.26.

3. 30/−30SG: $T^o = 20\,°C$; $T^t = T^b = 50\,°C$; $T^c = 20\,°C$; $N_y = 10\,N/mm$
As expected, the response to a change of uniform temperature of 30 K and the application of a force resultant $N_y = 10\,\text{N/mm}$ is the sum of the separate responses. For example $\varepsilon^o_x = -0.004395\%$ and $\varepsilon_2 = 0.113\%$.

4. 30/−30SG: $T^o = 20\,°C$; $T^t = 25\,°C$; $T^b = 50\,°C$; $T^c = 20\,°C$
This temperature gradient induces $\varepsilon^o_x = 0.01131\%$, $\varepsilon^o_y = 0.0385\%$; $\gamma^o_{xy} = 0$,

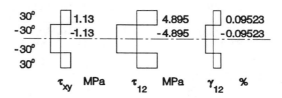

Figure 7.25 Response of 30/−30SG to all $T = 20\,°C$, $N_y = 10\,\text{N/mm}$.

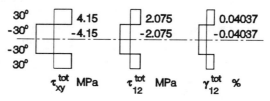

Figure 7.26 Response of 30/−30SG to $T^o = T^c = 20\,°C$, $T^t = T^b = 50\,°C$.

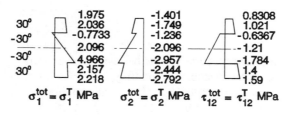

Figure 7.27 Response of 30/−30SG to $T^o = T^c = 20\,°C$, $T^t = 25\,°C$, $T^b = 50\,°C$.

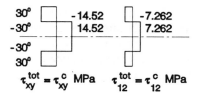

Figure 7.28 Response of $30/-30$SG to $T^\circ = T^t = T^b = 20\,^\circ\text{C}$, $T^c = 125\,^\circ\text{C}$.

$\kappa_x = 0.2323/\text{m}$, $\kappa_y = 0.5753/\text{m}$, and $\kappa_y = -0.2988/\text{m}$. The bending stress profiles in the principal directions are recorded in detail in Figure 7.27.

5. $30/-30$SG: $T^\circ = 20\,^\circ\text{C}$; $T^t = T^b = 20\,^\circ\text{C}$; $T^c = 125\,^\circ\text{C}$
The effect of cure at $125\,^\circ\text{C}$ is to induce at room temperature ($20\,^\circ\text{C}$) the shear stresses shown in Figure 7.28 together with $\sigma_1^c = -12.58\,\text{MPa}$ and $\sigma_2^c = +12.58\,\text{MPa}$.

Problem 7.26: Calculate the thermal expansion coefficients for laminate $30/-30$SG.

7.6.2 Balanced unsymmetric angleply laminate $30/-30$NSG

The following examples are based on the balanced unsymmetric angleply laminate $30/-30$NSG, which consists of two plies each 1 mm thick of prepreg

Table 7.3 Properties of balanced regular non-symmetric angle ply laminate $30/-30$NSG – each ply EG is 1 mm thick

A (kN/mm) B (kN)			B (kN) D (kNmm)		
63.54	17.53	0	0	0	-11.02
17.53	28.05	0	0	0	-4.351
0	0	21.83	-11.02	-4.351	0
0	0	-11.02	21.18	5.845	0
0	0	-4.351	5.845	9.349	0
-11.02	-4.351	0	0	0	7.276

a (mm/MN) h (/MN)			b (/MN) d (/MNmm)		
23.73	-10.21	0	0	0	29.83
-10.21	43.69	0	0	0	10.67
0	0	63	29.83	10.67	0
0	0	29.83	71.18	-30.62	0
0	0	10.67	-30.62	131.1	0
29.83	10.67	0	0	0	189

EG arranged as $(30°/-30°)_T$ and having the stiffness properties given in Table 7.3. The properties of individual plies are given in Table 7.1.

The balanced unsymmetric angleply laminate has non-zero terms B_{16} and B_{26}, which are responsible for twisting under either direct loads or under change of temperature.

1. 30/-30NSG: $T^o = 20\,°C$; $T^t = T^b = 20\,°C$; $T^c = 20\,°C$; $N_x = 10\,N/mm$
Under single direct loading, the midplane strains are $\varepsilon_x^\circ = 0.02373\%$, $\varepsilon_y^\circ = -0.01021\%$ and the twisting curvature is $\kappa_{xy} = 0.2983/m$. The stress profiles are shown in Figure 7.29 and the corresponding strain profiles in Figure 7.30. Strains ε_x and ε_y are uniform as expected.

2. 30/-30NSG: $T^o = 20\,°C$; $T^t = T^b = 50\,°C$; $T^c = 20\,°C$
Increasing the uniform temperature by 30 K induces $\varepsilon_x^\circ = 0.03177\%$, $\varepsilon_y^\circ = 0.07043\%$ and $\kappa_{xy} = 0.7843/m$, so that the non-symmetric angle ply laminate warps by twisting. Compare this twisting behaviour with that for a non-symmetric crossply laminate in Section 7.5.2, and the non-symmetric laminate based on isotropic plies.

Problem 7.27: Sketch the midplane strains and curvatures for laminate $30/-30NSG$ under the following conditions: $T^o = 20\,°C$, $T^t = T^b = 50\,°C$; $T^c = 20\,°C$; $N_x = 10\,N/mm$. What twisting moment would be necessary to just suppress the resulting twist?

Problem 7.28: How could you suppress the twisting curvature in Example 1?

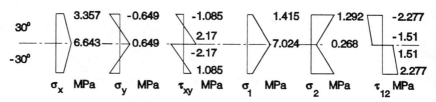

Figure 7.29 Stress response of $30/-30NSG$ to all $T = 20\,°C$, $N_x = 10\,N/mm$.

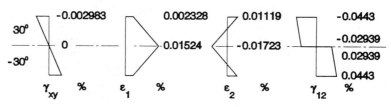

Figure 7.30 Strain response of $30/-30NSG$ to all $T = 20\,°C$, $N_x = 10\,N/mm$.

3. *30/−30NSG: $T^o = 20\,°C$; $T^t = 25\,°C$; $T^b = 50\,°C$; $T^c = 20\,°C$*
Application of this uniform temperature gradient results in a full set of midplane strains and curvatures: $\varepsilon^o_{x1} = 0.01853\%$, $\varepsilon^o_y = 0.04108\%$; $\gamma^o_{xy} = 0.01089\%$; $\kappa_x = 0.1324/m$, $\kappa_y = 0.2935/m$, $\kappa_{xy} = 0.4575/m$. The thermal stress and strain profiles are shown in Figures 7.31 and 7.32. The bending and shear strain profiles vary linearly with thickness as expected.

4. *30/−30NSG: $T^o = 20\,°C$; $T^t = T^b = 20\,°C$; $T^c = 125\,°C$*
Curing at 125 °C and cooling to 20 °C gives the stress distributions shown in Figure 7.33. It is interesting to note the significant value of transverse stress σ^c_2 at the interface.

(It is worth noting that the size effects mentioned in Section 7.5.2 for non-symmetric crossplies seem also to apply for non-symmetric angleply laminates. Rather than adopt the anticlastic curvature *in global co-ordinates* predicted

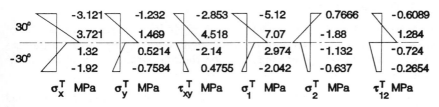

Figure 7.31 Stress response of 30/−30NSG to $T^o = T^c = 20\,°C$, $T^t = 25\,°C$, $T^b = 50\,°C$.

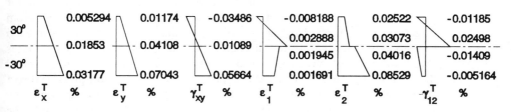

Figure 7.32 Strain response of 30/−30NSG to $T^o = T^c = 20\,°C$, $T^t = 25\,°C$, $T^b = 50\,°C$.

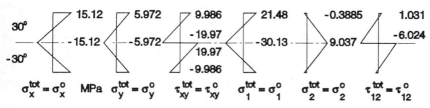

Figure 7.33 Response of 30/−30NSG to $T^o = T^t = T^b = 20\,°C$, $T^c = 125\,°C$.

above, large-scale non-symmetric angleply laminates hot-cured in flat moulds adopt after cooling a (usually) bistable single cylindrical curvature *in some principal direction* calculable using the transformation matrix (Section 3.3.10)).

7.7 EFFECT OF CHANGE OF TEMPERATURE ON STRENGTH OF LAMINATES

Earlier sections of this chapter have described how thermal or curing stresses (or both) are set up in laminates as the result of changes in temperature. This section describes some effects of these thermal and curing stresses on the strength of laminates under specific circumstances. The general principles are restated first, then examples are given which relate to changes of uniform temperature on the strength of balanced regular symmetric laminates.

7.7.1 The basic principles

1. A change of temperature in a free single ply induces thermal strains but not stresses. A change in temperature in a single ply not free to strain induces thermal stresses (and there may also be thermal strains, depending on the boundary conditions).

2. Plies in laminates built up from different isotropic materials bonded together are not free to expand in the plane, leading to thermal stresses in each ply, and hence to stress profiles through the thickness of the laminate.

3. Stresses in laminates made from bonded stacks of different plies, or similar plies aligned in different directions will induce stress profiles in the principal directions in each ply. In crossply laminates there are direct stresses in each principal direction within the plies and the laminates. In angleply laminates there are induced direct and shear stresses in both global and principal directions. In general there are developed both tensile and compressive direct stresses as well as positive and negative shear stresses. For a given laminate, values of thermal stresses in the stress profiles are directly proportional to the change of uniform temperature.

4. The stresses caused by changes of temperature contribute to the available strengths of the plies. When the combination of stresses in the principal directions in any one ply cause the Tsai–Hill equation to reach unity (Section 6.3.2), the ply is deemed to have failed. For a given thermal loading it is therefore possible to estimate from the (thermal) loading factor by how much the temperature differences must be multiplied in order that the laminate will fail, provided that there is no change of mechanism of failure or any dependence of properties on temperature and any other relevant factors.

5. As we have seen in Chapters 3 to 5, any mechanical loads will induce their own stress profiles in the global and principal directions. The profiles of

mechanical stress in the principal directions will add algebraically to the stresses induced by temperature change, thus changing the load factor to cause failure. Calculation of this combined load factor needs careful thought, as we shall discuss below. But the effect of the combination of thermal and mechanical stresses may well cause a change in mode and location of failure as well as a change of load factor needed to cause failure. It is very difficult to generalize the effect of change of temperature on the strength and mode of failure of a laminate, not least because the effects depend on the relative magnitudes of the thermal and mechanical loads. The following examples illustrate some representative patterns of behaviour in laminates based on the ply EG, for which data are given in Table 7.1.

7.7.2 Strength of laminates under mechanical or thermal loads

Crossply laminate 0/90SG

This balanced regular symmetric crossply laminate is based on four 0.5 mm plies laid up as $(0°/90°)_S$; the stiffness and strength of the cold-cured laminate was discussed in Section 6.5.1. Under an applied loads $N_x = 100$ N/mm, the stress profiles in the principal directions are given in Figure 6.5, and the Tsai–Hill criterion predicts first ply failure in the 90° plies at a load factor of 5.34.

Using the approach outlined in Section 7.5, the effect of increasing the (cold-cured) laminate temperature from 20 °C to 120 °C induces the thermal stresses in the principal directions of $\sigma_1^T = 18.09$ MPa, and $\sigma_2^T = -18.09$ MPa. Thus in the absence of mechanical stress, first ply failure is predicted at a notional thermal load factor of about 14, being controlled by the transverse compressive stresses. This is obviously not a practically useful conclusion, as epoxy resins do not survive sustained heating to 1420 °C, but the principle is important.

The effect of curing the laminate 0/90SG at 120 °C and cooling it to 20 °C induces the stresses $\sigma_1^c = -18.09$ MPa and $\sigma_2^c = +18.09$ MPa. These profiles relate to $\Delta T = -100$ K and therefore have the opposite signs to those thermal stress profiles caused by $\Delta T = +100$ K. Applying the Tsai–Hill criterion to the cure stresses alone suggests a thermal load factor of 5.51, with failure being dominated by the transverse tensile stress. The result means that if the laminate were cooled down by 551 K from the cure temperature, then it would crack up. Although in this example the result is not of practical interest, there are brittle high temperature resins used in polymer fibre composites for good reasons and which, if cured at inappropriate temperatures (very high ones by mistake), will destroy themselves during the cooling down after cure and before they even reach service. It is apparent that a change of sign in a temperature change has a marked effect on both the load factor to cause failure and the mode of failure.

Angleply laminate θ/ − θSG

This set of balanced regular symmetric angleply laminates is based on four
0.5 mm plies of EG. The strength behaviour of these room temperature
laminates under mechanical loads were discussed in Sections 6.6.1 and 6.6.3.
Under a direct load of $N_x = 100$ N/mm the laminate 30/− 30SG failed at a
load factor of 5.57 and the failure was dominated by the large shear stress in the
principal direction.

Raising the temperatures of the cold-cured laminates by 100 K to 120 °C
gives the stresses in the principal directions and the Tsai–Hill thermal load
factors shown in Table 7.4. It is clear that when a cold-cured angleply laminate
is heated during service, the mode and thermal load factor to cause failure
depends on the angle of the lay-up. In these examples at small angles from
above 0° to about 30° and from 60° towards 90° the failure is dominated by
shear. At 45°, because the shear is absent in the principal directions, the failure
of 45/ − 45SG is dominated by transverse compression.

When a hot-cured balanced regular symmetrical angleply laminate is
cooled down by 100 K to 20 °C, the mode and load factor to induce a notional
thermal cure failure also depends on angle. Representative data are shown in
Table 7.5. At angles up to 15° and beyond 75° the failure is shear controlled,
but from 30° to 60° the dominant stress/strength ratio is transverse tension.

Table 7.4 Thermal failure of $\theta/ - \theta$SG under increase of
temperature; $T^o = T^c = 20$ °C; $T^t = T^b = 120$ °C.
$[N] = 0$

θ	$\sigma_1/\sigma_{1\max}$	$\sigma_2/\sigma_{2\max}$	$\tau_{12}/\tau_{12\max}$	Tsai–Hill factor	Mode
15	0.003	0.012	0.07	14.04	S
30		0.044	0.087	10.14	S
45		0.061	0	13.94	TC

Table 7.5 Thermal cure failure of $\theta/ - \theta$SG after cooling
to room temperature; $T^o = T^t = T^b = 20$ °C; $T^c =$
120 °C. $[N] = 0$

θ	$\sigma_1/\sigma_{1\max}$	$\sigma_2/\sigma_{2\max}$	$\tau_{12}/\tau_{12\max}$	Tsai–Hill factor	Mode
15		0.032	0.07	12.9	S
30		0.12	0.087	6.75	TT + S
45		0.181	0	5.51	TT

7.7.3 Strength of laminates under combined thermal and mechanical loads

If in each ply the stresses in the principal directions at any thickness coordinate are $[\sigma_i^T + \sigma_i^m]$, and if the Tsai–Hill load factor is found to be F, then it is understood that failure will occur when each of the values of σ_i^T and σ_i^m are multiplied by F. This means that not only the mechanical loads are increased by F, but also that the temperature difference ΔT is multiplied by F. This would be a very unusual set of operating conditions.

It will now be apparent, therefore, that when a laminate carries both thermal (including curing) stresses and mechanical stresses, care must be taken in applying the Tsai–Hill (or other) failure criterion and interpreting the results.

A more common design question is to fix the thermal stresses (e.g. by curing hot and using cold, or by curing cold and using hotter or colder) and then determine the factor F^m by which the mechanical loads have to be multiplied to induce first ply failure. This means that the thermal stresses have to be separated out. The Tsai–Hill criterion then predicts failure will occur in the identified ply when

$$(\sigma_1^T + F^m \sigma_1^m)^2/(\sigma_{1\max})^2 - (\sigma_1^T + F^m \sigma_1^m)(\sigma_2^T + F^m \sigma_2^m)/(\sigma_{1\max})^2$$
$$+ (\sigma_2^T + F^m \sigma_2^m)^2/(\sigma_{2\max})^2 + (\tau_{12}^T + F^m \tau_{12}^m)^2/(\tau_{12\max})^2 = 1$$

This rather cumbersome expression can be reduced to a quadratic in F^m having the form $aF^{m2} + bF^m + c = 0$. It is worth noting that this gives only a crude estimate for F^m, because quite often there is some uncertainty about whether to use values for tensile or compressive strength when both are present in different proportions in the thermal/cure and mechanical profiles in the principal directions.

It is possible to adopt a similar approach when a laminate carries a given mechanical load and you wish to explore the effect of applying a range of thermal conditions. A further step involves looking at the effect of a range of thermal loads and a different (uncoupled) range of mechanical loads. The computer makes these awkward calculations possible very quickly, but our purpose is to understand and interpret the analysis first and then interpret the results.

In particular, the combination of even modest cure or thermal temperature changes can dramatically downgrade the load-bearing capabilities of the laminates. This is well known in principle and is described in the research literature, but it is unfortunate that documentation in most textbooks is at present rather sketchy.

7.8 HYGROSCOPIC EFFECTS

The discussion so far has ignored the effect of any fluid environment, although it is widely known that some polymers can absorb moisture and other fluids, which can change dimensions marginally or even substantially. We can begin to approach the problem of change in moisture content by using an analogy between temperature change and moisture concentration change, on the basis that they are both transport phenomena. We therefore introduce the concept of change of 'fluid' concentration, ΔC, and the coefficient of hygroscopic expansion, β, in order to account for change in dimension due to the uptake (or loss) of moisture. In this simple account we ignore the effect of moisture on the property under consideration, which is the same assumption we used throughout this chapter for change of temperature; the material is assumed to have stiffnesses and strengths which are independent of temperature or moisture concentration. We also assume here (for simplicity only) that the moisture content is in equilibrium throughout the material under consideration.

On this basis, the stress–strain relationships of an isotropic material take the form:

$$\varepsilon_1 = \sigma_1/E - v\sigma_2/E + \alpha\Delta T + \beta\Delta C$$

$$\varepsilon_2 = v\sigma_1/E + \sigma_2/E + \alpha\Delta T + \beta\Delta C$$

$$\gamma_{12} = \tau_{12}/G$$

The main points are that the solid usually swells on absorption of fluid, and shrinks when moisture is lost, and that changes in moisture content do not affect shear strain.

We shall not explore the application of the analysis in detail, but examples (assumed here to behave as isotropic sheets) where this kind of analysis explains observed phenomena include why a moist piece of cardboard will curl up when left on a flat surface exposed to the sun's heat, why a stale sandwich curls up at the edges, why toast made from moist bread by heating on one side only will adopt synclastic curvature.

For a unidirectional material, it is necessary to distinguish the coefficents of moisture expansion in the two principal directions, so that the stress–strain relationships become

$$\begin{pmatrix} \varepsilon_1 \\ \varepsilon_2 \\ \gamma_{12} \end{pmatrix} = \begin{pmatrix} S_{11} & S_{12} & 0 \\ S_{12} & S_{22} & 0 \\ 0 & 0 & S_{66} \end{pmatrix} \begin{pmatrix} \sigma_1 \\ \sigma_2 \\ 0 \end{pmatrix} + \begin{pmatrix} \alpha_1 \\ \alpha_2 \\ 0 \end{pmatrix} \Delta T + \begin{pmatrix} \beta_1 \\ \beta_2 \\ 0 \end{pmatrix} \Delta C$$

If we rely on the matrix to absorb the moisture rather than the fibres, it follows that $\beta_2 > \beta_1$. These types of data can be used with this simple analysis to explain why new wood shrinks more across the grain than along it, as it dries out to equilibrium moisture content and, by extension to laminate analysis, why plywood is dimensionally more stable than a similar wood modelled as a simple unidirectional block.

The argument can then be extended to account for the effect of moisture on strain in a unidirectional ply aligned at some reference direction θ to the principal directions. Transforming the moisture coefficients using the [T] matrix, as we did for thermal expansion coefficients in Section 7.3.4, we obtain the stress–strain relationships

$$
\begin{pmatrix} \varepsilon_x \\ \varepsilon_y \\ \gamma_{xy} \end{pmatrix} = \begin{pmatrix} S_{11}^* & S_{12}^* & S_{16}^* \\ S_{12}^* & S_{22}^* & S_{26}^* \\ S_{16}^* & S_{26}^* & S_{66}^* \end{pmatrix} \begin{pmatrix} \sigma_x \\ \sigma_y \\ \tau_{xy} \end{pmatrix} + \begin{pmatrix} \alpha_x \\ \alpha_y \\ \alpha_{xy} \end{pmatrix} \Delta T + \begin{pmatrix} \beta_x \\ \beta_y \\ \beta_{xy} \end{pmatrix} \Delta C
$$

FURTHER READING

The most readable introductions to change of uniform temperature in laminates in standard texts are by Jones, Datoo, and Agarwal and Broutman.

Agarwal, B.D. and Broutman, L.J. (1990) *Analysis and Performance of Fibre Composites*, Wiley.

Datoo, M.H. (1991) *Mechanics of Fibrous Composites*, Elsevier.

Hyer, M.W. (1983) Non-linear effects of elastic coupling in unsymmetric laminates, in Z. Hashin and C.T. Herakovich (eds) *Mechanics of Composite Materials: recent advances*, Pergamon Press, Oxford.

Hyer, M.W. (1981) Calculations of the room-temperature shapes of unsymmetric laminates, *J. Composite Materials*, **15**, 296.

Hyer, M.W. (1982) The room-temperature shapes of four-layer crossply laminates, *J. Composite Materials,* **16**, 318.

Jones, R.M. (1975) *Mechanics of Composite Materials*, McGraw-Hill.

Schapery, J. (1968) Thermal expansion coefficients of composite materials based on energy principles, *J. Composite Materials* **2**, 380–404.

8	# Stiffness of thin-walled structures

8.1 INTRODUCTION

The earlier chapters of this book have examined in detail an introduction to how flat panels behave under in-plane loads or moments. The thin flat plate, and (to an approximation) the thin singly-curved panel, is a two-dimensional structural element. It is clearly the fundamental building block for composite structures, and it is therefore essential that the behaviour of the flat plate is well-understood before looking at more complicated structures. Even with the detail studied in the previous chapters, we have not covered all possible types of loading situations – transverse loading of plates is obviously missing – but to cover this would require a much deeper knowledge of the theory of plates and shells than we have space to cover within the scope of Chapter 2 for isotropic materials and in the rest of this book for composites: for further details the reader is advised to consult more advanced texts cited in the bibliography.

The purpose of this chapter is to examine how some simple one-dimensional structures behave under a modest range of representative loading situations. Most of these structures are taken as long and slender, with thin walls, and examples of them include skis, driveshafts, golf club shafts, vaulting poles, pipes or hose, and wind turbine blades. For illustration only we can also use the following approach to understand the elementary mechanics of parts of thin-walled structures such as pressure vessels and aeroplane wings.

Our objective is to develop a method of approach to selected stiffness problems involving laminated construction. Not all situations will be analysed to give exact solutions, but the nature of the main approximations will be identified. The treatment will be indicative rather than comprehensive, and some simple problems are given for the reader to try out. Some of these problems indicate the types of approach needed if quantitative design is to be valid for polymer/fibre composites. The end result is a basis for tackling more advanced structures having more complicated shapes.

Some of the following ideas are developed in the more advanced texts on composites, although usually in a rather abstract form. It would be very satisfactory – though rather ambitious – if this volume provided the reader with a framework for persuing with some confidence the more advanced ideas laid out in the detailed texts and research papers.

Inevitably in a short book we cannot deal with all the complicated situations which arise in practice. In this chapter we shall be concerned with stiffness behaviour; parallel arguments can be developed for strength.

8.1.1 Thin-walled structures

The first question we need to answer is what we mean by a thin-walled structure. Clearly the thickness is much less than the other two dimensions. This includes all the plates we have discussed in the previous chapters. In this chapter we shall impose the further restriction that our thin-walled structures are usually long and slender, so the transverse dimensions are small compared to the length. Examples of these slender structures include pipe and hose, and vaulting poles. We shall also look at some thin-walled structures which are obviously not long and slender, such as pressure vessels. In addition, we shall look briefly at slender structures which consist of thin plates joined at their long edges to form profiles such as I, T, [, hollow square or rectangle, all of which can be efficient in bending. A selection of these is shown in Figure 8.1.

For isotropic materials we have already studied in Chapter 2 the load-deformation behaviour of long slender structures. At that elementary stage we did not dwell on the special problems of thin-walled structures. Thin-walled structures are attractive to the designer because, by proper design, they can offer good stiffness per unit mass, particularly in bending. In addition to the behaviour mentioned in Chapter 2, thin-walled structures offer the following challenges, which call upon more advanced concepts:

1. Shear stresses and strains can assume great importance, and warping can occur.
2. Not only can the slender thin-walled structures buckle overall, but the thin walls can be prone to local buckling under compressive loads. The structures are also sensitive to the nature and magnitude of initial imperfections introduced during manufacture.

We can only comment in outline on these advanced concepts of structural mechanics: they are covered in great detail by Murray (1984).

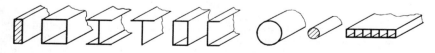

Figure 8.1 Slender profiles.

To these challenges even for isotropic materials, we bring the added challenges of the use of directional composite materials.

8.1.2 Construction of thin-walled structures

The thin-walled structures mentioned above can be made in several ways, depending on their shape, and can have many different internal structures.

Most unreinforced thermoplastics materials can be readily extruded to make isotropic structural elements such as pipes, rod, curtain rails, window frames, rainwater goods, sealing strips, fence posts, and replacements for corrugated card (with or without flat faces).

In composite materials, hollow cross-sections can be filament wound, braided, or pultruded. Other profiles can be pultruded. We shall limit our discussions to straight profiles, but it is possible to pultrude some sections with single curvature along their length. It is also possible to make complicated profiles using prepreg and appropriate moulds or mandrels. Although thin-walled structures can be made by the contact moulding processes, the dimensional accuracy is usually insufficient when substantial loads are to be carried and the imprecision of construction increases the likelihood of buckling under compressive stress.

The details of construction can be varied over a wide range to match the service requirements, and most of the stacking arrangements discussed in previous chapters can be used, though sometimes the full range of effects is

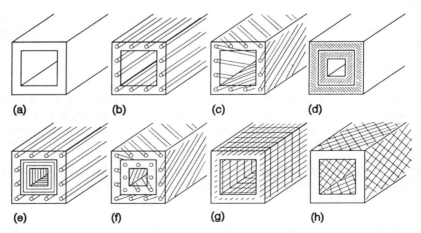

Figure 8.2 Some simple profile constructions: (a) isotropic wall; (b) unidirectional ply along length; (c) unidirectional ply at angle to length; (d) laminated walls of dissimilar plies; (e) laminated walls at 0° and 90° to length; (f) laminated walls at $\pm\theta$ to length; (g) woven cloth plies aligned with axis; (h) woven plies at $\pm45°$ to length.

suppressed by the constraints of the cross-sectional shape. Some simple examples are shown in Figure 8.2, and discussed below.

We shall first discuss the behaviour of isotropic materials, then look at elements based on a single unidirectional ply where the fibres are aligned along the axis of the section (the most obvious achievement in basic pultrusion) or at an angle to it. The next step is to review the behaviour of structures based on laminates using isotropic plies: this will enable us to revise and develop our understanding of laminates and the use of the language of laminates, for both symmetric and unsymmetric laminates, without becoming bogged down by directional effects. Finally, we shall examine the various forms of laminate used in thin-walled structures, including both cross-ply and angle-ply laminates, and some special constructions.

8.1.3 Co-ordinate systems

In the discussion of the behaviour of plates, we used the x direction as a reference (Figure 8.3), and all ply angles were measured with reference to it. In Section 2.5 on beam bending we used y as the longitudinal direction, with z as the depth. In this discussion of thin-walled structures, we shall use x as the longitudinal direction, but it will become apparent (especially in bending) that sometimes we have to use the y direction to denote thickness instead of the z direction.

For example in an I-section (Figure 8.4), the flange A carries bending stresses along the length, x, of the beam, which vary through the thickness, z. In

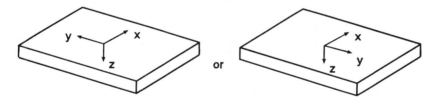

Figure 8.3 Co-ordinate systems for a thin flat plate.

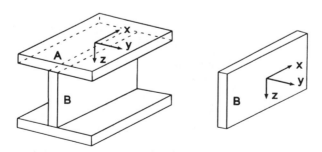

Figure 8.4 Coherent co-ordinate system for elements of an I beam.

the web, B, needed to transmit shear under a moment M_x, the depth of the beam becomes z and the thickness y in the beam co-ordinate system.

This is an unwelcome situation, because using y for thickness confuses earlier discussions. On this account the reader is strongly urged to define carefully the co-ordinates used, and equally carefully to be clear about the significance of directions implied in the stiffness matrices [ABD], and the related compliance matrices [abhd].

8.1.4 Loading conditions and end constraints

In this chapter we shall look at four loading situations: axial tension and compression, internal pressure of round enclosed hollow cylinders, applied torques, and bending of beams. In particular, and recalling remarks made in Section 2.8, we need to beware of the likelihood of buckling, which may occur where compressive stresses are present, e.g. under axial compression of struts (Figure 8.5), in the compression part of a beam, especially with wide flanges or open webs (Figure 8.6), under torsion of tubes where the shear can be decomposed into equal orthogonal tensile and compressive stresses, and in the shear web of a deep beam under bending (Figure 8.7).

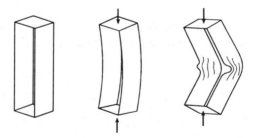

Figure 8.5 Euler and local buckling of hollow slender channel.

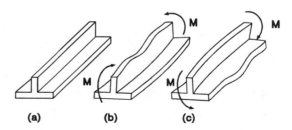

Figure 8.6 Buckling of a T beam under an applied moment; (b) web in compression; (c) flange in compression.

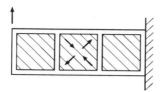

Figure 8.7 Shear buckling in deep thin-walled beam under bending.

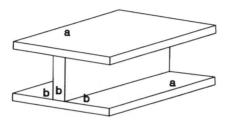

Figure 8.8 Edge conditions in beam elements: (a) free edge; (b) built-in (fixed) at the join.

Our discussion will focus on axial loads, torques and moments acting about the centroid of the section. Transverse loads in beams may be either concentrated at a point, or distributed along a (part) length of the beam.

End conditions need to be identified with some care. We have the usual end conditions for beams: simply-supported or built-in (fixed) ends. In structures made from plate elements fastened at the long edges (Figure 8.8), we usually assume that the joined edges are fixed, and it is this condition which contributes much to the bending stiffness and buckling stability. In wide flanges the free edge is free to deform, and the wider the flange the lower the stiffness becomes – this is especially important in estimates of the local buckling capability of a thin-walled structure where the stresses are compressive.

8.2 IN-PLANE STIFFNESS OF THIN-WALLED STRUCTURES

There are two special cases of practical interest: long structures under axial tension or compression (without buckling), and hollow structures under internal pressure. The effect of both being applied together can be calculated by superposition, unless buckling is likely.

8.2.1 Axial tension of unidirectional materials

In this discussion the structure has a constant cross-section and wall thickness. It is assumed that the axial load acts through the centroid, with the load uniformly distributed over the complete cross-section, and that the longitudinal strain(s) are constant along the length, with stresses free to vary through the thickness in laminate constructions. In slender structures we usually ignore changes in transverse dimensions in elementary analysis of behaviour of isotropic materials, so the problem becomes one-dimensional. In many composite slender structures we may need to consider both transverse and shear deformations, as well as longitudinal deformations, especially where coupling effects occur in laminates.

The deformation caused by axial loads involves calculation of the product of the effective longitudinal modulus, E_{eff}, and the area of cross-section, A. Thus the change in length ΔL_x of a rod of initial length L_{xo} under a load P_x is

$$\Delta L = P_x L_{xo} / E_{eff} A_y \tag{8.1}$$

Calculation of E_{eff} is trivial for an isotropic material, or a single unidirectional ply. It is worthwhile recasting the familiar Equation (8.1) in terms of the lamina(te) compliances [a] – this may seem excessively long-winded for an isotropic material, but it becomes increasingly useful as the complexity of construction of the cross-section develops.

For a unidirectional material having its principal directions aligned with the axis of the structure (Figure 8.9), we can relate the strains to the applied stresses, and the wall thickness (assumed constant) using

$$[\varepsilon^\circ] = [a][N] = [a][\sigma]h \tag{8.2}$$

Under an axial load acting alone, we therefore have

$$E_x = E_1 = E_{eff} = \sigma_x / \varepsilon_x^\circ = [a_{11} h]^{-1} \tag{8.3}$$

and, for a width b, this then leads to

$$E_{eff} A_y = a_{11}^{-1} h^{-1} \cdot bh = b a_{11}^{-1} \tag{8.4}$$

We can also calculate the transverse strain $\varepsilon_2 = a_{12} N_x$, and there are no shear strains because $a_{16} = 0$ for a cross-section where fibres are aligned in the direction of the applied load.

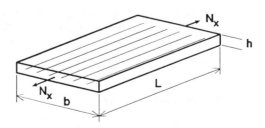

Figure 8.9 A flat plate loaded in its principal directions.

The stress in the axially loaded tiebar or strut is uniform, and simply calculated as P_x/A_y, or for a uniform wall thickness h and contour perimeter c as P_x/ch.

In general it seems straightforward for composites to use the more formalized approach of Equation (8.4) than the traditional approach of (8.1). We shall therefore persue the formalized approach for laminates.

Where a thin-walled slender structure is made from a unidirectional material which is aligned at θ to the axial direction, we know from Section 3.3 that the extensional compliance is fully populated. For such as off-axis unidirectional structure the behaviour is therefore complicated because of the shear which accompanies the application of axial tension. This is not conceptually troublesome for a circular cross-section, because the shear path is neat and continuous round the circumference: the tube merely twists along its axis. An example of such a structure is sold for light-duty hose, and consists of an open coil spring enclosed is a plasticized clear PVC compound. Experiments with a long length of this held vertically at one end, and axially loaded at the free end, confirm that if the mass of the tube is small the strains conform with $\varepsilon_x = a_{11}N_x$, $\varepsilon_y = a_{12}N_x$ and $\gamma_{xy} = a_{16}N_x$.

Where the ends of a thin walled structure are not free to twist or contract laterally, then the constraint must be taken into account. This is especially important in realizing practical designs in composites, or in testing specimens or products.

For an axially-loaded short annular tube with fibres at an angle θ to the axis, suppression of axial rotation by suitable end conditions requires that $\gamma_{xy} = 0$. This is achieved by applying a suitable internal torque N_{xy}, which is calculated from the relevant components of Equation 8.2: $\gamma_{xy} = 0 = a_{16}N_x + a_{66}N_{xy}$, i.e. $N_{xy} = (-a_{16}/a_{66})N_x$, so that the axial strain is now given by $\varepsilon_x = (a_{11} - a_{16}^2/a_{66})N_x$.

For a long annular tube, the constraint at the grips can be treated as localized and can be ignored to a first approximation. This clearly has implications for the size of test specimens to obtain meaningful data, because of the compliance factor $(a_{11} - a_{16}^2/a_{66})$ distinguishing axial strain in long and short specimens.

For a square section tube based on a single ply at an angle to the length direction (Figure 8.10), axial tension in any one side would induce shear if the edges were unbonded. Because edges are unbonded, internal shear forces are needed to maintain structural integrity (the principle of compatibility), and the structure warps out of plane as well as twisting. Open channel sections (Figure 8.11) also show this tendency to warp out of plane.

Axial stiffness of thin-walled laminates structures

The approach closely follows that outlined above, but the stress profile across the wall thickness changes where plies have different values of modulus. Plies

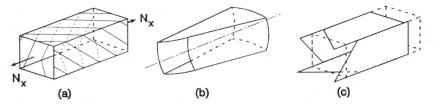

Figure 8.10 Deformation of hollow square tube under axial tension: (a) wound off-axis; (b) warping with bonded edges; (c) shear if edges not bonded.

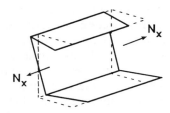

Figure 8.11 Warping of unbalanced angleply out-of-plane under axial tension acting through centroid.

are bonded together and, as we have seen in Chapters 4 to 7, there can be considerable constraint exercised in each ply. It is necessary to calculate the axial and transverse stress profiles because they are the basis for calculating the strength of the structure.

For a balanced symmetric laminate based on isotropic plies, the procedure is to calculate the [A] and [a] matrices using the procedures of Chapter 4. The axial strain response, which is uniform across the wall thickness, is then calculated using Equation 8.2. For a non-symmetric laminate based on isotropic plies (Figure 8.12) under axial tension, the structure will try to bend as

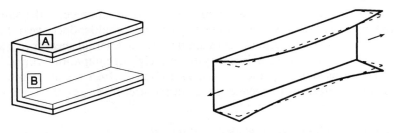

Figure 8.12 Bending at free edges of channel of non-symmetric laminate based on isotropic plies.

well as extend and contract: this is most noticeable at the free edges, and is suppressed where the plate elements are joined together.

There is the special case of a thin-walled tube, made from an unsymmetric laminate, which is under axial loading. The bending stiffness of the tube completely suppresses any bending in the axial and the transverse directions (always provided the tube does not have a very large radius).

For thin-walled structures based on laminates made from unidirectional plies (or woven cloth), the procedures are similar to those outlined above, with the various stiffnesses and compliances outlined in Chapter 5. For angleply constructions there are circumstances where, in flat plates, axial tension may induce twisting as well as bending curvatures. In closed slender thin-walled structures all three curvatures may be suppressed, and this must be allowed for in predicting performance.

There are likely to be no major problems with open sections such as I or channel forms, when under axial tension, provided the wall is made from an isotropic ply, or a unidirectional ply aligned with the axis. However, when using a balanced symmetric crossply or angleply laminate there may be some risk of splitting at the free edges between plies at different angles, because the flanges can be regarded as narrow plates. We shall comment on the use of non-symmetric laminates further in Section 8.2.2.

> **Problem 8.1:** Compare the strain response to an axial tensile load of $N_x = 1$ N/mm of two thin-walled tubes each having a wall thickness of 1 mm and a diameter of 40 mm, and sketch the axial stress profiles. Each tube wall has a total thickness of 0.5 mm steel and 0.5 mm aluminium. Tube A is made from the symmetric laminate SAL4S for which data are given in Table 4.3. Tube B is based on the non-symmetric laminate SAL2NS, for which data are in Table 4.6.

> **Problem 8.2:** A tension member is to be made in the form of a channel section. Which laminate would be best to use: $(0°/90°)_s$ or $(90°/0°)_s$?

8.2.2 Axial compression

We should distinguish the behaviour of bulky low aspect ratio products from those having thin walls. The bulky low aspect ratio products – thick short sections – should not buckle and where this is so the techniques for analysing the effects of compressive loads are the same as those described for tensile loads in Section 8.2.1.

But the reader knows that many structures made from fibre/polymer composite laminates are thin-walled, and we must therefore make some assessment of the likelihood of buckling collapse either by Euler buckling of slender structures or by local buckling.

Simple structures made from a unidirectional material with the fibres aligned in the direction of the compressive load are relatively easy to deal with using the value of modulus relevant to the load direction. Structures made from a unidirectional ply loaded off-axis can in principle be handled in the same way, but the designer should look at end conditions carefully: any constraints could promote a local shearing and perhaps introduce an unexpected eccentricity.

Structures of curved closed cross-sections made from symmetric laminates can also be handled in the same way because the closed form will suppress any out-of-plane bending from coupling effects, remembering that buckling is essentially a bending phenomenon, even though in-plane loads are applied. Carefully-made symmetric crossply laminates should not cause too much difficulty when used in open sections, but the associated twist in open-section angleply laminates may cause structures to buckle under lower loads than those predicted here.

For an unsymmetric laminate, the procedure is in general much more complicated. We have already seen that flat plates based on unsymmetric laminates will show bending/extensional coupling because the [B] matrix is populated. Where elements of flat strips are bonded together at the long edges, e.g. in an I section, then the curvature is suppressed at the joint, but there may be some bending at the free edge of flanges. This may not have too much effect on behaviour under axial tension, but the bending does introduce an eccentricity which could promote a considerably greater chance of initial local buckling under axial compression.

It is therefore good practice to avoid non-symmetric lay-ups where open-edged thin-walled structures are used in compression.

Problem 8.3: A straight vertical pultruded section 100 mm long, 4 mm wide and 2 mm thick, is made from a unidirectional composite having the properties of EG. It is to be tested under axial longitudinal compression in a rig which ensures that the ends are fixed (built-in) with a test length of 50 mm. From Table 6.2 the compliance coefficient for this laminate is $a_{11} = 10.96$ mm/MN. How will it buckle and at what strain?

Problem 8.4: Two identical tubes made from $30/-30$SG with a total wall thickness of 2 mm are 0.1 m long with a mean diameter of 150 mm. Their ends are carefully machined to be exactly perpendicular to the axis of the tube. They are to be tested in axial tension and in axial compression. Data for this laminate are given in Tables 6.31 and 6.32. Estimate the loads at which you would expect failure to occur.

Problem 8.5: Straight circular thin-walled tubes are to be made from symmetric laminates 0/90SG and $+45/-45$SG, for which compliance data are given in Table 5.13. The tubes are 0.5 mm thick, 50 mm diameter and 1 m long. Compare the buckling loads for these tubes when under uniaxial compresive load.

8.2.3 Internal pressure

The procedures outlined in Section 8.2.1 for uniaxial loading of thin-walled sections also apply for thin-walled tubing, the only major difference being that under internal pressure acting alone, the hoop and axial stresses are tensile, and there is no risk of buckling.

We have already discussed in Chapters 3 and 5 the behaviour of some simple thin-walled tubes modelled as flat plates. We looked at special tubes made from a single isotropic or unidirectional ply in Chapter 3, and mentioned the behaviour of tubes made from symmetric laminates in Chapter 5. We shall now introduce the issues which it is necessary to take into account when looking at tubes having nonsymmetric crossply or angleply construction.

Problem 8.6: A thin-walled tube is to be made by a special lamination process from pre-preg based on EG, for which the ply elastic constants are $E_1 = 45.6\,\text{GPa}$, $E_2 = 10.73\,\text{GPa}$, $G_{12} = 5.14\,\text{GPa}$, and $v_{12} = 0.274$. The tube wall is to be modelled as an angleply laminate, and is to be made in two arrangements, each using four 0.25 mm plies: (a) $(30°/-30°)_S$; and (b) $(30°/30°/-30°/-30°)_T$, where angles are measured with respect to the longitudinal axis of the tube. The ABD matrices and their inverses are given below. Compare the stiffness behaviour of both tubes under internal pressure.

(a) Stiffnesses of $(30°/-30°)_S$ based on 0.25 mm EG plies

A (kN/mm)			D (kNmm)		
31.77	8.767	0	2.648	0.7306	0.6887
8.767	14.02	0	0.7306	1.169	0.272
0	0	10.91	0.6887	0.272	0.9095

a (mm/MN)			d (/MNmm)		
38.04	−23.78	0	533.9	−257.7	−327.2
−23.78	86.18	0	−257.7	1044	−117.1
0	0	91.62	−327.2	−117.1	1382

(b) Stiffnesses of $((30°)_2/(-30°)_2)_T$ based on 0.25 mm plies of EG

A (kN/mm) B (kN)			B (kN) D (kNmm)		
31.77	8.767	0	0	0	−2.755
8.767	14.02	0	0	0	−1.088
0	0	10.91	−2.755	−1.088	0
0	0	−2.755	2.648	0.7306	0
0	0	−1.088	0.7308	1.169	0
−2.755	−1.088	0	0	0	0.9095

a (mm/MN) h (/MN)			b (/MN) d (/MNmm)		
47.45	− 20.41	0	0	0	119.3
− 20.41	87.39	0	0	0	42.69
0	0	126	119.3	42.69	0
0	0	119.3	569.4	− 245	0
0	0	42.69	− 245	1049	0
119.3	42.69	0	0	0	1512

Problem 8.7: An experimental design of thin-walled hose for temporary water supply is intended to be made from four 1 mm thick plies of undirectional cord-reinforced rubber having the stacking sequence $(30°/-30°)_s$, where the angle is relative to the hoop direction.

By gross incompetence the stacking sequence of the manufactured hose is found to be $(30°/-30°/30°)_T$. The mean diameter is 120 mm. The transformed compliance matrix [a] for the actual manufactured hose is

$$[a] = \begin{pmatrix} 7.383 & -19.87 & -0.444 \\ -19.87 & 59.31 & 0.091 \\ -0.444 & 0.091 & 2.178 \end{pmatrix} \text{mm/kN}$$

One of the tests for the hose consists of holding a 10 m length of hose vertically by the top end, with the lower end sealed and free to move. An internal pressure is applied to the top end of the hose and changes in deformation of the hose are recorded. Calculate the deformations in the hose and changes in leading dimensions under an internal air pressure of 0.4 MPa, and sketch the likely visual response in a schematic diagram, clearly indicating the alignment of the cords in the hose.

What would be the deformation response at the top of the hose if the hose were pressurised with water to 0.4 MPa?

8.3 TORSION LOADING OF THIN-WALLED STRUCTURES

8.3.1 Closed circular cross-sections

We have already examined in Chapter 2 the behaviour under torsion of rods and thin-walled tubes made from isotropic materials. The governing concept for stiffness was shown to be the torsional stiffness GJ, where G is the shear modulus and $J = 2\pi R_m^3 h$, the polar second moment of area. There is therefore the temptation to transfer this concept into the analysis of stiffness behaviour of thin-walled structures based on a single ply or a laminated construction. However, we need to cover the full behaviour of composites, so it is better to calculate all the strain responses, which then takes account of any interactions not present in isotropic materials.

Using thin walls, we assume that under an applied torque, T, the average shear stress across the wall is constant. For a tube of mean radius R and thickness h, we found in Chapter 2 that the shear stress τ_{xy} is given by $\tau_{xy} = T/(2\pi R^2 h)$, and hence the force resultant N_{xy} is given by

$$N_{xy} = \tau_{xy} h \qquad (8.5)$$

Hence the strain responses can be calculated using

$$\begin{pmatrix} \varepsilon^\circ \\ \kappa^\circ \end{pmatrix} = \begin{pmatrix} a & b \\ h & d \end{pmatrix} \begin{pmatrix} N \\ M \end{pmatrix} \qquad (8.6)$$

For isotropic materials, or orthotropic materials aligned with the axis of the tube, there is no coupling between shear stress and direct strains. But we have already seen that for a unidirectional ply loaded in-plane off-axis, or for unbalanced angleply laminates, there is shear-extension coupling, which needs to be taken into account. For non-symmetric lay-ups there are the challenges of suppression of bending-extension coupling.

Once the strains, $[\varepsilon]$, have been determined, the associated stresses $[\sigma]_f$ in each ply can be calculated for the usual way using

$$[\sigma]_f = [Q]_f [\varepsilon] \qquad (8.7)$$

Again it cannot be overemphasized that all the stresses need to be calculated if estimates of strength are to be made: failure may well not coincide with the highest stress calculated because of the severe anisotropy of the plies.

Problem 8.8: A vertical tube, suspended from the top end, is made from the non-symmetric laminate C + 4NS. The laminate is 0.5 mm thick, 400 mm long, and has a mean diameter of 50 mm. In the following experiments the gauge length is set at 300 mm, and the 50 mm ends are to be ignored. Calculate the change in length and diameter of the gauge length when the tube is under: (a) a uniaxial longitudinal load N_y; (b) an internal pressure p represented by $N_x = 2N_y$; (c) What is the angle of twist induced when the free ends of the tube react to opposing couples N_{xy}?

Problem 8.9: A tube 50 mm mean diameter, 800 mm long and 0.5 mm thick is made from four 0.125 mm plies of HM-CARB. Tube A is laid up as $(-45°/+45°)_S$, and tube B as $(-45°/-45°/+45°/+45°)_T$, for which the compliance matrices are as follows:

Tube A:

[a] mm/kN			[d]/kNmm		
110.3	−89.67	0	5738	−3862	1321
−89.67	110.3	0	−3862	5738	1321
0	0	43.49	1321	1321	3947

Tube B:

	a (mm/MN) h (/kN)			b (/kN) d (/kNmm)	
127.9	−72.07	0	0	0	−313.3
−72.07	127.9	0	0	0	−313.3
0	0	116.6	−313.3	−313.3	0
0	0	−313.3	6141	−3459	0
0	0	−313.3	−3459	6141	0
−313.3	−313.3	0	0	0	5595

Compare the stiffness response of each tube when a torque of 30 Nm is applied, and calculate the angle of twist.

8.3.2 Non-circular sections

When a torque is applied to the ends of a rectangular hollow section, warping occurs, that is, a plane cross-section no longer remains plane. This is a geometrical effect which is independent of the type of material used: it occurs even in isotropic materials where the coupling effects of composite mechanics are not present.

8.3.3 Torsion of open sections

Where the shear path is not closed, the shear completes its path by flowing along the long open edge, as shown in Figure 8.13. The result is a shear between the long open faces. The shear stiffness of the open section is much less than that of a similar but closed section, as handling the profiles used to bind reports or to hang picture posters will confirm. Detailed analysis for torsion of open sections is given by Murray for isotropic materials and by Datoo (1991) for composite materials.

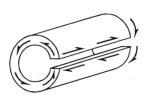

Figure 8.13 Development of shear in open section under applied torque.

8.4 BENDING OF THIN-WALLED BEAM STRUCTURES

8.4.1 Introduction

We have already discussed elementary aspects of stresses and deflections in beams made from isotropic materials in Chapter 2. What we shall now do is to introduce some of these ideas using the language of composites. This will provide some familiarity with the simple concepts before we later discuss the behaviour of beams made from certain classes of composite materials. The emphasis throughout will be to introduce an approximate theory on the basis of which the performance of beams can usefully be estimated; more exact theory will be found in books such as Datoo (1991) and Tsai (1988).

It is worth re-reading the main assumptions made in Chapter 2 about simple bending theory. Beams are long slender structures where the width, b, and the depth, h, are small compared with the length, L. In this discussion beams will be loaded in the (x, z) plane, so that curvature arises in the (x, z) plane. Loads and moments are assumed to act through or about the centroid of the beam.

8.4.2 Isotropic materials

Many of the simplest beam problems involve relating bending stresses and bending deflections to applied moments or transverse loads. This section revises elementary beam theory with the objective of introducing the terminology of laminates for later use.

Solid rectangular cross-section

Following the methods outlined in Chapter 2, we can express bending stresses in terms of applied full bending moments $M_b(x)$ and the second moment of area about the centroid of the beam, I:

$$\sigma_b(z) = M_b(x)z/I$$

Clearly the maximum value of bending stress in an isotropic material always occurs at the axial co-ordinate of maximum value of bending moment at the surfaces of the beam, $z = \pm h/2$:

$$\sigma_{bmax}(\pm h/2) = \pm M_{bmax}(x)h/2I$$

Bending deflections $w_b(x)$ are related to the applied moment $M_b(x)$ or the transverse load $F_z(x)$ by

$$w_b(x) = k_m M_b(x)L^2/EI \quad \text{or} \quad w_b(x) = k_F F_z(x)L^3/EI$$

where k_m and k_F are (different) dimensionless constants which depend on end conditions and the nature and variation with x of the moment or load (Section 2.7.1).

For geometrical reasons, the strain profile through the depth of the beam is linear $(\varepsilon_x(z) = z\kappa_x)$, and within any given material the stress profile varies linearly with depth co-ordinate because $\sigma_x(z) = zE\kappa_x$. There will however be a discontinuity in stress at the interface between materials having different moduli.

The stiffness of a beam is given by EI and, to relate the concepts of composite mechanics to beam design, we seek relationships for both E and EI. Ignoring the width (y) dependence of stiffness terms, we have for a symmetric beam $\varepsilon_x = a_{11} N_x$ and $\kappa_x = d_{11} M_x$. Because we ignore the dependence of terms on y, we may also use $N_x = A_{11}\varepsilon_x$ and $M_x = D_{11}\kappa_x$, so that in this special case of a symmetric beam $A_{11} = a_{11}^{-1}$ and $D_{11} = d_{11}^{-1}$. (This is not true for symmetric plates.)

Thus for a isotropic beam $E = \sigma_x/\varepsilon_x = (ha_{11})^{-1} = A_{11}/h$. Because the full bending moment M_b is related to the bending moment per unit width M_x by $M_b = bM_x$, we have $M_b = bD_{11}\kappa_x = EI\kappa_x$, so $EI = bD_{11}^{-1}$. In terms of laminate concepts we find the bending deflection profile can be written as

$$w_b(x) = k_F F_z(x)L^3/bD_{11} = k_F F_z(x)L^3 d_{11}/b$$

For a cantilever loaded at the free end, $k_F = \tfrac{1}{3}$ gives the maximum deflection at the free end: $w_{bmax} = F_z L^3 d_{11}/3b$.

Note the D_{11} and d_{11} taken from the stiffness and compliance matrices for full isotropic plates will be based on $Eh^3/(1 - v^2)$ and will introduce an error factor of $(1 - v^2)$ when used in beam calculations. In laminated composites based on unidirectional plies having a high anisotropy ratio, this problem is minimized because $v_{12} \gg v_{21}$, so $(1 - v_{12}v_{21}) \approx 1$ to a good approximation.

Where a shear modulus in beam co-ordinates is required, the application of the in-plane shear stress τ_{xz} alone will give $\gamma_{xz} = ha_{66}\tau_{xz}$, and hence $G_{xz} = (ha_{66})^{-1}$.

It cannot be overemphasized that the co-ordinate system must be identified with extreme care: it is most unwise to rush into calculations without checking meticulously that the stiffnesses [A] and [D], or the compliances [a] and [d], correspond with the intended co-ordinates.

Simple laminated beams of rectangular cross-section

The dimensions of the cross-section of one simple laminated beam are shown in Figure 8.14: For this three-ply beam, which is made from a core B and two skins A laid parallel to the midplane of bending, the bending stiffness $EI = E_A I_A + E_B I_B$ is given by

$$E_A I_A = E_A [b(h_A + h_B)^3/12 - bh_B^3/12]$$
$$E_B I_B = E_B bh_B^3/12$$

The strain profile $\varepsilon_x = z/R_x$ is linear and continuous, and the stress profiles

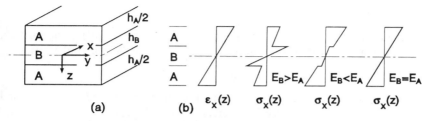

(a)   (b) $\varepsilon_x(z)$ $\sigma_x(z)$ $\sigma_x(z)$ $\sigma_x(z)$

Figure 8.14 Stress and strain profiles in beam made from isotropic plies.

$\sigma_x = zE/R_x$ are linear but have a discontinuity at the interface between materials A and B if the materials have different values of modulus, as shown in Figure 8.14.

For a composite beam the steps for the calculation of maximum bending stress are necessarily more involved than those for an isotropic material. In particular, having seen how stresses in laminated flat plates can vary dramatically through the thickness, the most damaging highest values of stresses (and combinations of stresses) are not necessarily at the surface (i.e. furthest from the midplane). For laminates of isotropic plies, see the example of $\sigma_1(z)$ for laminate SAL2NS under $N_1 = 20\,\text{N/mm}$ in Section 4.6, and see some of the stress profiles in the principal directions for angleply laminates.

The first step towards calculation of stress profiles is to determine the maximum curvature in the beam using $\kappa_{max} = M_{max}/EI = M_b d_{11}/b = M_b/bD_{11}$. The second step is to calculate the stress in each ply using $\sigma_{xf} = [Q_{11}^*]_f [\kappa_{max}]z$, where z is measured from the centroid of the laminate (which may not be the midplane).

A different arrangement of the three ply beam is shown in Figure 8.15, where plies are laid up perpendicular to the plane of bending. The stiffness of the section about the y axis is $EI = E_A b_A h^3/12 + E_B b_B h^3/12$, and the bending strain and stress profiles are (in the z direction) linear and continuous, as sketched in Figure 8.15.

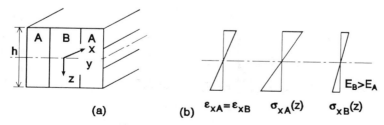

(a)   (b) $\varepsilon_{xA} = \varepsilon_{xB}$ $\sigma_{xA}(z)$ $\sigma_{xB}(z)$

Figure 8.15 Stress and strain profiles in beam made from isotropic plies.

Hollow rectangular section

For the horizontal hollow rectangular section (Figure 8.16(a)) made from an isotropic material, the second moment of area about the centroid $z = 0$ is

$$I_y = bh^3/12 - (b - 2h_2)(h - 2h_1)^3/12$$

The stiffness of the section is $E_x I_y$ and the stress σ_x caused by an applied moment M_x (or vertical force F_z) is given by $\sigma_x = M_x z/I_y$.

An alternative approach, which is necessary for composite materials, is to analyse the elements shown separately in Figure 8.16(b). The end result for the second moment of area about the original centroid is the same of course, but looks less elegant. Using the parallel axes theorem, we find:

$$I_y = 2bh_1^3/12 + 2bh_1(h - h_1)^2/4 + 2h_2(h - 2h_1)^3/12$$

Now let us suppose (Figure 8.16(b)) that the horizontal walls of the section are made from material A and the vertical walls from material B. The second moment of the section is the same as for a single material, but the stresses and curvatures are different. The two elements have the second moments about the y axis (composite) centroid

$$I_{yA} = 2bh_1^3/12 + bh_1(h - h_1)^2/4 \quad \text{and} \quad I_{yB} = 2h_2(h - 2h_1)^3/12$$

The strain is linear through the depth of the beam:

$$\varepsilon_x = z/R_x$$

The stress profile is linear within each material because

$$\sigma_x = zE/R$$

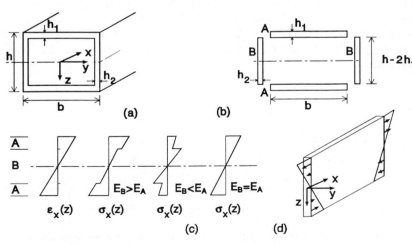

Figure 8.16 Decomposition of hollow section and stress profiles.

but there is a discontinuity in stress at the interface between materials A and B if there is a change of modulus, as shown in Figure 8.16(c). If the beam is made from only one material, then the stress profile will be linear and continuous. The argument will be similar for an I-section beam.

The reader will notice that the bending stress profile in the web may be represented in plate terms by an in-plane stress $\sigma_x(z)$, as shown in Figure 8.16(d).

8.4.3 Beams based on unidirectional plies

Beam of unidirectional material

For a unidirectional composite beam such as an I section made by extrusion, with fibres aligned along the axis of the beam, we may use the customary design formula for calculating bending stiffness, where the longitudinal modulus E_1 is appropriate for bending stress and the in-plane shear modulus G_{12} for shear. The global co-ordinates (x, y, z) for the beam are shown in Figure 8.17.

For a beam where the unidirectional material is aligned at some angle θ to the longitudinal axis, the deflections under a transverse load depend not only on the modulus along the beam axis, but also on the direction of the load. A load transverse to the thickness of the laminate will also induce twisting curvature, whereas a transverse load in the plane of the laminate will not twist.

Problem 8.10: A horizontal cantilever beam 30 mm wide, 2 mm deep and 100 mm long is made from laminate 00G, with fibres aligned along the beam. The bending compliance (Table 6.32) is $d_{11} = 32.89$ /MNmm. Calculate the vertical deflection under a vertical end load of 1 N passing through the centroid of the cross-section.

Problem 8.11: A long slender beam of uniform rectangular cross-section is to be made from a unidirectional array of glass fibres bonded together with a thermosetting resin. Estimate the optimum volume fraction of fibres

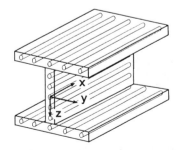

Figure 8.17 An I beam made from unidirectional material.

necessary to achieve the best bending stiffness per unit weight of composite. Outline the assumptions you make in your analysis.

Problem 8.12: A horizontal cantilever 10 mm wide, 2 mm deep and 100 mm long is made from laminate 30G with fibres aligned in the plane at 30° to the axis of the beam. In-plane and bending compliances are given in Tables 6.31 and 6.32. Calculate the leading deformations in the beam when a load of 1 N is applied at the free end through the centroid of the cross-section: (a) vertically; (b) horizontally.

Balanced symmetric crossply and angleply laminates

Deformations in beams made from flat narrow sections of symmetric crossply or angleply laminates can be analysed in a similar manner to that outlined in the previous subsection. If the beam thickness is in the laminate z direction, and the beam length is in the x direction then we can make the following observations. In a symmetric crossply laminates, load F_z and F_y will induce deflection but no twist. In a symmetric angleply laminate F_z will induce deflection w and twist per unit length ϕ, and load F_y will induce deflection v only (if shear is ignored).

For beams aligned as above but made from non-symmetric laminates the behaviour is rather more complicated. In a non-symmetric crossply, under load F_z the deflection w is accompanied by an elongation of the neutral axis; and under F_y there is deflection v with a small amount of shear. In a non-symmetric angleply, under F_z we expect curvature to achieve deflection w but the beam also seeks to shear in its plane because of b_{16}; under F_y we also develop deflection v and some twisting curvature caused by h_{26}.

Design of built-up laminated beams

A sandwich beam (Figure 8.18(a)) may consist of a core (C), two facing skins (A), and two vertical sides (B). The core may be a uniform foam. The skins and side walls may be a multi-ply balanced symmetric laminate construction made by pultrusion or filament winding.

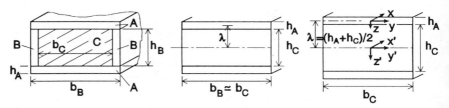

Figure 8.18 (a) Built-up sandwich beam; (b) sandwich without vertical side walls; (c) co-ordinate system.

The co-ordinate directions chosen for either the flanges or the web are likely to have different nomenclature from those customary in plate theory, as already explained.

To a first approximation, the bending stiffness may be calculated neglecting the small contribution from the sidewalls. Model the beam as a symmetrical sandwich ACA (Figure 8.18(b)), where the centroid of the skins, a, is at a distance $\lambda = (h_A + h_c)/2$ from the midplane.

Let us look first at the bending stiffness of the core (Figure 8.18(c)): this is simply given by $E_c I_y = E_c b h_c^3/12$.

Now we look at the skin, where the arguments are more subtle. The top skin, using its co-ordinate system about its midplane, is under a stress $\sigma_x(z)$, so ignoring shear stress we have $\varepsilon_x = h a_{11} \sigma_x$, so $E_x = (h a_{11})^{-1}$, and $\kappa_x = d_{11} M_x$, so $E_x I_y/b = (d_{11})^{-1}$. The bending stiffness of any one facing skin is $E_x I_y = b/d_{11}$. The bending stiffness of one skin about the neutral axis of the built up beam at $z' = 0$ is

$$E_x I_{y'} = E_x I_y + E_x \lambda^2 A_A = b/d_{11} + \lambda^2 b/a_{11}$$

The total bending stiffness of the top and bottom skins is therefore

$$(E_x I_{y'})_A = 2b/d_{11} + 2\lambda^2 b/a_{11}$$

The first term represents the bending stiffness of the skin about its own midplane, and the second term the bending stiffness of the skin about the midplane of the built-up beam. Clearly the greater the separation of the skins, the greater the domination by the second term.

We can now compare the approaches we have described. Consider a sandwich beam 10 mm wide consisting of two layers of mild steel each 0.2 mm thick separated by a foam core 0.6 mm thick. The elastic constants are $E_s = 208\,\text{GPa}$ and $E_f = 70\,\text{MPa}$.

Using conventional solid body mechanics we calculate the bending stiffness as

$$
\begin{aligned}
EI &= b(E_s h^3 - E_s h_f^3 + E_f h_f^3)/12 \\
&= (0.01)(208 \times 10^9 (10^{-9} - 2.16 \times 10^{-10}) + 7 \times 10^7 \times 2.16 \times 10^{-10})/12 \\
&= 0.1359\,\text{Nm}^2
\end{aligned}
$$

Using the approach of laminate analysis we need to find

$$EI = b(2/d_{11} + 2\lambda^2/a_{11} + E_f h_f^3/12)$$

For a mild steel sheet 0.2 mm thick, $E_1 = E_2 = 208\,\text{GPa}$, $v_{12} = 0.28$, $G_{12} = 81.3\,\text{GPa}$, and hence the [ad] matrix is

[a] mm/kN			[d] /kNmm		
24.04	−6.731	0	7212	−2019	0
−6.731	24.04	0	−2019	7212	0
0	0	61.5	0	0	18450

We see that for a steel sheet 0.2 mm thick we have

$a_{11} = 24.04 \times 10^{-9}$ m/N; $d_{11} = 7.212$ /Nm

$EI = (0.01)(2/7.212 + 2 \times 16 \times 10^{-8}/24.04 \times 10^{-9} + 7 \times 10^{7} \times 2.16 \times 10^{-10}/12)$
$\quad = 0.1359$ Nm^2

As expected both methods give the same result.

Problem 8.13: A mild steel tube 20 mm mean diameter and 1 mm wall thickness satisfies service requirements for bending and torsional stiffness. There are some other characteristics, particularly environmental resistance and maintenance, which encouraged the designer to undertake a partial feasibility study for replacing the steel by a composite tube. In this book he finds sufficient data to estimate leading dimensions for tubes made from HM-CARB high modulus carbon fibre/epoxy prepreg and EG prepreg. Bot prepregs are available 0.125 mm thick. What lay-up would you consider for each replacement composite?

Data for steel are in Table 4.1, for EG (2 mm thick) in Table 6.31, and for HM-CARB (0.5 mm thick) in Table 5.13.

8.4.4 Shear stresses and deflections

For isotropic materials we found in Chapter 2 that the shear stress in a rectangular cross-section beam under some transverse load has a parabolic profile through the thickness from zero at the upper and lower surfaces, with a maximum at the neutral plane. The shear deflection of a beam under transverse load has the general form

$$w_s = k_s F_z L/bhG$$

where k_s is a dimensionless constant which takes account of the end conditions and the nature and variation with x of the applied load. Values of k_s for some common situations are given in Section 2.7.3.

For a rectangular section beam the ratio of deflections in shear and in bending is

$$w_s/w_b \propto (E/G)(h^2/L^2)$$

The conclusion for isotropic materials is that shear deflections in long slender beams are usually negligible compared with bending deflections.

Whereas for isotropic materials E is typically 2.5 to 3 times G, for composite materials there are many practical circumstances where $E_x \gg G_{xy}$, so that for the same beam dimensions and loading, w_s may even be comparable with w_b. It is good practice therefore to check shear deflections in beams made from composite materials, especially when the span/depth ratio is small.

Problem 8.14: Show that $E_x/G_{xy} = a_{66}/a_{11}$, and compare values of this ratio for the ply data in Chapter 3, and the laminate data in Chapter 5.

FURTHER READING

Datoo, M.H. (1991) *Mechanics of Fibrous Composites*, Elsevier.

Murray, N.W. (1984) *Introduction to the Theory of Thin-walled Structures*, Oxford University Press. (Isotropic materials only)

Tsai, S.W. (1988) *Composites Design* (4th edn), Think Composites.

Vinson, J.R. and Sierakowski, R.L. (1986) *The behaviour of Structures Composed of Composite Materials*, Nijhoff.

Appendix: outline answers to selected problems

CHAPTER 1

Outline 1.2: The area of one fibre end is $A_e = \pi D_f^2/4$; the cylindrical area is $A_c = \pi D_f L_f$. We seek $2A_e = 0.01A_c$: $L_f/D_f = 50$.

Outline 1.3: We may assume that the surface of the ends of the glass fibres is negligible compared with the cylindrical area along the length. A mass of glass M occupies a volume $V = M/\rho$, and the length of the fibres is $V = \pi D^2 L/4$. The area of the cylindrical surface of the fibres is $A = \pi D L$. Combining these expression, we find the surface area per unit mass is $A/M = 4/\rho D = 157\,\text{m}^2/\text{kg}$, and hence the surface area of 20 g is about $3.14\,\text{m}^2$. The volume of the glass fibres is a mere $V = M/\rho = 0.02\,\text{kg}/(2540\,\text{kg/m}^3) = 7.874 \times 10^{-6}\,\text{m}^3 = 7.874\,\text{ml}$.

Outline 1.4: For hexagonal packing $V_f = \pi\sqrt{3}/6 = 90.7\%$. For square packing, $V_f = \pi/4 = 78.5\%$.

Outline 1.5: It is common in production to express quantities in terms of weight fractions w_i, whereas in analysis and design the term volume fraction V_i is used. For the composite as a whole, or for any individual constituent, the two quantities are related by density: for a weight of material W_i of density ρ_i in a composite mass W_c, the weight fraction of i is $w_i = W_i/W_c$. If the volume of i is V_i, then $W_i = \rho_i V_i$ and the volume fraction is defined as $V_i = V_i/V_c$.

Thus we have $w_f = W_f/W_c = W_f/(W_f + W_b + W_{cc})$, and hence

$$V_f = (w_f/\rho_f)/(w_f/\rho_f + w_p/\rho_p + w_{cc}/\rho_{cc}) = 0.139$$

Outline 1.6: (a) regular balanced non-symmetric; (b) regular symmetric unbalanced; (c) no terms apply; (d) symmetric for purposes of analysis; (e) no terms apply.

Outline 1.7: Resin-rich interstices in cloth; bending of rovings as they interlace each other reduces strength because of the off-axis loading effect.

Outline 1.8: (a) The precise answer depends on the number of fibres through the thickness of the prepreg tape. Where this is large, typically 10–12, laying prepreg crossply or parallel will sensibly give the same volume fraction of fibres in the laminate as in the prepreg. Where fibres are crossed in single layers, $V_f = 3\pi/16 = 58.9\%$.

(b) An element of plain woven cloth is shown in Figure A1.1. The volume of the element is $4\alpha^2 r^3$. By Pythagoras $R^2 = \alpha^2 r^2/4 + (R + r)^2$, so $R = (r/2)(1 + \alpha^2/4) = \cos^{-1}(R - r)/R$. The volume of a complete toroid is $v = 2\pi^2 R r^2$, so the volume of one fibre AB in the element is $(2\theta/2\pi)2\pi^2 R r^2$. There are two such fibres in the element. Hence the volume fraction of fibres is $V_f = \theta\pi R/\alpha^2 r$. For three diameters spacing, $\alpha = 6$, $R = 5r$ and $V_f = 0.281$.

For the closest possible plain weave based on single radially stiff rigid filaments, $\alpha = 2\sqrt{3}$, $\theta = \pi/4$, hence $V_{f\max} = 0.411$, but such a fabric (if it could be made at all) would behave more like an isotropic sheet than a fabric, and would squeak if any forces were applied to it.

Outline 1.9: Figure A1.2 shows the tightest possible plane-knit structure assuming the fibres remain circular with radius r. There is symmetry about O_1O_2 of the loop having the centreline A through J, which interlaces with an identical loop.

In the plane of the large diagram angle $FGO_1 = \phi = \cos^{-1}(2/3) = 48.2°$. Hence $O_1O_2 = 2(3r \sin \phi) = 4.47r$. So the volume of the parallelopiped cell is $4r\cdot 8r\cdot(8r + 4.47r) = 399r^3$. Angle $FO_1E = 180° - 2 \times 48.2° = 83.6°$.

The volume occupied by the two fibre loops in the unit cell is not easy to estimate precisely. The roughest (under-) estimate assumes that the half-loop CH is only located in the plane and subtends an obtuse angle $CO_1H = 4 \times 48.2° + 83.6° = 276.4°$. The volume of four such half-loops is then $4 \times (276.4/360) \times 2\pi^2 (3r)r^2 = 182r^3$, giving a volume fraction of fibres $V_f = 182r^3/399r^3 = 0.456$.

This is an underestimate because segments of the type FG are actually longer because they spiral out-of-plane. The in-plane straight length of FG is $3r$ and

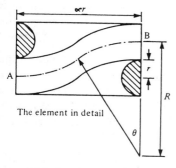

The element in detail

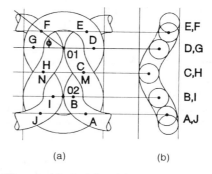

(a)   (b)

Figure A1.1 An element of woven cloth.   **Figure A1.2** A tight plain-knit structure.

G is displaced out-of-plane by r from the plane of F, so very crudely we may calculate the out-of-plane straight length of F as $\sqrt{(10)}r$. So a toroidal element in the plane FGO_1 subtends the angle $FO_1G = 2\sin^{-1}((\sqrt{10})/6) = 63.6°$, and hence has a part-toroidal volume $V_{FG} = (63.6/360)\cdot 2\pi^2\cdot 3r^2 = 10.46r^3$. More straightforwardly, the toroidal element EF has a volume $V_{EF} = (83.6/360)\cdot 2\pi^2\cdot 3r^3 = 13.75r^3$. Within the unit cell, which contains two loops, we have four segments EF and sixteen segments FG. Hence the volume of the fibres is more closely represented as $4V_{EF} + 16V_{FG} = 222r^3$. The maximum volume fraction in this structure, which is locked rigid, is $222r^3/399r^3 = 0.556$.

A conventional loose-knit structure stretches readily under modest load because the curved loops unbend, as we shall discuss in Section 5.2.3. The model structure in this problem is so tight that it cannot unbend. So the volume fraction of fibres in a normal knitted structure is much less than that calculated here. However most knitted structures are based on bundles of fibres, the cross-section of which distorts by flattening and spreading at contact points between adjacent loops, and this dramatically alters the maximum achievable volume fraction of fibres.

Outline 1.10: (a) Regular balanced antisymmetric; (b) regular balanced antisymmetric; (c) regular unbalanced symmetric; (d) regular balanced antisymmetric; (e) regular balanced symmetric; (f) balanced symmetric (regular only if the 90° material is made from two separate plies of thickness $h/4$); (g) regular balanced symmetric; (h) regular balanced non-symmetric; (i) regular balanced non-symmetric.

CHAPTER 2

Outline 2.4: $p = \rho gh = 2500 \times 9.81 \times 20 = 500\,\text{kPa}$ (compressive).

Outline 2.5: The area of the wall A is $30\,\text{m}^2$ so the force $F = pA = 30 \times 1000 = 30\,\text{kN} = 3$ tonnes force uniformly distributed. Even though the pressure is very small, the force is large, so the ducting (and its supports) will have to be able to resist this force, especially at the edges of the sheeting. Bending deflections in thin-walled ducting can be very large unless well-supported at suitable intervals.

Outline 2.6: Regard the sphere as thin-walled, thickness h and mean radius R_m. Cut the sphere across a diameter and blank off one cap face with a flat plate. The force on the flat plate from the pressure p inside the half-sphere is approximately $F = \pi R_m^2 p$. The force exerted by the stress in the wall of the sphere is $F = \sigma\cdot 2\pi R_m h$. For equilibrium the forces must balance, hence $\sigma = pR_m/2h$. Substituting numerical values, $\sigma = (1 \times 100/2 \times 10) = 5\,\text{MPa}$.

Outline 2.7: (a) if the shear forces are applied to the large faces of the block, then the shear stress in the plane is the same in both blocks: $\tau = 100/(0.1 \times 0.1) =$

$10000 \, \text{N/m}^2$. (b) if the shear forces are applied to narrow opposite faces, then $\tau_1 = 100/(0.1 \times 0.01) = 100\,000 \, \text{N/m}^2$ and $\tau_2 = 100/(0.1 \times 0.02) = 50\,000 \, \text{N/m}^2$.

Outline 2.8: No. The analysis in Section 2.3.4 shows that the forces are related to the length of the edge of the element over which they are applied, and hence $F_y/dy = F_z/dz$.

Outline 2.9: Faces ABFE and DCGH carry stresses due to the torque. AEHD and BFGC carry complementary shear stresses. ABCD and EFGH are stress-free.

Outline 2.10: Shear stresses on ABC and DEF are caused by the applied torque. ACFD and ABED carry complementary shear stresses. CBEF is stress free.

Outline 2.11: Those faces which carry stress do not (in principle) distort, and those faces which do distort are stress free.

Outline 2.12: CBEF is an unloaded face and distorts when the rod is twisted about its longitudinal axis. ABED, ACFD, ABC and DEF carry stress and are notionally unstrained.

Outline 2.13: $90° = 1.571^c$. $\quad \gamma_{max} = \pm R\theta/L = 0.01 \times \pm 1.571/0.4 = \pm 0.0393$, i.e. $\gamma_{max} = \pm 3.93\%$. We assume that the neck does not buckle during this large twist, and the shear strains are small so that $\tan \gamma = \gamma$.

Outline 2.14:
(a) $F_x = 50 \, \text{kN}$, hence $\sigma_x = 50 \times 10^3/(1 \times 8 \times 10^{-3}) = 6.25 \, \text{MPa}$

$$\varepsilon_x = \sigma_x/E = 6.25 \times 10^6/3 \times 10^9 = 2.08 \times 10^{-3} = 0.208\%$$
$$\varepsilon_y = -v\varepsilon_x = -0.0729\%$$
$$L_x = L_o(1 + \varepsilon_x) = 1(1 + 0.00208) = 1002.08 \, \text{mm}$$
$$L_y = L_o(1 + \varepsilon_y) = 1(1 - 0.000729) = 999.27 \, \text{mm}$$

(b) $F_{xy} = 30 \, \text{kN}$, hence $\tau_{xy} = 30\,000/(1 \times 8 \times 10^{-3}) = 3.75 \, \text{MPa}$

$$\gamma_{xy} = \tau_{xy}/G = 3.75 \times 10^6/1.11 \times 10^9 = 0.00338 = 0.338\%$$

L_o remains unchanged at $1000 \, \text{mm}$
Change of angle is 0.00338 radians, i.e. about $0.194°$.
(c) $F_x = F_y = 50 \, \text{kN}$. $\sigma_x = \sigma_y = 6.25 \, \text{MPa}$

$$\varepsilon_x = \sigma_x/E - v\sigma_y/E = 0.65 \times 6.25 \times 10^6/3 \times 10^9 = 0.001354 = 0.1354\%$$

$L_x = L_o(1 + \varepsilon_x) = 1001.354 \, \text{mm} = L_y$

Outline 2.15: The shear stress is independent of the direction of loading: $\tau = F/A = 6000/(0.1 \times 0.15) = 4 \times 10^5 \, \text{N/m}^2$. $\gamma = \tau/G = 0.1333$. The in-plane displacement $u \approx \gamma h = 0.1333 \times 0.01 = 1.333 \, \text{mm}$.

Outline 2.16: (a) False. The hoop strain describes fractional increase in diameter. Using Hooke's law we find $\varepsilon_H = \sigma_H/E - v\sigma_A/E = (pR/Eh)(1 - v/2)$, from which we can see that doubling the pressure doubles the hoop strain but not the diameter, because the new diameter is $2R(1 + \varepsilon_H)$. (b) True, because $\sigma_H = pR/h$. (c) True, see (a). (d) False. The original enclosed volume is approximately $V_o = \pi R^2 L$.

The new volume V is $V = \pi R^2 (1 + \varepsilon_H)^2 L (1 + \varepsilon_A)$. Assuming small strains so we can neglect products of small quantities, the fractional increase in volume of the vessel is $\varepsilon_v = \varepsilon_A + 2\varepsilon_H$. From Hooke's law we find $\varepsilon_A = (pR/Eh)(\frac{1}{2} - v)$, hence $\varepsilon_v = (pR/2Eh)(5 - 4v)$. (e) False for almost all materials. $\varepsilon_H/\varepsilon_A = (2 - v)/(1 - 2v)$. This expression only has the same value as σ_H/σ_A when $v = 0$, which is true for some low density flexible foams.

Outline 2.17: (a) Assume here that the polymer behaves elastically. The mean diameter is $D_m = 0.775$ m. The cross-sectional area $A = \pi D_m h = \pi \times 0.775 \times 0.025 = 6.09 \times 10^{-2}$ m². The tensile stress is $\sigma = F/A = 4 \times 10^5/6.09 \times 10^{-2} = 6.57 \times 10^6$ Pa. The tensile strain is $\varepsilon = \sigma/E = 6.57 \times 10^6/7 \times 10^8 = 9.386 \times 10^{-3}$. The change in length is $\Delta L = \varepsilon L_o = 9.386 \times 10^{-3} \times 1 \times 10^3 = 9.386$ m.

(b) The hoop stress $\sigma_H = pD/2h = 0.32 \times 10^6 \times 0.775/0.05 = 4.96 \times 10^6$ Pa. $\sigma_A = 2.48 \times 10^6$ Pa. The strains may be calculated from

$$\begin{pmatrix} \varepsilon_H \\ \varepsilon_A \\ \gamma_{AH} \end{pmatrix} = \begin{pmatrix} 1/E & -v/E & 0 \\ -v/E & 1/E & 0 \\ 0 & 0 & 1/G \end{pmatrix} \begin{pmatrix} \sigma_H \\ \sigma_A \\ \tau_{AH} \end{pmatrix} = \begin{pmatrix} 1.43 \times 10^{-9} & -5.72 \times 10^{-10} & 0 \\ -5.72 \times 10^{-10} & 1.43 \times 10^{-9} & 0 \\ 0 & 0 & 4 \times 10^{-9} \end{pmatrix} \begin{pmatrix} \sigma_H \\ \sigma_A \\ \tau_{AH} \end{pmatrix}$$

$$\varepsilon_H = 1.43 \times 10^{-9} \times 4.96 \times 10^6 - 5.72 \times 10^{-10} \times 2.48 \times 10^6 = 5.674 \times 10^{-3}$$

$$\Delta D_m = \varepsilon_H D_m = 5.674 \times 10^{-3} \times 0.775 = 4.4 \times 10^{-3} = 4.4 \text{ mm}$$

$$\varepsilon_A = -5.72 \times 10^{-10} \times 4.96 \times 10^6 + 1.43 \times 10^{-9} \times 2.48 \times 10^6 = 7.094 \times 10^{-4}$$

$$\Delta L = \varepsilon_A = 7.094 \times 10^{-4} \times 1 \times 10^3 = 0.709 \text{ m}.$$

In practice polyethylene will creep under load, the strain will increase with time under load. This book does not discuss this pattern of behaviour, which is very important for polymers; there is a full discussion of this in the author's *Engineering with Polymers*.

Outline 2.18: From the outline to Problem 2.16 we have $\Delta V/V_o = (pR/2Eh)(5 - 4v) = (5 \times 10^5 \times 62.5/2 \times 3 \times 10^9 \times 5)(5 - 1.4) = 0.00375$. We assume the pipe is thin-walled and that the restraint on radial expansion at each end of the tube can be neglected.

Outline 2.19: (a) In a thin-walled pipe the hoop stress σ_H is twice the axial stress, σ_A, so the strain response is given by

$$\begin{pmatrix} \varepsilon_H \\ \varepsilon_A \\ \gamma_{AH} \end{pmatrix} = \begin{pmatrix} 1/E & -v/E & 0 \\ -v/E & 1/E & 0 \\ 0 & 0 & 1/G \end{pmatrix} \begin{pmatrix} \sigma_H \\ \sigma_H/2 \\ \tau_{HA} \end{pmatrix}$$

In the hoop direction the strain is $(v/2)$ less than if the pipe were only under hoop stress. In the axial direction there will be no axial strain at all under internal pressure if $v = 0.5$. This is exploited in the rubber inner tubes for pneumatic bicycle tyres, for which $v = 0.5$. The closed tube does change its diameter on inflation, which is desirable to achieve a snug fit under the tyre carcase, but does not change its length, which is desirable to avoid creasing and damage within the tyre which is relatively undeformable in the 'length' direction.

(b) In a thin-walled sphere under internal pressure, the hoop stress in any direction is the same. So in any two perpendicular directions we have identical hoop stresses, σ, giving

$$\begin{pmatrix} \varepsilon_x \\ \varepsilon_y \\ \gamma_{xy} \end{pmatrix} = \begin{pmatrix} 1/E & -v/E & 0 \\ -v/E & 1/E & 0 \\ 0 & 0 & 1/G \end{pmatrix} \begin{pmatrix} \sigma \\ \sigma \\ \tau_{xy} \end{pmatrix}$$

This means that the restraining effect is the same as that described in the text above.

Outline 2.20:

S2	$\sigma_y; \ -\sigma_x; \ -\sigma_x + \sigma_y$
S9	$+\tau_{xy}$
S10	$-\tau_{xy}$
S11	$\sigma_x + \tau_{xy}; \ -\sigma_y + \tau_{xy}; \ \sigma_x - \sigma_y + \tau_{xy}$
S12	$\sigma_x - \tau_{xy}; \ -\sigma_y - \tau_{xy}; \ \sigma_x - \sigma_y - \tau_{xy}$
S20	$\sigma_x + \sigma_y - \tau_{xy}$
S25	$-\sigma_x - \sigma_y + \tau_{xy}$

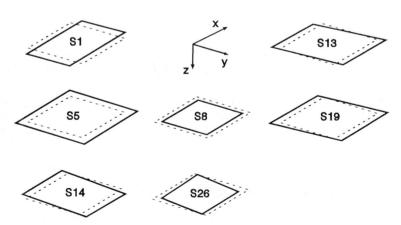

Figure A2.1 Deformations of isotropic sheet under several in-plane loading conditions.

Missing (Figure A 2.1) are the following:

S1	$\sigma_x; \ -\sigma_y; \ \sigma_x - \sigma_y$
S5	$\sigma_x + \sigma_y$
S8	$-\sigma_x - \sigma_y$
S13	$\sigma_y + \tau_{xy}; \ -\sigma_x + \tau_{xy}; \ -\sigma_x + \sigma_y + \tau_{xy}$
S14	$\sigma_y - \tau_{xy}; \ -\sigma_x - \tau_{xy}; \ -\sigma_x + \sigma_y - \tau_{xy}$
S19	$\sigma_x + \sigma_y + \tau_{xy}$
S26	$-\sigma_x - \sigma_y - \tau_{xy}$

Outline 2.21: We know $T = (G\theta/L)\cdot(\pi R^4/2)$, hence $\theta = 2LT/(G\pi R^4) = 2 \times 4000 \times 45\,000/(80 \times 10^9 \times 3.14 \times (0.0625)^4) = 94.29$ radians, i.e. about 15 complete revolutions. The shear stress at the surface of the shaft is $\tau_{xy} = G\theta R/L = 80 \times 10^9 \times 94.29 \times 0.0625/4000 = 11.79\,\text{MPa}$, which is well within the capability of the material, which would typically be able to withstand some $80\,\text{MPa}$.

Outline 2.22: $\theta = 4° = 4/57.29 = 0.0698^c$

$$T = 2\pi R^3 hG\theta/L = 2\pi \cdot (0.05)^3 \times 6 \times 10^{-3} \times 1.4 \times 10^9 \times 0.0698/1.1 = 418.6\,\text{Nm}$$

$$\tau = G\theta R/L = 1.4 \times 10^9 \times 0.0698 \times 0.05/1.1 = 4.44\,\text{MPa}$$

$$\gamma = \tau/G = 4.44 \times 10^6/1.4 \times 10^9 = 3.17 \times 10^{-3} = 0.317\%$$

Outline 2.23: (a) Using the $x'x'$ base in Figure 2.37, the first moment of an elemental strip distant z from $x'x'$ of thickness dz and constant width b is $bz\,dz$. Hence we confirm formally that the centroid is at the co-ordinate

$$z^* = (b\textstyle\int_0^h z\,dz)/bh = (bh^2/2)/bh = h/2$$

Taking the second moment of the same elemental strip about the centroid axis xx we have

$$I = b \int_{-h/2}^{+h/2} z^2\,dz = (b/3)[(+h/2)^3 - (-h/2)^3] = bh^3/12$$

(b) First we must locate the centroid. Taking first moments of the flange and web elements separately we have $[bh_1 + b_1(h-h_1)]z^* = bh_1(h - h_1/2) + b_1(h-h_1)^2/2$, which leads to $z^* = [bh_1(2h - h_1) + b_1(h - h_1)^2]/[2\{bh_1 + b_1(h-h_1)\}]$. We can now calculate the second moment of area of the complete Tee section by calculating the second moment of area of the flange and web elements about their own centroids and achieving the second moment of these elements about z^* using the parallel axis theorem:

$$I_{x^*x^*} = bh_1^3/12 + b_1h(h-h_1)^3/12 + bh_1(h - h_1/2 - z^*)^2 + b_1(h-h_1)(z^* - (h-h_1)/2)^2$$

Both stress and strain profiles are linear with depth.

Outline 2.24: Let the total depth of the section be h_1, total breadth b, and wall thickness h_2. We can calculate the second moment of area of the complete beam, I_x, in terms of the second moments of the flanges (I_{fx}) and the web (I_{wx}), all terms about the centroid of the complete beam.

$$I_{wx} = h_2(h_1 - 2h_2)^3/12 \quad \text{and} \quad I_{fx} = 2(bh_2^3/12 + bh_2(h_1 - h_2)^2/4)$$

hence

$$I_{wx}/I_x = [h_2(h_1 - 2h_2)^3]/[b_1h_1^3 - (b_1 - h_2)(h_1 - 2h_2)^3]$$

Putting $b = h_1/2$ and assuming $h_2 \ll h_1$, we find $I_{wx}/I_x \approx (h_1 - 6h_2)/4h_1$, and if $h_2 = 0.1h_1$, $I_{wx}/I_x \approx 0.1$.

Outline 2.25: $I = bh^3/12 = 12 \times 125 \times 10^{-12}/12 = 1.25 \times 10^{-10} \text{m}^4$

$$\kappa_x = M/EI = 0.1/1 \times 10^9 \times 1.25 \times 10^{-10} = 0.8/\text{m}$$

$$R_x = 1/\kappa_x = 1.25 \text{ m}$$

At the top surface of the beam $z = -2.5 \text{ mm}$

$$\varepsilon_x = z\kappa_x = -2.5 \times 10^{-3} \times 0.8 = -2 \times 10^{-3} = -0.2\%$$

$$\sigma_x = E\varepsilon_x = -2 \times 10^{-3} \times 9 \times 10^9 = -18 \text{ MPa}$$

Outline 2.26: For circular corrugations the periodic width is $w_c = 4R_m$, with wall thickness h and depth h_c (Figure 2.36). The second moment of area about the centroid of the corrugations is the same as that of a circle, i.e. $I_{c1} \approx \pi R_m^3 h$. Along the length, the second moment of area for a length w_c is $I_{L1} = w_c h^3/12 = R_m h^3/3$. The anisotropy ratio for bending stiffness is $R_1 = I_{c1}/I_{L1} = 10(R_m/h)^2$. If $R_m = 5h$, then $R_1 \sim 250$.

For square corrugations, (ignoring the taper needed in practice, which has a small effect on the essence of these calculations), the second moment of area along the sheet about its centroid is $I_{L2} = w_c h^3/12$. Across the width of the sheet we have, per unit corrugation, $I_{c2} = [(w_c/2 + h)h_c^3 - (w_c/2 - h)(h_c - 2h)^3]/12$. (It is sensible to check extremes. If $w_c = 2h$, we have a solid sheet of width $w_c = 2h$ and depth h_c confirmed by $I_c = hh_c^3/6$. If $w_c \gg h$ and $h_c \gg h$, and noting that $(h_c - 2h)^3 \sim h_c^3 - 6h_c^2 h$ if products of small quantities are neglected, $I_c \sim w_c h_c^2 h/4$.)

So for a real sheet the ratio of stiffnesses for equal lengths is given by the ratio $R_2 = I_{c2}/I_{L2} \sim 3(h_c/h)^2$. If $h_c \sim 10h$, $R_2 \sim 300$.

The bending stiffness per unit weight along the corrugations may be represented by I/A. Hence the ratio of longitudinal bending stiffnesses per unit weight is given by

$$(I/A)_{sq}/(I/A)_{circ} = (h_c^2/8)/(h_c^2/8) = 1$$

Outline 2.27:

(a) $D_{11} = D_{22} = D = Eh^3/[12(1 - v^2)] = 3 \times 10^9 \times 8 \times 10^{-9}/[12(0.8775)] = 2.279 \text{ Nm}$

$$D_{12} = -vD = -0.7979 \text{ Nm}$$

$$\kappa_x = M_x/[D(1 - v^2)] = 1/[2.279 \times 0.8775] = 0.500/\text{m}$$

$$\kappa_y = -v\kappa_x = -0.175/\text{m}$$

To suppress anticlastic curvature we have to ensure that $\kappa_y = 0$, so that $\kappa_y = (-vM_x + M_y)/[D(1 - v^2)]$, i.e. $M_y = +vM_x = 0.35$ N, and hence

$$\kappa_x = [M_x - vM_y]/[D(1 - v^2)] = M_x/D = 1/2.279 = 0.4388 \text{ /m}.$$

(b) $\kappa_x = (M_x - vM_y)/[D(1 - v^2)] = 0.65 \times 1/[2.279 \times 0.8775] = 0.325/\text{m} = \kappa_y$
The application of equal moments M_x and M_y reduces the curvature κ_x from 0.5/m (from M_x acting alone) and results in synclastic rather than anticlastic curvature.

(a) $\qquad \sigma_x = [Ez/(1 - v^2)](\kappa_x + v\kappa_y)$

$$= [(3 \times 10^9 \times -1 \times 10^3)/0.8775](0.5 + 0.35 \times -0.175) = -1.5 \text{ MPa}$$

$$\varepsilon_x = (h/2)\kappa_x = -10^{-3} \times 0.5 = -5 \times 10^{-4} = -0.05\%$$

$$\varepsilon_y = (h/2)\kappa_y = 0.0175\%$$

The strain profiles are shown in Figure 2.42.

Outline 2.28:

M1	M_x; $-M_y$; $M_x - M_y$
M5	$M_x + M_y$
M8	$-M_x - M_y$
M9	M_{xy}
M10	$-M_{xy}$
M13	$M_y + M_{xy}$; $-M_x + M_{xy}$; $-M_x + M_y + M_{xy}$
M14	$M_y - M_{xy}$; $-M_x - M_{xy}$; $-M_x + M_y - M_{xy}$
M19	$M_x + M_y + M_{xy}$
M26	$-M_x - M_y - M_{xy}$

Missing (Figure A2.2) are M2 (M_y; $-M_x$; $-M_x + M_y$). M11 ($M_x + M_{xy}$; $-M_y + M_{xy}$; $M_x - M_y + M_{xy}$). M12 ($M_x - M_{xy}$; $-M_y - M_{xy}$; $M_x - M_y - M_{xy}$). M20 ($M_x + M_y - M_{xy}$). M25 ($-M_x - M_y + M_{xy}$).

Outline 2.29: To sketch the deformed sheet we need to calculate the midplane strains and curvatures. We can use expressions such as $\varepsilon_x = z\kappa_x$, and $\varepsilon_x^\circ = (\varepsilon_x^t + \varepsilon_x^b)/2$:

$$\varepsilon_x^\circ = (0.3121 + 0.3046)/2 = 0.30835\%$$

$$\varepsilon_y^\circ = (-0.2848 - 0.2335)/2 = -0.25415\%$$

$$\gamma_{xy}^\circ = (-0.3266 - 0.1239)/2 = -0.22525\%$$

From $\varepsilon_x^t = (-h/2)\kappa_x$ and $\varepsilon_x^b = (+h/2)\kappa_x$, we have $\kappa_x = -(\varepsilon_x^t - \varepsilon_x^b)/h$

$$\kappa_x = -(0.3046 - 0.3121) \times 0.01/4 \times 10^{-3} = 0.01875/\text{m}$$

$$\kappa_y = -(-0.2748 + 0.2335) \times 0.01/4 \times 10^{-3} = 0.10325/\text{m}$$

$$\kappa_{xy} = -(-0.3266 + 0.1239) \times 0.01/4 \times 10^{-3} = 0.5068/\text{m}$$

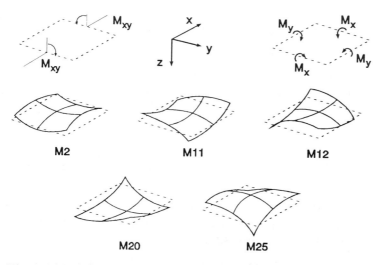

Figure A2.2 Deformations of flat sheet under different bending moments.

The midplane stresses are related to the midplane strains by

$$\begin{pmatrix} \sigma_x \\ \sigma_y \\ \tau_{xy} \end{pmatrix} = \begin{pmatrix} E/(1-v^2) & vE/(1-v^2) & 0 \\ vE/(1-v^2) & E/(1-v^2) & 0 \\ 0 & 0 & G \end{pmatrix} \begin{pmatrix} \varepsilon_x^\circ \\ \varepsilon_y^\circ \\ \gamma_{xy}^\circ \end{pmatrix}$$

from which we find that $\sigma_x/\sigma_y = (\varepsilon_x^\circ + v\varepsilon_y^\circ)/v\varepsilon_x^\circ + \varepsilon_y^\circ)$, i.e. only v is needed to calculate the required stress ratios.

Outline 2.30: See Figure A2.3.

Outline 2.31: The profiles are shown in Figure A 2.4. The bending stress is zero at the neutral axis, and has maximum values at upper and lower surface. The shear stress is greatest at the neutral axis and zero at top and bottom surfaces.

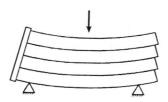

Figure A2.3 Shear in paperback book in three-point bending.

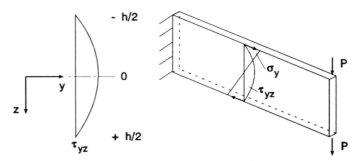

Figure A2.4 Shear stress in rectangular section beam under shear force Q_z.

Outline 2.32: Under a pure bending moment M_y, the bending moment is constant along the length of the beam and hence $dM_y/dy = 0$. Hence the shear force is zero and thus the shear stress profile is uniformly zero.

Outline 2.33: The second moment of area of the beam (from Section 2.5.6) is

$$I = bh^3/12 = 0.005 \times 0.01^3/12 = 4.167 \times 10^{-10} \, m^4$$

(a) Referring to Figure A 2.5, the maximum bending moment occurs at the built-in end, $M_o = -WL = -1.5 \, Nm$. The maximum compressive stress occurs on the underside of the beam at $z = +h/2$, with the value calculated in accordance with Section 2.5.5 as $\sigma_c = Mz/I = -1.5 \times 0.005/4.167 \times 10^{-10} = -18 \, MPa$.

(b) We need to find the general expression for slope and deflection in the end-loaded cantilever. Taking the origin at the built-in end (Figure A 2.5), then the moment at any point y is $M = -M_o + R_o y = -W(L - y)$. Hence $d^2w/dy^2 = -M/EI = +(W/EI)(L - y)$, hence $dw/dy = (W/EI)(Ly - y^2/2) + C_1$; at $y = 0$, $dw/dy = 0$, so $C_1 = 0$. $w(y) = (W/EI)(Ly^2/2 - y^3/6) + C_2$, and at $y = 0$, $w = 0$, so $C_2 = 0$. Hence the slope at $y = 0.105 \, m$ is $0.0164^c \approx 0.94°$, and the deflection is 1.292 mm.

(c) The mass of the beam is $m = 1600 \times 0.005 \times 0.01 \times 0.15 = 0.012 \, kg$, which corresponds to a uniformly distributed load of $9.81 \times 0.012 = 0.118 \, N$.

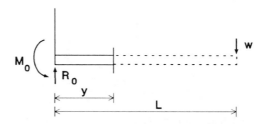

Figure A2.5 Cantilever beam under end load.

The load per unit length is therefore $w_1 = 0.118\,\text{N}/0.15\,\text{m} = 0.787\,\text{N/m}$. From Figure 2.62, and using the principle of superposition, the deflection at the free end, taking self-weight loading into account, is $W_{max} = (L^3/EI)$ $(W/3 + w_1 L/8) = 1.8 + 0.008 = 1.808\,\text{mm}$. In this example the effect of the self-weight loading is small.

Outline 2.34: $\sigma_c = P_c/A = \pi^2 EI/AL^2 = \pi^2 \sigma_c I/\varepsilon_c AL^2$, hence $\varepsilon_c = (\pi^2 I/AL^2)$

Outline 2.35: The Euler buckling condition is $P_c \propto EI/L^2 \propto E\pi R^4/L^2$. The load exerted by the mass of the tree is $P_c \propto \rho g \pi R^2 L$, so $\rho g R^2 L \propto ER^4/L^2$, and hence for same material $L \propto R^{2/3}$, which is a statement of Kleiber's law. We can refine the analysis to take account of the non-uniform cross-sectional area of the tree trunk, and that the load is acting non-uniformly axially along the trunk rather than concentrated at the ends, but the basic shape of the result still stands, although the constant of proportionality is slightly different.

Outline 2.36: We need to find the radius of gyration for each section. (a) For the thin-walled tube $r \approx \sqrt{\{\pi R_m^3 h/4)/(2\pi R_m h)\}} = R_m/2$. Hence $L = 60r = 30R_m$, so a tube behaves as if it were more slender than it really is. (b) The minimum second moment of area is $I = h^4/4$, so $I/A = 3h^2/4$, leading to $L = 30h\sqrt{3}$.

Outline 2.37: (a) Ignoring any restraint exerted by surrounding fibres, the second moment of area of one fibre is $I_f = \pi R_f^4/4 = 4.91 \times 10^{-22}\,\text{m}^4$, so for the complete bundle $P_{ca} \approx 1000\pi^2 EI_f/L^2 = 3.44\,\text{mN}$. (b) If the fibres are close-packed and bonded together with no voids, then as a rough approximation the radius of the circular bundle is about $R_b = 1.66 \times 10^{-4}\,\text{m}^2$, with a cross-sectional area $A_b = 8.631 \times 10^{-8}\,\text{m}^2$ and $I_b = 5.93 \times 10^{-16}\,\text{m}^4$. Hence $P_{cb} \approx \pi^2 EI_b/L^2 = 4.15\,\text{N}$, and the slenderness ratio of the bundle is $(L/r)_b \approx 120$. The ratio of the buckling loads is therefore $P_{cb}/P_{ca} = 1207$.

Outline 2.38: (a) Crossover occurs when $\varepsilon_c = kh/D = \pi^2/(L/r)^2$, and hence (b) $(L/r) \propto \sqrt{(D/h)}$.

Outline 2.39: $\lambda = 4.9R\sqrt{(R/h)} = 2\,\text{m}$. The critical buckling pressure is predicted as $p_c = E(h/R)^3/[4(1-v^2)] = 10 \times 10^9 (6/100)^3/[4 \times 0.91] = 0.6\,\text{MPa}$. This is substantially below the value suggested for failure under internal pressure. The calculation assumes no value of safety factor, and it would be wise to down-rate the external applied pressure by a factor of perhaps 2 in a practical design.

CHAPTER 3

Outline 3.1: $1 - v^2 = 0.8911$; $S_{11} = 1/E = 1.429 \times 10^{-11}\,\text{m}^2/\text{N}$, and so on:

$$[S] = \begin{pmatrix} 14.29 & -4.714 & 0 \\ -4.714 & 14.29 & 0 \\ 0 & 0 & 38.46 \end{pmatrix} \times 10^{-12}\,\text{m}^2/\text{N}$$

$$[Q] = \begin{pmatrix} 78.55 & 25.92 & 0 \\ 25.92 & 78.55 & 0 \\ 0 & 0 & 26 \end{pmatrix} \times 10^{-9} \, \text{N/m}^2$$

Outline 3.2: The thickness co-ordinate in the plate is measured from the midplane, positive downwards. For the general case of a stress varying linearly with thickness, $\sigma_x(z) = a + bz$, where $a = (\sigma_x^b + \sigma_x^t)/2$ and $b = (\sigma_x^b - \sigma_x^t)/h$. Hence

$$N_x = \int_{-h/2}^{+h/2} \sigma_x(z)dz = ah, \text{ and so } N_x = 1.5 \, \text{N/mm}^2 \times 2 \, \text{mm} = 3 \, \text{N/m}$$

$$M_x = \int_{-h/2}^{+h/2} \sigma_x(z)z \, dz = bh^3/12 : M_x = (\sigma_x^b - \sigma_x^t)h^2/12 = 3 \times 2^2/12 = 1\text{N}.$$

Outline 3.4: The determinant of the matrix $[A]$ is $|A| = (A_{11}A_{22} - A_{12}^2)A_{66}$ which is positive. The cofactor a_{12}, by the rules for matrix inversion, is $-A_{11}A_{66}/|A|$, and both A_{11} and A_{66} are positive. The same procedure applies for the sign of d_{12}.

Outline 3.5: Because the rubber is assumed linear elastic, the response is simply the sum of each effect alone. The result is $\varepsilon_1^\circ = 5\%$, $\varepsilon_2^\circ = -2.45\%$, $\kappa_1 = 1.50/\text{m}$ and $\kappa_2 = -0.735/\text{m}$, and is shown diagramatically in Figure A3.1(a). Stress and strain profiles are also additive, as shown in Figure A3.1(b). These profiles can be checked by calculation, e.g. $\sigma_1 = Q_{11}(\varepsilon_1 + z\kappa_1) + Q_{12}(\varepsilon_2 + z\kappa_2)$.

Outline 3.6: We can determine the curvatures using $\kappa_1 = d_{12}M_2 = 3.0/\text{m}$, $\kappa_2 = d_{22}M_2 = -1.47/\text{m}$, and $\kappa_{12} = d_{66}M_{12} = 9.009/\text{m}$, and these are shown schematically in Figure A3.2 together with strain profiles.

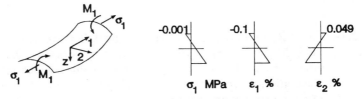

Figure A3.1 Deformation of NR under $N_1 = 0.1 \, \text{N/mm}$ and $M_1 = 0.1 \, \text{N}$.

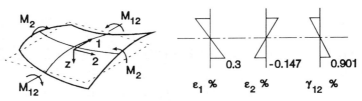

Figure A3.2 Deformation of NR under $M_2 = M_{12} = 0.002 \, \text{N}$.

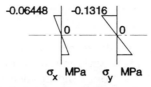

Figure A3.3 Stress profiles in NR under $\kappa_1 = 100/m$.

Outline 3.7: Assuming $\kappa_1 = 0$ we have $M_1 = D_{12}\kappa_2 = (0.0004299 \text{ kNmm}) \times 1.00/m = 0.0004299 \text{ N}$. $M_2 = D_{22}\kappa_2 = 0.0008773 \text{ N}$. Stress profiles are given in Figure A3.3.

Outline 3.8:

$S_{11} = 1/E_1 = 7.463 \times 10^{-12} \text{ Pa}^{-1}$; $S_{12} = v_{12}S_{11} = -2.090 \times 10^{-12} \text{ Pa}^{-1}$

$S_{22} = 1/E_2 = 112 \times 10^{-12} \text{ Pa}^{-1}$; $S_{66} = 1/G_{12} = 196.1 \times 10^{-12} \text{ Pa}^{-1}$

$v_{21} = v_{12}E_2/E_1 = 0.28 \times 8.9/134 = 0.0186$; $J = (1 - v_{12}v_{21}) = 0.995$

$Q_{11} = E_1/J = 134.7 \text{ GPa}$; $Q_{12} = v_{12}E_2/J = 2.504 \text{ GPa}$

$Q_{22} = E_2/J = 8.945 \text{ GPa}$; $Q_{66} = G_{12} = 5.1 \text{ GPa}$

$A_{11} = Q_{11}h = 134 \times 10^9 \times 1.25 \times 10^{-4} = 16.84 \times 10^6 \text{ N/m} = 16.84 \text{ kN/mm}$

$A_{12} = 0.313 \text{ kN/mm}$; $A_{22} = 1.118 \text{ kN/mm}$; $A_{66} = 0.6375 \text{ kN/mm}$

$D_{11} = Q_{11}h^3/12 = 21.92 \times 10^{-3} \text{ Nm}$

$D_{12} = 407.6 \times 10^{-6} \text{ Nm}$

$D_{22} = 1.456 \times 10^{-3} \text{ kNmm}$; $D_{66} = 8.301 \times 10^{-4} \text{ kNmm}$

To calculate the compliance coefficients, we must invert the [A] and [D] matrices.

$$|A| = (A_{12}A_{22} - A_{12}^2)A_{66} = 11.94$$

$$a_{11} = A_{22}A_{66}/|A| = 1.118 \times 0.6375/11.94 = 0.05969 \text{ mm/kN}$$

$$a_{12} = -A_{12}A_{66}/|A| = -0.01672 \text{ mm/kN}$$

$$a_{22} = A_{11}A_{66}/|A| = 0.8989 \text{ mm kN}$$

$$a_{66} = (A_{11}A_{22} - A_{12}^2)/|A| = 1.569 \text{ mm/kN}$$

Similarly $d_{11} = 45.85/\text{mmkN}$; $d_{12} = -12.84/\text{mmkN}$; $d_{22} = 690.3/\text{mmkN}$; $d_{66} = 1205/\text{mmkN}$.

Outline 3.9(a): From Section 1.3.1 we noted that $E_1 = V_f E_f + V_m E_m$. Because $\varepsilon_f = \varepsilon_1$, we can write $\sigma_f = E_f \cdot \varepsilon_1$, so the load taken by the fibres, P_f can be expressed as $P_f = A_f \sigma_f = A_f E_f \sigma_1/E_1 = A_f E_f P_1/A_1 E_1 = V_f E_f P_1/E_1$ and hence $P_f/P_1 = 1/(1 + V_m E_m/V_f E_f)$. Substituting numerical values for V_f (say, 0.3, 0.5, 0.7) and E_m/E_f (say 0.1 and 0.01) leads to the following conclusions: Fibre load-bearing efficiency increases as the proportion of fibres increases and increases as E_f/E_m increases. This only applies when the load is applied along the fibre direction.

Outline 3.9(b): The resulting curvatures are $\kappa_1 = d_{11}M_1 = 1.648/\text{m}$ and $\kappa_2 = d_{12}M_1 = -0.593/\text{m}$ and the anticlastic curvature is schematically shown in Figure A3.4(a). As expected from bending theory the ratio of curvatures corresponds to the major Poisson's ratio $\nu_{12} = -\kappa_2/\kappa_1$, when only moment M_1 is applied. The stress and strain profiles vary linearly with thickness co-ordinate, as shown in Figure A3.4(b).

Outline 3.10: The strains are $\varepsilon_2^\circ = 6.906\%$ and $\varepsilon_1^\circ = -0.1978\%$, as shown diagrammatically in Figure A3.5. The nylon cords have little stiffening effect when the sheet is loaded transversely, but the longitudinal stiffening does substantially reduce the longitudinal contraction. The minor Poisson's ratio is $\nu_{21} = -\varepsilon_1^\circ/\varepsilon_2^\circ = 0.002864$, and is much smaller than the major Poisson's ratio, because of the high anisotropy ratio of this sheet, $E_1/E_2 = 910/7.24 = 125.7$.

Outline 3.11: $N_1 = A_{11}\varepsilon_1 + A_{12}\varepsilon_2 = (1.822 + 0.005218)\,\text{kN/mm} \times 0.05 = 91.35\,\text{N/mm}$; $N_y = A_{12}\varepsilon_1 + A_{22}\varepsilon_2 = 0.9857\,\text{N/mm}$.

Outline 3.12:

$$E_1 = 1.74\,\text{GPa}; \quad E_2 = 14.1\,\text{MPa}; \quad \nu_{12} = 0.547; \quad G_{12} = 2.5\,\text{MPa}$$
$$\nu_{21} = (E_2/E_1)\nu_{12} = (0.0141/1.74) \times 0.547 = 4.433 \times 10^{-3}$$

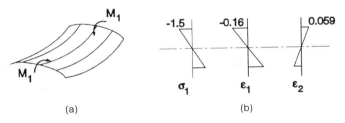

(a)   (b)

Figure A3.4 Deformation, and stress and strain profiles, for NE0 under $M_1 = 1\,\text{N}$.

Figure A3.5 Deformation of NE0 under $N_2 = 1\,\text{N/mm}$.

$$S_{11} = 1/E_1 = 5.747 \times 10^{-10} \, m^2/N$$
$$S_{12} = -v_{12}/E_1 = -3.144 \times 10^{-10} \, m^2/N$$
$$S_{22} = 1/E_2 = 709.2 \times 10^{-10} \, m^2/N$$
$$S_{66} = 1/G_{12} = 4000 \times 10^{-10} \, m^2/N$$

$$S = \begin{pmatrix} 5.747 & -3.144 & 0 \\ -3.144 & 709.2 & 0 \\ 0 & 0 & 4000 \end{pmatrix} \times 10^{-10} \, m^2/N$$

$$\varepsilon_2 = 0 = S_{12}\sigma_1 + S_{22}\sigma_2 + S_{66} \cdot 0$$

Hence

$$\sigma_1/\sigma_2 = -S_{22}/S_{12} = -709.2/-3.144 = 225.6$$

The obvious procedure for calculating the strains (a) is to use the compliances [S]. For illustrative purposes, and to prepare the ground for laminate analysis, the extensional stiffnesses are also calculated (b) and inverted.

(a)

$$\varepsilon_1 = S_{11}\sigma_1 + S_{12}\sigma_2 + 0/\tau_{12} = 0.011274 = 1.1274\%$$
$$\varepsilon_2 = S_{12}\sigma_1 + S_{22}\sigma_2 + 0.\tau_{12} = 0.04336 = 4.336\%$$
$$\gamma_{12} = 0.\sigma_1 + 0.\sigma_2 + S_{66}\tau_{12} = 0.04 = 4\%$$

(b) Writing $J = (1 - v_{12}v_{21}) = 1 - 0.547 \times 0.004 = 0.9978$

$$Q_{11} = E_1/J = 1.7438 \, GPa; \quad Q_{12} = v_{12}E_1/J = 7.7303 \, MPa$$
$$Q_{22} = E_2/J = 14.13 \, MPa; \quad Q_{66} = G_{12} = 2.5 \, MPa$$

Hence

$$A_{11} = hQ_{11} = 1.9 \times 10^{-3} \times 1.7438 \times 10^9 = 3.313 \, MPam = 3.313 \, kN/mm$$
$$A_{12} = hQ_{12} = 14.688 \, kPam; \quad A_{22} = hQ_{22} = 26.847 \, kPam;$$
$$A_{66} = hQ_{66} = 4.75 \, kN/mm$$

Input loads are

$$N_1 = \sigma_1 h = 20 \times 10^6 \times 1.9 \times 10^{-3} = 38 \, kN/m$$
$$N_2 = \sigma_2 h = 7 \times 10^5 \times 1.9 \times 10^{-3} = 1.33 \, kN/m$$

The inverse extensional stiffness matrix [a] is calculated as

$$[A]^{-1} = \begin{pmatrix} A_{22}/J' & -A_{12}/J' & 0 \\ -A_{12}/J' & A_{11}/J' & 0 \\ 0 & 0 & 1/A_{66} \end{pmatrix} = \begin{pmatrix} 0.30258 & -0.16554 & 0 \\ -0.16554 & 373.39 & 0 \\ 0 & 0 & 2105.3 \end{pmatrix} \times 10^{-6} \, m/N$$

where $J' = A_{11}A_{22} - A_{12}^2$. Hence $\varepsilon_1 = a_{11}N_1 + a_{12}N_2 + a_{16}N_{12} = 0.30258 \times 10^{-6} \times 38 \times 10^3 - 0.16554 \times 10^{-6} \times 1.33 \times 10^3 + 0 = 0.01127$ which agrees with that calculated by the much simpler method.

Outline 3.13:

$$[S] = \begin{pmatrix} 1/E_1 & -v_{12}/E_1 & 0 \\ -v_{21}/E_2 & 1/E_2 & 0 \\ 0 & 0 & 1/G_{12} \end{pmatrix}$$

$$|S| = \frac{1}{E_1 E_2 G_{12}} - \frac{(-v_{12})(-v_{21})}{E_1 E_2 G_{12}} = \frac{(1 - v_{12}v_{21})}{E_1 E_2 G_{12}}$$

$$[S^c] = [S^c]^T = \begin{pmatrix} \dfrac{1}{E_2 G_{12}} & \dfrac{+v_{21}}{E_2 G_{12}} & 0 \\ \dfrac{v_{12}}{E_1 G_{12}} & \dfrac{1}{E_1 G_{12}} & 0 \\ 0 & 0 & \dfrac{(1 - v_{12}v_{21})}{E_1 E_2} \end{pmatrix}$$

$$[Q] = [S]^{-1} = \frac{[S^c]^T}{|S|} = \begin{pmatrix} \dfrac{E_1}{1 - v_{12}v_{21}} & \dfrac{v_{21}E_1}{1 - v_{12}v_{21}} & 0 \\ \dfrac{v_{12}E_2}{1 - v_{12}v_{21}} & \dfrac{E_2}{1 - v_{12}v_{21}} & 0 \\ 0 & 0 & G_{12} \end{pmatrix}$$

Outline 3.15: The following detailed analysis formalizes what common sense suggests, and provides a useful check on concepts.

Longitudinal windings (Figure A3.6(a))
Under internal pressure $\tau_{AH} = 0$ and $\sigma_H = 2\sigma_A$, hence $\gamma_{AH} = \tau_{AH}/G_{12} = 0$. $\varepsilon_A = (\sigma_A/E_1)(1 - v_{12})$, which is small because E_1 is large. $\varepsilon_H = -v_{12}\sigma_A/E_1 + 2\sigma_A/E_2$, which is large because E_2 is small and the second term dominates.

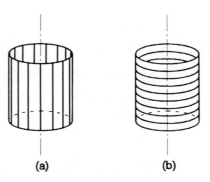

(a)   (b)

Figure A3.6 (a) Longitudinal windings; (b) hoop windings.

Under opposing torques, shear strain is the result of applied shear stress; $\varepsilon_H = \varepsilon_A = 0$.

Under axial tensile load, σ_A is defined by P/A, and $\sigma_H = \tau_{AH} = 0$. $\varepsilon_A = \sigma_A/E_1$, and $\varepsilon_H = -v_{12}\sigma_A/E_1$, both direct strains being small because E_1 is large.

Hoop windings (Figure A 3.6(b))
Under internal pressure, $\gamma_{AH} = 0$; $\varepsilon_A = \sigma_A/E_2 - 2v_{12}\sigma_A/E_1$, large because E_2 is small; $\varepsilon_H = -v_{12}\sigma_A/E_1 + 2\sigma_A/E_1$, small because E_1 is large.

Under opposing torques $\gamma_{AH} = \tau_{AH}/G_{12}$, the same as for fibres along the axis.

Under axial load, $\gamma_{AH} = 0$. $\varepsilon_A = \sigma_A/E_2$, large because E_2 is small; $\varepsilon_H = \sigma_A/E_1$, small because E_1 is large.

Outline 3.16: Using the [T] matrix Equation (3.43) we have, for a lamina of thickness h:

$$h\tau_{12} = h\sigma_x(-\sin\theta\cos\theta) + h\sigma_y(\sin\theta\cos\theta)$$

If $\sigma_x = \sigma_y$, then $\tau_{12} = 0$ for any value of θ, and it does not matter whether applied force resultants are tensile or compressive. Indeed this is true for any material, not just unidirectional composites.

Outline 3.17: By inspection $\sigma_x = N_x/h = \sigma_y = 2.5\,\text{N/mm}^2$, and $\tau_{xy} = 0.5\,\text{N/mm}^2$. Using Equation (3.34) we have

$$\begin{pmatrix}\sigma_a \\ \sigma_b \\ \tau_{ab}\end{pmatrix} = [T(60°)]\begin{pmatrix}\sigma_x \\ \sigma_y \\ \tau_{xy}\end{pmatrix} = \begin{pmatrix}0.25 & 0.75 & 0.866 \\ 0.75 & 0.25 & -0.866 \\ -0.433 & +0.433 & -0.5\end{pmatrix}\begin{pmatrix}\sigma_x \\ \sigma_y \\ \tau_{xy}\end{pmatrix}$$

Hence $\sigma_a = 0.25 \times 2.5 + 0.75 \times 2.5 + 0.866 \times 0.5 = 2.933\,\text{N/mm}^2$. $\sigma_b = 2.067\,\text{N/mm}^2$, $\tau_{xy} = -0.25\,\text{N/mm}^2$.

Outline 3.18:

$$\begin{pmatrix}\varepsilon_1 \\ \varepsilon_2 \\ \gamma_{12}/2\end{pmatrix} = [T(30°)]\begin{pmatrix}\varepsilon_x \\ \varepsilon_y \\ \gamma_{xy}/2\end{pmatrix} = \begin{pmatrix}0.75 & 0.25 & 0.866 \\ 0.25 & 0.75 & -0.866 \\ -0.433 & 0.433 & 0.5\end{pmatrix}\begin{pmatrix}0.005 \\ 0.007 \\ 0.001\end{pmatrix} = \begin{pmatrix}0.006366 \\ 0.005634 \\ 0.001366\end{pmatrix}$$

Hence $(\varepsilon_1, \varepsilon_2, \gamma_{12}) = (0.006366, 0.005634, 0.002732)$.

Outline 3.19: We have $\sigma_x = -3.5\,\text{MPa}$, $\sigma_y = +7\,\text{MPa}$ and $\tau_{xy} = -1.4\,\text{MPa}$. From Equation (3.10).

$$\begin{pmatrix}\sigma_1 \\ \sigma_2 \\ \tau_{12}\end{pmatrix} = \begin{pmatrix}0.25 & 0.75 & 0.866 \\ 0.25 & 0.75 & -0.866 \\ -0.433 & 0.433 & 0.5\end{pmatrix}\begin{pmatrix}\sigma_x \\ \sigma_y \\ \tau_{xy}\end{pmatrix}$$

Hence $\sigma_1 = 0.25 \times -3.5 + 0.75 \times 7 + 0.866 \times -1.4 = 3.1626\,\text{MPa}$

$$\sigma_2 = 0.3374\,\text{MPa}, \quad \tau_{12} = 5.2465\,\text{MPa}$$

Using Equation (3.33), $S_{11} = 7.4128 \times 10^{-11}\,m^2/N$, $S_{12} = -2.857 \times 10^{-11}\,m^2/N$, $S_{22} = 28.57 \times 10^{-11}\,m^2/N$, and $S_{66} = 23.81 \times 10^{-11}\,m^2/N$. So from Equation (3.63):

$$S_{11} = (7.1428 \times 0.625 + 28.57 \times 0.5625 + 18.096 \times 0.1875) \times 10^{-11}$$
$$= 199.1 \times 10^{-12}\,m^2/N$$

and hence the coefficients of the transformed compliance matrix are:

$$[S] = \begin{pmatrix} 199.1 & 4.464 & -130.9 \\ 4.464 & 91.96 & -54.64 \\ -130.9 & -54.64 & 370.2 \end{pmatrix} \times 10^{-12}\,m^2/N$$

The required strains are

$$\varepsilon_x = [199.1 \times -3.5 \times 10^6 + 4.464 \times 7 \times 10^6 + (-130.9) \times (-1.4 \times 10^6)] \times 10^{-12}$$
$$= -4.823 \times 10^{-4} = -0.04823\%, \quad \varepsilon_y = 0.07046\%, \quad \gamma_{xy} = -0.04426\%$$

Outline 3.20: We can recast the expression for $1/E_x$ in Equation (3.64) in terms of $\cos\theta$, using $\sin^2\theta = 1 - \cos^2\theta$, to give a quadratic in $\cos^2\theta$:

$$(1/E_1 + 1/E_2 + 2v_{12}/G_{12})\cos^4\theta + (1/G_{12} - 2v_{12}/E_1 - 2/E_2)\cos^2\theta + (1/E_2 - 1/E_x) = 0$$

We seek the values of θ_1 for $E_x = 0.99E_1$ and θ_2 for $E_x = 0.95E_1$. Substituting numerical values and solving the quadratic in $\cos^2\theta$ gives $\theta_1 = 1°$, and $\theta_2 = 2.275°$. To obtain accurate values of longitudinal modulus it is necessary to align the fibres precisely in the direction of test. A 2° misalignment produces nearly 5% error.

Outline 3.21: We can recast the expression for $1/G_{xy}$ in Equation (3.64) as a quadratic in $\cos^2\theta$, and substitute $G_{xy} = 0.95\,G_{12}$ to give $2.667\cos^4\theta + 133.3\cos^2\theta - 10.526 = 0$, from which $\theta = 79°$, i.e. 11° misalignment.

Outline 3.22: We can combine v_{xy} and E_x in Equation (3.64) as $v_{xy} = [A(\sin^4\theta + \cos^4\theta) + B(\sin^2\theta\cos^2\theta)]/[C\cos^4\theta + D\sin^4\theta + E\sin^2\theta\ \cos^2\theta]$, and then express this equation in terms of $\sin\theta$, which has the form: $v_{xy} = [A + M\sin^2\theta + N\sin^4\theta]/[C + P\sin^2\theta + Q\sin^4\theta]$. We can now find the maximum value of v_{xy} by setting $dv_{xy}/d\theta = 0$. Using $s = \sin\theta$ and $c = \cos\theta$ gives $dv_{xy}/d\theta = sc[(C + Ps^2 + Qs^4)(2M + 4Ns^2) - (A + Ms^2 + Ns^4)(2P + 4Qs^2)]/[C + Ps^2 + Qs^4]$. So provided $C + Ps^2 + Qs^4 \neq 0$, the problem reduces to $\sin\theta = 0$, or $\cos\theta = 0$, (which are both legitimate solutions, but neither is a maximum), or, after some rearrangement, we obtain $\sin^2\theta = -(CM - AP)/[2(CN - AQ)]$. Going back over all the changes of symbol, we can now substitute values of elastic constants: $A = v_{12}/E_1$, $B = -(1/E_1 + 1/E_2 - 1/G_{12})$, $C = 1/E_1$, $D = 1/E_2$, $E = (1/G_{12} - 2v_{12}/E_1)$, $M = B - 2A$, $N = -M$, $P = E - 2C$, and $Q = C + D - E$.

For a high modulus carbon fibre epoxy ply having $E_1 = 208\,GPa$, $E_2 = 7.6\,GPa$, $G_{12} = 4.8\,GPa$ and $v_{12} = 0.3$, we find v_{xy} has its maximum value when $\sin^2\theta = 0.239$, i.e. $\theta = \pm 29.25°$.

For a rayon cord reinforced rubber ply having $E_1 = 1.74\,\text{GPa}$, $E_2 = 14.1\,\text{MPa}$, $v_{12} = 0.547$, $G_{12} = 2.5\,\text{MPa}$, we find v_{xy} reaches a maximum at $\theta = \pm 37°$.

Outline 3.23: From the expression for S_{12}^* we have, for $\theta = 45°$

$$4/E_{45} = 1/E_1 + 1/E_2 + (1/G_{12} - 2v_{12}/E_1)$$

We can therefore eliminate G_{12} and v_{12} from the general expression:

$$1/E_x = \cos^4\theta/E_1 + \sin^4\theta/E_2 + (4/E_{45} - 1/E_1 - 1/E_2)\sin^2\theta\cos^2\theta$$

leading to $E_{60} = 10.04\,\text{GPa}$.

Outline 3.24: We re-express Q_{16}^* in terms of multiple angles to give

$$Q_{16}^*(\theta) = -(1/4)(Q_{11} - Q_{22})\sin 2\theta$$
$$- (1/8)(Q_{11} + Q_{22} - 2Q_{12} - 4Q_{66})\sin 4\theta$$
$$= A\sin 2\theta + B\sin 4\theta$$
$$\therefore dQ_{16}^*(\theta)/d\theta = 2A\cos 2\theta + 4B\cos 4\theta = 0$$

for a maximum or minimum.

Rewriting in terms of $\cos 2\theta$, we have $4B\cos^2 2\theta + A\cos 2\theta - 2B = 0$ i.e. $\cos 2\theta = [-A \pm \sqrt{(A^2 + 32B^2)}]/8B$. Thus for a high modulus carbon fibre epoxy composite having stiffnesses $Q_{11} = 180.7\,\text{GPa}$, $Q_{12} = 2.41\,\text{GPa}$, $Q_{22} = 8.032\,\text{GPa}$, $Q_{66} = 5\,\text{GPa}$, we find $\cos 2\theta = -1.0465$ or $+0.4777$: $\theta = +30.73°$ is one practicable solution.

Outline 3.25: Using $[\kappa]_x = [T]^{-1}[\kappa]_1$ from Equation (3.44) $\kappa_x = -\kappa_{12}\sin\theta\cos\theta$. $d\kappa_x/d\theta = -\kappa_{12}(\cos^2\theta - \sin^2\theta) = 0$, i.e. $\theta = \pm 45°$.

Outline 3.26: $\kappa_x = \kappa_1\cos^2\theta - 2\kappa_{12}\sin\theta\cos\theta$. $d\kappa_x/d\theta = -2\kappa_1\sin\theta\cos\theta - 2\kappa_{12}(\cos^2\theta - \sin^2\theta) = 0$, and expressing in terms of 2θ, $\tan 2\theta = -2\kappa_{12}/\kappa_1$.

Outline 3.27: $\kappa_x = \kappa_1\cos^2\theta + \kappa_2\sin^2\theta = \kappa_1\cos^2\theta - v\kappa_1\sin^2\theta$, so $\tan\theta = \sqrt{(1/v)}$; for $v = 0.3$, $\theta = 61.3°$. $\kappa_y = \kappa_1\sin^2\theta - v\kappa_1\cos^2\theta = 0$, so $\tan\theta' = \sqrt{v}$, and for $v = 0.3$. $\theta' = 28.7°$.

Outline 3.28: Under N_x acting alone, and assuming that the specimen is free to contract transversely and to take up any shear strain, we find $\varepsilon_x = a_{11}N_x = a_{11}h\sigma_x$, hence $\sigma_x/\varepsilon_x = E_x = 1/(a_{11}h) = 1/(56630\,\text{mm/MN} \times \text{mm}) = 8.829\,\text{N/mm}^2$. $v_{xy} = -a_{12}/a_{11} = --39160/56630 = +0.691$. Under σ_y alone, $E_y = 1/(a_{22}h) = 5.5\,\text{N/mm}^2$. Under τ_{xy} alone, $G_{xy} = 1/(a_{66}h) = 1/(121900 \times 10^{-6} \times 2) = 4\,\text{N/mm}^2$. Note that the value of shear modulus G_{xy} under off-axis in-plane shear stress is substantially larger than that for the same ply under shear stress in its principal directions, this is quite normal but the magnitude of the increase depends on the material.

Outline 3.29: This shear force resultant induces midplane strains $\varepsilon_x° = a_{16}N_{xy} = -7.465\%$, $\varepsilon_y° = 1.532\%$ and $\gamma_{xy}° = 12.19\%$ as shown in Figure A3.7(a). From Equation (3.9) $\tau_{xy} = N_{xy}/h = (1\,\text{N/mm})/2\text{mm} = 0.5\,\text{MPa}$.

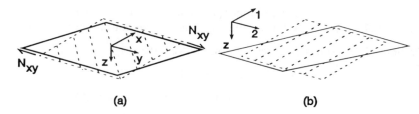

Figure A3.7 Deformation of NE30 under $N_{xy} = 1$ N/mm.

The stresses and strains in the principal directions are

$$\sigma_1 = 0.433 \text{ MPa}; \; \sigma_2 = -0.433 \text{ MPa} \text{ and } \tau_{12} = 0.25 \text{ MPa}; \text{ and}$$
$$\varepsilon_1 = 0.06471\%; \; \varepsilon_2 = -5.998\%; \text{ and } \gamma_{12} = 13.89\%$$

The strains in the principal directions are shown in Figure A3.7(b).

Outline 3.30: The boundary conditions are $\varepsilon_x = 0.05$, $N_y = N_{xy} = 0$. We find $N_x = \varepsilon_x/a_{11} = (0.05/56.63) \text{ kN/mm} = 0.8829 \text{ N/mm}$, so that $\varepsilon_y = a_{12}N_x = -3.457\%$, $\gamma_{xy} = a_{66}N_x = -6.591\%$.

Outline 3.31: Boundary conditions $\gamma_{xy} = 0.05$. $\varepsilon_x = \varepsilon_y = 0$ give $N_x = A_{16}\gamma_{xy} = 0.5874 \text{ kN/mm} \times 0.05 = 29.37 \text{ N/mm}$ $(\sigma_x = 14.68 \text{ MPa})$, $N_y = A_{26}\gamma_{xy} = 9.76 \text{ N/mm}$ $(\sigma_y = 4.88 \text{ MPa})$, $N_{xy} = A_{66}\gamma_{xy} = 17.16 \text{ N/mm}$ $(\tau_{xy} = 8.582 \text{ MPa})$. By use of the stress transformation matrix $[T]$ we find $\sigma_1 = 19.67 \text{ MPa}$, $\sigma_2 = -0.1004 \text{ MPa}$, $\tau_{12} = +0.045 \text{ MPa}$.

Outline 3.32: The resulting curvatures are $\kappa_x = d_{12}M_y = -11.75/\text{m}$, $\kappa_y = d_{22}M_y = 27.27/\text{m}$ and $\kappa_{xy} = d_{26}M_y = 4.595/\text{m}$, as shown schematically in Figure A3.8. The minor Poisson's ratio is $v_{yx} = -\kappa_x/\kappa_y = 11.75/27.27 = 0.4309$ (as expected from the response to the transverse load N_y).

Outline 3.33: (a) $\varepsilon_x^\circ = 5.663\%$, $\varepsilon_y^\circ = -3.916\%$, $\gamma_{xy}^\circ = -7.465\%$, $\kappa_x = -8.934/\text{m}$, $\kappa_y = 26.69/\text{m}$, $\kappa_{xy} = 0$. To achieve suppression of twisting, $M_{xy} = -0.01256 \text{ N}$. (b) $\varepsilon_x^\circ = 5.663\%$, $\varepsilon_y^\circ = -3.916\%$, $\gamma_{xy}^\circ = -7.465\%$, $\kappa_x = 0$, $\kappa_y = 2.368/\text{m}$, $\kappa_{xy} = 0$. To suppress the curvatures we need to apply $M_x = 0.2723 \text{ N}$ and $M_{xy} = 0.1541 \text{ N}$.

Outline 3.34: The free development of twisting curvature implies that $M_{xy} = 0$, hence $\kappa_{xy} = -(D_{16}/D_{66})\kappa_x = -17.11/\text{m}$, leading to $M_x = (D_{11} - D_{16}^2/D_{66})\kappa_x = 0.0838 \text{ N}$ and $M_y = (D_{12} - D_{16}D_{26}/D_{66})\kappa_x = 0.0361 \text{ N}$.

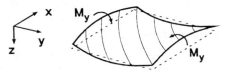

Figure A3.8 Deformation of NE30 under $M_y = 0.1$ N.

Outline 3.35: By matrix algebra we can construct a table of strain responses to unit force resultants (Table A3.1):

Outline 3.36: By matrix algebra we can construct a table of curvature responses to unit force resultants (Table A3.2):

Outline 3.37: For off-axis loading of a lamina we can use either

$$\begin{pmatrix} \varepsilon_x \\ \varepsilon_y \\ \gamma_{xy} \end{pmatrix} = \begin{pmatrix} S_{11}^* & S_{12}^* & S_{16}^* \\ S_{12}^* & S_{22}^* & S_{26}^* \\ S_{16}^* & S_{26}^* & S_{66}^* \end{pmatrix} \begin{pmatrix} \sigma_x \\ \sigma_y \\ \tau_{xy} \end{pmatrix}$$

or the form to be developed for laminate analysis, which we use here:

$$\begin{pmatrix} \varepsilon_x^\circ \\ \varepsilon_y^\circ \\ \gamma_{xy}^\circ \end{pmatrix} = \begin{pmatrix} a_{11} & a_{12} & a_{16} \\ a_{12} & a_{22} & a_{26} \\ a_{16} & a_{26} & a_{66} \end{pmatrix} \begin{pmatrix} N_x \\ N_y \\ N_{xy} \end{pmatrix}$$

Table A3.1

Loading	$\varepsilon_x^\circ \%$	$\varepsilon_y^\circ \%$	$\gamma_{xy}^\circ \%$
$+N_x$	$+0.01488$	-0.004572	-0.007127
$-N_x$	-0.01488	$+0.004572$	$+0.007127$
$+N_x+N_y$	$+0.01031$	$+0.01031$	-0.01425
$-N_x+N_y$	-0.01946	$+0.01946$	0
$+N_x-N_y$	$+0.01946$	-0.01946	0
$-N_x-N_y$	-0.01031	-0.01031	$+0.0142$
$+N_x+N_{xy}$	$+0.007756$	-0.0117	$+0.0183$
$-N_x+N_{xy}$	-0.02201	-0.002555	$+0.003256$
$+N_x-N_{xy}$	$+0.02201$	$+0.002555$	-0.003256
$-N_x-N_{xy}$	-0.007756	$+0.0117$	-0.0183
$+N_x+N_y+N_{xy}$	$+0.003184$	$+0.003184$	$+0.01118$
$-N_x+N_y+N_{xy}$	-0.02658	$+0.01233$	$+0.02543$
$+N_x-N_y+N_{xy}$	$+0.01233$	-0.02658	$+0.02543$
$+N_x+N_y-N_{xy}$	$+0.01744$	$+0.02744$	-0.03968
$-N_x-N_y+N_{xy}$	-0.01744	-0.02744	$+0.03968$
$-N_x-N_y-N_{xy}$	-0.003184	-0.003184	-0.01118
$+N_y$	-0.004572	$+0.01488$	-0.007127
$-N_y$	$+0.004572$	-0.01488	$+0.007127$
$+N_x+N_{xy}$	-0.0117	$+0.007756$	$+0.0183$
$-N_x+N_y$	-0.002555	-0.02201	$+0.03256$
$+N_y-N_{xy}$	$+0.002333$	$+0.02201$	-0.03256
$-N_y-N_{xy}$	$+0.0117$	-0.007756	-0.0183
$+N_{xy}$	-0.007127	-0.007127	$+0.02543$
$-N_{xy}$	$+0.007127$	$+0.007127$	-0.02543
$-N_x+N_y-N_{xy}$	-0.01233	$+0.02658$	-0.02543
$+N_x-N_y-N_{xy}$	$+0.02658$	-0.01233	-0.02543

Table A 3.2

Loading	κ_x/m	κ_y/m	κ_{xy}/m
$+M_x$	$+7.144$	-2.195	-3.421
$-M_x$	-7.144	$+2.195$	$+3.421$
$+M_x + M_y$	$+4.949$	$+4.949$	-6.842
$-M_x + M_y$	-9.339	$+9.339$	0
$+M_x - M_y$	$+9.339$	-9.339	0
$-M_x - M_y$	-4.949	-4.949	$+6.842$
$+M_x + M_{xy}$	$+3.723$	-5.615	$+8.785$
$-M_x + M_{xy}$	-10.56	-1.226	$+15.63$
$+M_x - M_{xy}$	$+10.56$	$+1.226$	-15.63
$-M_x - M_y$	-3.723	5.615	-8.785
$+M_x + M_y + M_{xy}$	$+1.528$	$+1.528$	5.364
$-M_x + M_y + M_{xy}$	-12.76	$+5.918$	$+12.21$
$+M_x - M_y + M_{xy}$	$+5.918$	-12.76	$+12.21$
$+M_x + M_y - M_{xy}$	$+8.37$	$+8.37$	-19.05
$-M_x - M_y + M_{xy}$	$+8.37$	-8.37	$+19.05$
$+M_x - M_y - M_{xy}$	-1.528	-1.528	-5.364
$+M_y$	-2.195	$+7.144$	-3.421
$-M_y$	$+2.195$	-7.144	$+3.421$
$+M_x + M_{xy}$	-5.615	$+3.723$	$+8.785$
$-M_y + M_{xy}$	-1.226	-10.56	$+15.63$
$+M_y - M_{xy}$	$+1.226$	10.56	-15.63
$-M_y - M_{xy}$	$+5.615$	-3.723	-8.785
$+M_{xy}$	-3.421	-3.421	$+12.21$
$-M_{xy}$	$+3.421$	$+3.421$	-12.21
$+M_x + M_y - M_{xy}$	-5.918	$+12.76$	-12.21
$+M_x - M_y - M_{xy}$	$+12.76$	-5.918	-12.21

(a) Let x be the hoop direction and y the axial direction in the tube. Under axial tension, N_y, we have

$$\varepsilon_x^{\circ} = -3.92\,N_y \quad \text{hoop contraction}$$

$$\varepsilon_y^{\circ} = +9.09\,N_y \quad \text{axial extension}$$

$$\gamma_{xy} = 1.53\,N_y \quad \text{small positive shear (positive twist)}$$

(b) Under internal pressure, p, we know that $N_x = 2\,N_y$, and hence

$$\varepsilon_x^{\circ} = (2a_{11} + a_{12})N_y = 9.4\,N_y$$

$$\varepsilon_y^{\circ} = (2a_{12} + a_{22})N_y = 1.25\,N_y$$

$$\gamma_{xy}^{\circ} = (2a_{16} + a_{26}) = -13.41\,N_y$$

We see that the tube increases in length and diameter, and develops large negative shear.

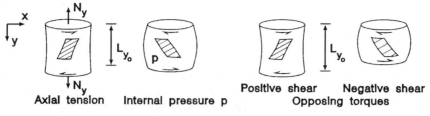

Figure A3.9 Deformations of thin-walled tube made from NE30.

(c) Under opposing torques N_{xy} at each end:

$$\varepsilon_x^\circ = a_{16} N_{xy} = -7.47 N_{xy}$$
$$\varepsilon_y^\circ = a_{26} N_{xy} = 1.53 N_{xy}$$
$$\gamma_{xy}^\circ = a_{66} N_{xy} = -12.2 N_{xy}$$

If N_{xy} is positive the tube contracts its diameter, extends slightly along its axis and twists in a positive sense. If N_{xy} is negative the tubes expands in diameter, decreases in length and twists in a negative sense.

The four responses are shown diagrammatically in Figure A3.9.

CHAPTER 4

Outline 4.1: $[\varepsilon^\circ]_L$ and $[\kappa]_L$ refer to the laminate as an entity and are obtained from externally applied force and moment resultants using the stiffnesses [A], [B] and [D] for the complete laminate. $[\varepsilon(z)]_f$ refers to the fth ply and is calculated as $[\varepsilon(z)]_f = [\varepsilon^\circ]_L + z[\kappa]_L$; for small curvatures we assume that $[\kappa]_f = [\kappa]_L$. Given $[\varepsilon(z)]_f$, we can then calculate the internal stress profiles using $[\sigma(z)]_f = [Q]_f [\varepsilon(z)]_f = [Q]_f [\varepsilon_L^\circ + z\kappa_L]$. It is commonplace to omit the subscripts f and L as the context always makes it clear which is intended.

Outline 4.2: $[D] = (1/3)\Sigma[Q]_f (h_f^3 - h_{f-1}^3)$ hence $D_{66} = (1/3)Q_{66}((+h/2)^3 - (-h/2)^3) = G_{12}h^3/12$, but $G = E/[2(1+v)] = E(1-v)/[2(1-v^2)]$, hence $D_{66} = Eh^3(1-v)/[12 \times 2(1-v^2)] = (1/2)D(1-v)$.

Outline 4.3: The thickness co-ordinates from the top surface of the laminate are $-h/2, -h/4, 0, +h/4$ and $+h/2$. We therefore can write: $[A] = [Q_a](-h/4 - (-h/2)) + [Q_b](h/4 - (-h/4)) + [Q_a](h/2 - h/4) = (h/2)\{[Q_a] + [Q_b]\}$ i.e. the extensional stiffness comes half from material a and half from material b.
$[D] = (1/3)[Q_a((-h/4)^3 - (-h/2)^3) + Q_b((h/4)^3 - (-h/4)^3) + Q_a((h/2)^3 - (h/4)^3)] = (h^3/96)(7Q_a + Q_b)$. Clearly the outer plies make a major contribution to the bending stiffness, and this confirms the analysis given in Section 2.5.6.

Outline 4.4: The strain response is a uniform $\gamma_{12} = a_{66}N_{12} = 0.03728\%$ in all layers, because they are bonded together. The stresses in the two materials are different because of the different moduli. Thus we have $\tau_{12}(\text{Al}) = Q_{66}(\text{Al})\gamma_{12}^\circ = 26\,\text{GPa} \times 0.0003728 = 9.692\,\text{MPa}$, and $\tau_{12}(\text{S}) = Q_{66}(\text{S})\gamma_{12}^\circ = 81.3\,\text{GPa} \times 0.0003728 = 30.31\,\text{MPa}$, as shown in Figure A4.1.

Outline 4.5: $E_1 = \sigma_1/\varepsilon_1 = 1/(a_{11}h_L) = 1/(7.19 \times 10^{-6} \times 1) = 139.1\,\text{GPa}$ $(= E_2$ because $a_{22} = a_{11})$. Using the formal definition $v_{12} = -\varepsilon_2/\varepsilon_1$ under a single uniform stress σ_1, $\varepsilon_1 = a_{11}N_1$, and $\varepsilon_2 = a_{12}N_1$, so $v_{12} = -a_{12}/a_{11} = -(-2.106)/7.19 = +0.293$. As the plies are isotropic, $v_{21} = v_{12}$. $G_{12} = 1/a_{66}h_L) = 1/(18.64 \times 10^{-6} \times 1) = 53.65\,\text{GPa}$.

Outline 4.6: It is helpful to distinguish two situations: (a) $\varepsilon_x^\circ = 0$; and (b) ε_x° free to contract.

(a) Applying $\varepsilon_x^\circ = 0$, $\varepsilon_y^\circ = 0.1\%$ induces $N_x = A_{12}\varepsilon_y^\circ = 44.56\,\text{N/mm}$ and $N_y = A_{22}\varepsilon_y^\circ = 152.1\,\text{N/mm}$, with the stress profiles shown in Figure A4.2(a).

(b) Applying $\varepsilon_y^\circ = 0.1\%$ with ε_x° free to contract induces $N_x = 0 = A_{11}\varepsilon_x^\circ + A_{12}\varepsilon_y^\circ$, so $\varepsilon_x^\circ = -A_{12}\varepsilon_y^\circ/A_{11} = -0.0293\%$, and $N_y = A_{12}\varepsilon_x^\circ + A_{22}\varepsilon_y^\circ = 139.1\,\text{N/mm}$, with the stress profiles (e.g. $\sigma_{xS} = Q_{12S}\varepsilon_y$) shown in Figure A4.2(b).

Outline 4.7: For $R_m = 0.02\,\text{m}$, $h = 0.001\,\text{m}$, $p = 2.5\,\text{MPa}$, $T = 100\,\text{Nm}$ we have $\sigma_H = pR_m/h = 50\,\text{MPa}$, $N_H = N/\text{mm}$, $N_A = 25\,\text{N/mm}$, $\tau_{AH} = T/(2\pi R_m^2 h) = 39.8\,\text{MPa}$, $N_{AH} = 39.8\,\text{N/mm}$. Let the axis of the tube correspond to the x direction in laminate analysis. $\varepsilon_A = a_{11}N_A + a_{12}N_H = 0.007445\%$, $\varepsilon_H = a_{12}N_A + a_{22}N_H = 0.03069\%$, and $\gamma_{AH} = a_{66}N_{AH} = 0.07418\%$. The change in enclosed volume is $\pi R_i^2 L[1 - (1 + \varepsilon_H)^2(1 + \varepsilon_A)] \approx \pi R_i^2 L(2\varepsilon_H + \varepsilon_A) = 0.54 \times 10^{-6}\,\text{m}^3$. The stress profiles are shown in Figure A4.3.

Outline 4.8: The resulting twisting curvature is given by

$$\kappa_{12} = d_{66}M_{12} = (161.3/\text{mmMN}) \times 2\,\text{N} = 0.3226/\text{m}$$

Figure A4.1 Response of SAL4S to $N_{12} = 20\,\text{N/mm}$.

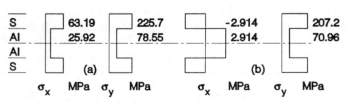

Figure A4.2 Response of SAL4S to: (a) $\varepsilon_x = 0$, $\varepsilon_y = 0.1\%$; (b) $\varepsilon_y = 0.1\%$.

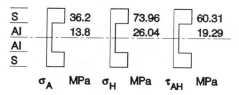

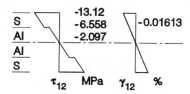

σ_A MPa σ_H MPa τ_{AH} MPa

Figure A4.3 Response of SAL4S tube to $p = 2.5$ MPa. $T = 100$ Nm.

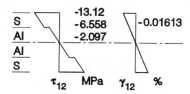

τ_{12} MPa γ_{12} %

Figure A4.4 Response of SAL4S to $M_{12} = 2$ N.

The strain profile varies linearly with thickness co-ordinate, $\gamma_{12}(z) = z\kappa_{12}$, and the stress profile shows a discontinuity at the interface between steel and aluminium because of the change of modulus (Figure A4.4). At the interface above the midplane, $h = -0.25$ mm, and we find

$$\tau_{12}(S) = Q_{66}(S) \times -0.25 \times 10^{-3} \times 0.3226 = -6.558 \text{ MPa}$$
$$\tau_{12}(Al) = Q_{66}(Al) \times -0.25 \times 10^{-3} \times 0.3226 = -2.097 \text{ MPa}$$

Outline 4.9: Allowing no transverse curvature to develop $(\kappa_x = 0)$, and applying $\kappa_y = 0.1$/m induces $M_y = \kappa_y/(d_{22} - d_{12}^2/d_{11}) = 1.728$ N and $M_x = -d_{12}M_y/d_{11} = 0.4875$ N with stress profiles shown in Figure A4.5(a).

Allowing transverse curvature κ_x to develop we find $M_x = 0$, $M_y = (D_{22} - D_{12}^2/D_{11})\kappa_y = 1.59$ N and $\kappa_x = -D_{12}\kappa_y/D_{11} = -0.02824$/m, with stress profiles in Figure A4.5(b).

Outline 4.10: The shear strain response, $\gamma_{12}^\circ = a_{66}N_{12} = 0.06147\%$, is uniform through the thickness, leading to stresses of $\sigma_1(S) = Q_{11}(S)\gamma_{12}^\circ = 49.98$ MPa and $\sigma_1(F) = 0.01482$ MPa.

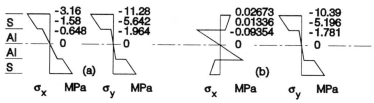

σ_x MPa σ_y MPa σ_x MPa σ_y MPa

Figure A4.5 Stress profiles in SAL4S: (a) $\kappa_2 = 0.1$/m, $\kappa_1 = 0$; (b) $\kappa_2 = 0.1$/m.

Outline 4.11: The curvatures are $\kappa_1 = d_{11}M_1 = 0.1472/\text{m}$ and $\kappa_2 = d_{12}M_1 = -0.04121/\text{m}$, as shown diagrammatically in Figure A 4.6(a). The stress and strain profiles are shown in Figure A 4.6(b), and show clearly that the foam acts merely as a separator: its role in bending is to prevent shear between the inside surfaces of the steel plies.

Outline 4.12: $A_{11} = A_{22} = 90.4 \text{ kN/mm}$, $A_{12} = 25.33 \text{ kN/mm}$, $A_{66} = 32.64 \text{ kN/mm}$. $D_{11} = D_{22} = 610.8 \text{ kNmm}$, $D_{12} = 171.1 \text{ kNmm}$, $D_{66} = 220.2$ kNmm. In-plane stiffnesses are not affected because increasing the foam from 0.6 to 5 mm contributes negligible in-plane stiffness. The bending stiffnesses are dramatically increased because the steel is now much further from the midplane of the laminate.

Outline 4.13: Taking steel as the top ply, $B_{11} = (1/2)[Q_{11S}(0 - (-h/2)^2) + Q_{11A}((h/2)^2 - 0)] = (Q_{11A} - Q_{11S})(h^2/8) = (78.55 - 225.7) \times 10^3 \times 1/8 = -18.394 \text{ kN}$. Other values of B_{ij} are given in Table 4.6. If the top layer were aluminium, then $B_{11} = +18.394 \text{ kN}$.

Outline 4.14: Consider a laminate of thickness h, made from two materials a and b (Figure A4.7), having thicknesses $h_a = (h/2)(1 - \alpha)$ and $h_b = (h/2)(1 + \alpha)$. $[B] = (1/2)[[Q]_a((-\alpha h/2)^2 - (-h/2)^2) + [Q]_b((h/2)^2 - (-\alpha h/2)^2)] = (h^2/8)[[Q]_a(\alpha^2 - 1) + ([Q]_a/[Q]_b)(1 - \alpha^2)]$. By differentiation $[B]$ has its maximum value when $\alpha = 0$ (i.e. plies have the same thickness) and when $[Q]_b \gg [Q]_a$.

Outline 4.15: For two plies of materials a and b (Figure A4.8) $[B] = (1/2)([Q]_a(0 - (-h/2)^2 + [Q]_b((h/2)^2 - 0)) = (h^2/8)(-[Q]_a + [Q]_b)$. So if $[Q]_a > [Q]_b$ then $[B]$ will be negative. Note that although B_{ij} may be positive

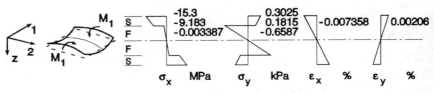

Figure A 4.6 Curvatures and profiles for MSF232S under $M_1 = 2$ N.

Figure A 4.7 Thickness co-ordinates for non-regular non-symmetric laminate.

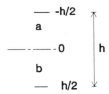

Figure A4.8 Thickness co-ordinates for regular nonsymmetric laminate.

or negative, [b] and [h] may contain terms of either sign when the full ABD matrix is inverted.

Outline 4.16: The deformation responses are $\varepsilon_1^\circ = 0.00263\%$, $\varepsilon_2^\circ = -0.000915\%$, $\kappa_1 = 0.212/m$ and $\kappa_2 = -0.0656/m$. The midplane strains are small, and the transverse curvature is negligible in this example. The strain profiles are linear through the thickness as expected, and are shown in Figure A4.9. The longitudinal stress profile is linear through the thickness of each ply, but there is a discontinuity at the interface between the aluminium and the steel caused by the different values of modulus. The transverse stress profile is calculated from

$$\sigma_2 = Q_{12}(\varepsilon_1^\circ + z\kappa_1) + Q_{22}(\varepsilon_2^\circ + z\kappa_2) \tag{A4.1}$$

using appropriate materials constants for each ply.

Outline 4.17: For two plies the surface co-ordinates from the top are $-h/2(a)$, 0, $(b) + h/2$. $2[B] = [Q_a](0 - (-h/2)^2) + [Q_b](h/2)^2 - 0)$, hence $[B] = (h^2/8)$ $([Q_a] - [Q_b])$.

For four plies $a/b/a/b$, the reader will find $[B] = (h^2/16)([Q_b] - [Q_a])$; and in general for n plies alternating in pairs a/b, $[B] = (h^2/4n)([Q_b] - [Q_a])$. As the number of alternating plies increases, the laminate behaves increasingly as if it were more symmetric.

Outline 4.18: We shall discuss three different responses.

(a) Application of $\varepsilon_y^\circ = 0.1\%$, with no imposed restraints. We realize that $N_x = M_x = M_y = 0$ so from Equation (4.22) we find $N_y = \varepsilon_y^\circ/a_{22} = 113.3$ N/mm, $\varepsilon_x^\circ = a_{12}N_y = -0.031\%$, $\kappa_x = h_{12}N_y = -0.5183/m$ and $\kappa_y = 1.489/m$. The stress profiles are in Figure A4.10(a).

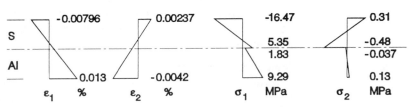

Figure A4.9 Response of SAL2NS to $M_1 = 2$ N.

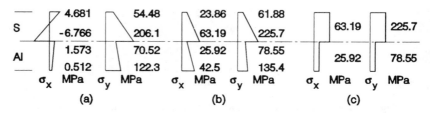

Figure A4.10 Responses of SAL2NS to: (a) $\varepsilon_y^\circ = 0.1\%$; (b) $\varepsilon_x^\circ = 0$, $\varepsilon_y^\circ = 0.1\%$; (c) $\varepsilon_y^\circ = 0.1\%$, $\varepsilon_x^\circ = \kappa_x = \kappa_y = 0$.

(b) Application of $\varepsilon_y^\circ = 0.1\%$, with ε_x° prevented. Putting $\varepsilon_x^\circ = 0 = M_x = M_y$ gives $N_y = \varepsilon_y/(a_{22} - a_{12}^2/a_{11}) = 125.4$ N/mm and $N_x = -a_{12}N_y/a_{11} = 38.87$ N/mm, so that $\kappa_x = h_{11}N_x + h_{12}N_y = -0.06285/$m and $\kappa_y = 1.469/$m. Stress profiles in Figure A4.10(b).

(c) Application of $\varepsilon_y^\circ = 0.1\%$, with $\varepsilon_x^\circ = \kappa_x = \kappa_y = 0$. To obtain this heavily restrained response we need to apply Equation (4.19), giving $N_x = A_{12}\varepsilon_y^\circ = 44.56$ N/mm, $N_y = A_{22}\varepsilon_y^\circ = 152.1$ N/mm, $M_x = B_{12}\varepsilon_y^\circ = -4.659$ N and $M_y = B_{22}\varepsilon_y^\circ = -18.39$ N. Stress profiles are shown in Figure A4.10(c).

Outline 4.19: Both have the same amount of steel and of aluminium. [D] is a function of $(h_f^3 - h_{f-1}^3)$ from Equation (4.13), so whether the contribution of any ply is above or below the midplane at the same location $|h(f)|$ is not important for laminates made from isotropic plies. For SAL2NS the ply surface co-ordinates are $-h/2$, 0, $+h/2$, so $D_{11} = (1/3)\{Q_{11S}(0 - (-h/2)^3 + Q_{11A}((+h/2)^3 - 0)\} = (h^3/24)(Q_{11S} + Q_{11A})$. For SAL4NS, ply surfaces are $-h/2$, $-h/4$, 0, $+h/4$, $+h/2$, hence $D_{11} = (1/3)\{Q_{11S}((-h/4)^3 - (-h/2)^3 + (+h/4)^3 - 0) + Q_{11A}(0 - (-h/4)^3 + (+h/2)^3 - (+h/4)^3)\} = (h^3/24)(+h^3/24)(Q_{11S} + Q_{11A})$.

CHAPTER 5

Outline 5.1: From Section 3.3.7, the axial stiffness of the coil is $(F/x)_c = Ed^4 \cos \theta/[16D^3n(1 + v \cos \theta)]$, and using $v = 0.33$ as a typical value, $(F/x)_c = (Ed\sqrt{3})/200\,000$. Ignoring any straight ends, the length of yarn in five furns is $L_y = n\pi D/\cos \theta = 100\pi d/\sqrt{3}$. The axial stiffness of the straight yarn is $(F/x)_y = EA/L = (Ed\sqrt{3})/400$. The ratio of the stiffnesses is $R = (F/x)_y/(F/x)_c = 500$. More turns or a larger coil diameter would increase the ratio dramatically. This example merely serves to show how large the effect of curling, twisting or coiling can be.

Outline 5.2: (b) Each filament is 4 denier; 9 km of filament has a mass of 4 g, so 40 g of filament is 90 km long. The volume of filaments is $V = 40$ g/$(1350$ kg/m$^3) = 2.96 \times 10^{-5}$m^3, and hence the radius of the filament is given by $\pi R^2 L = V$, so that $R = 10$ μm and the fibre diameter is 20 μm.

Outline 5.3: For a symmetric four-ply stack $(0°/90°)_s$ we find $D_{11}/D_{22} = (7Q_{11} + Q_{22})/(Q_{11} + 7Q_{22})$. For an eight-ply stack $(0°/90°/0°/90°)_s$ of the same total thickness, $D_{11}/D_{22} = (11Q_{11} + 5Q_{11})/(5Q_{11} + 11Q_{22})$. Clearly convergence to $D_{11} \approx D_{22}$ is rapid as the number of plies increases.

Outline 5.4: $A_{11} = 2Q_{11}(0)h_f + 2Q_{11}(90)h_f = 2 \times 0.125(180.7 + 8.032) = 47.19 \, \text{kN/mm}$. For calculation of D_{12} it is better to use the formal ply co-ordinates.

$$D_{12} = (\tfrac{1}{3})\{Q_{12}(0)((-h_L/4)^3 - (-h_L/2)^3 + (h_L/2)^3 - (h_L/4)^3)$$
$$+ Q_{12}(90)((+h_L/4)^3 - (-h_L/4)^3)\}$$
$$= (h_L^3/12)Q_{12} = (0.5^3/12) \times 2.41 = 0.0251 \, \text{kNmm}$$

Outline 5.5: (a) Under N_x alone the longitudinal strain profile is uniform. If the plies were unbonded, the $0°$ plies would contract laterally much more $(v_{12} = 0.3)$ than the $90°$ plies $(v_{21} = 0.3 \times 8/180 = 0.0133)$. To achieve uniform contraction, an internal tension is applied to the $0°$ plies and an internal compression to the $90°$ plies.

(b) The discontinuity in bending stress $\sigma_y(z)$ at the interface arises from a difference in modulus in the plies either side of the interface, using the same kind of argument based on v_{12} and v_{21}.

Outline 5.6: Under a single load N_x and using the data in Table 5.2, we find $E_x = 1/(a_{11}h_L) = 1/(21.21 \times 10^{-6} \times 0.5) = 94.3 \, \text{GPa}$; $v_{12} = -a_{12}/a_{11} = -(-0.5414)/21.21 = +0.0255$. For this crossply laminate $v_{21} = v_{12}$. Note that v_{12} for the laminate is substantially less than the major ratio for the single ply but greater than the minor ratio. $G_{xy} = 1/(a_{66}h_L) = 1/(400 \times 10^{-6} \times 0.5) = 5 \, \text{GPa}$. The result for the shear modulus is expected because both the $0°$ and the $90°$ plies are under the same shear stress and achieve the same shear strain.

Outline 5.7: The strain response is $\gamma_{xy} = 0.4\%$ under a uniform shear stress $\tau_{xy} = 20 \, \text{MPa}$. $\tau_{xy}(z)$ is uniform because the transformed reduced shear stiffness Q_{66} is the same in plies oriented at $0°$ and $90°$ to the reference x direction. The values of $Q_{11}(0°)$ and $Q_{11}(90°)$ are different, hence the discontinuity at the interfaces between differently oriented plies.

Outline 5.8: The resulting curvatures are $\kappa_x = d_{11}M_x = 0.604/\text{m}$ and $\kappa_y = d_{12}M_x = -0.0491/\text{m}$. These curvatures correspond to the stress and strain profiles in Figure A5.1.

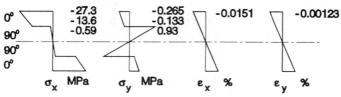

Figure A5.1 Response of $C + 4S$ to $M_x = 1 \, \text{N}$.

Outline 5.9: (a) Assuming ε_x° is free to contract we find $\varepsilon_y^\circ = -\nu_{12}\varepsilon_x^\circ = -0.0255 \times 0.02 = -0.00051\%$. Hence $N_x = A_{11}\varepsilon_x^\circ + A_{12}\varepsilon_y^\circ = 9.432\,\text{N/mm}$. $N_y = A_{12}\varepsilon_x^\circ + A_{22}\varepsilon_y^\circ = 0$. The stress in the top 0° ply is $\sigma_x = Q_{11}(0^\circ)\varepsilon_x^\circ + Q_{12}(0^\circ)\varepsilon_y^\circ = 36.13\,\text{MPa}$. In the 90° ply care must be taken to identify the appropriate coefficients of $[Q]$. We calculate the stress σ_x in the 90° ply as $\sigma_x = Q_{11}(90^\circ)\varepsilon_x^\circ + Q_{12}(90^\circ)\varepsilon_y^\circ = 1.594\,\text{MPa}$.

(b) Assuming $\varepsilon_x^\circ = 0.02\%$ and $\varepsilon_y^\circ = 0$ we find $N_x = A_{11}\varepsilon_x^\circ = 9.438\,\text{N/mm}$ and $N_y = 0.241\,\text{N/mm}$, leading to the stress profiles in Figure A5.2.

Outline 5.10: Assuming the 0° corresponds to the axis of the drum, we have $\kappa_x = 0$, $\kappa_y = 0.5/\text{m}$. Hence $M_x = D_{12}\kappa_y = 0.01255\,\text{N}$ and $M_y = D_{22}\kappa_y = 0.1543\,\text{N}$, giving the stress profiles shown in Figure A5.3.

Outline 5.11: For the unbalanced regular symmetric crossply laminate FSP stressed in its principal directions, $A_{11} \neq A_{22}$, and hence the midplane direct strains under N_y are different from those under N_x. In detail we find $\varepsilon_x^\circ = a_{12}N_y = -1.787\,\text{mm/MN} \times 10\,\text{N/mm} = -0.001783\%$; and $\varepsilon_y^\circ = a_{22}N_y = 0.06251\%$. the midplane shear strains are the same under both N_x and N_y because A_{16} (and a_{16}) are the same. The stress profiles (Figure A5.4) reflect the differences in A_{11} and A_{22}.

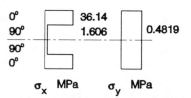

Figure A5.2 Response of C + 4S to $\varepsilon_y^\circ = 0$ and $\varepsilon_x^\circ = 0.02\%$.

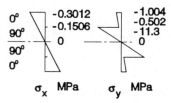

Figure A5.3 Response of C + 4S to $\kappa_x = 0$ and $\kappa_y = 0.5/\text{m}$.

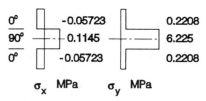

Figure A5.4 Response of FSP to $N_y = 10\,\text{N/mm}$.

Outline 5.12: $\kappa_x = d_{11}M_x = 13.78\,/\text{mmMN} \times 1\,\text{N} = 0.01378/\text{m}$; $\kappa_y = d_{12}M_x = -0.003729/\text{m}$.

Outline 5.13: Under $N_x = 10\,\text{N/mm}$, $v_{xy} = 0.001787/0.0329 = 0.05419$. Under $N_y = 10\,\text{N/mm}$, $v_{yx} = 0.001787/0.06251 = 0.02859$.

$E_x = 1/(a_{11}h_L) = 1/(32.91 \times 10^{-6} \times 4.5) = 6.752\,\text{GPa}$. $E_y = 1/(a_{22}h_L) = 3.555\,\text{GPa}$. $G_{xy} = 1/(a_{66}h_L) = 0.364\,\text{GPa}$.

Outline 5.14: If we designate in the two-ply laminate the top ply as a and the bottom ply as b, we find from $B = (\frac{1}{2})\Sigma Q_f(h_f^2 - h_{f-1}^2)$ that

$$B_{11} = (\tfrac{1}{2})[Q_{11a}(0°)(0 - (-h/2)^2 + Q_{11b}(90°)((h/2)^2 - 0)]$$

but plies a and b have the same properties but orientated differently, so we can write $Q_{11a}(0°) = Q_{11}$, and $Q_{11b}(90°) = Q_{22}$, hence

$$B_{11} = (\tfrac{1}{2})[Q_{11}(-h^2/4) + Q_{22}(+h^2/4)] = (h^2/8)[Q_{22} - Q_{11}]$$

Similarly

$$B_{22} = (\tfrac{1}{2})[Q_{22}(0 - (-h/2)^2) + Q_{11}((h/2)^2 - 0)] = (h^2/8)[Q_{11} - Q_{22}] = -B_{11}$$

For a non-symmetric arrangement $(0°/90°/0°/90°)_T$ of total thickness h:

$$B_{11} = (\tfrac{1}{2})[Q_{11}((-h/4)^2 - (-h/2)^2 + Q_{22}(0 - (-h/4)^2) + Q_{11}((h/2)^2 - 0)$$
$$+ Q_{22}((h/2)^2 - (h/4)^2)]$$

from which $B_{11} = (h^2/16)[Q_{22} - Q_{11}]$, and hence as the number of plies increases, the size of B decreases at constant thickness of laminate, and hence the non-symmetric character of the laminate diminishes.

Outline 5.15: By symmetry $Q_{12} = Q_{21}$, and hence $B_{12} = (\tfrac{1}{2})[Q_{12}((-h/2)^2 - 0) + Q_{12}(0 - (h/2)^2)] = 0$, and so the corresponding inverse coefficient $b_{12} = 0$. $\kappa_y = b_{12}N_x = 0$. $N_x = 0$.

Outline 5.16: It would seem essential to apply only a moment M_x to eliminate the curvature κ_x, but further thought will confirm that this will also induce an anticlastic curvature κ_y. To suppress both curvatures it is necessary to apply moments M_x and M_y via edge clamps, which may be calculated from the simultaneous equations $\varepsilon_x = a_{11}N_x + b_{11}M_x + b_{12}M_y$; $\kappa_x = b_{11}N_x + d_{11}M_x + d_{12}M_y$; $\kappa_y = b_{12}N_x + d_{12}M_x + d_{22}M_y$. Noting that $\kappa_x = 0 = \kappa_y = b_{12}$, we find $M_y = -(d_{12}/d_{22})M_x = -b_{11}N_x/(d_{11} - d_{12}^2/d_{22})$. Substituting numerical values, $M_x = -1.144\,\text{N}$, $M_y = -0.02922\,\text{N}$, leading to $\varepsilon_x = 0.02121\%$ and $\varepsilon_y = -0.0005414\%$.

Outline 5.17: Taking $N_x = N_y = 10\,\text{N/mm}$, we can calculate $\varepsilon_x = \varepsilon_y = 0.05558\%$, $\kappa_x = 3.131/\text{m}$ and $\kappa_y = -3.131/\text{m}$. For $N_{xy} = 10\,\text{N/mm}$ we obtain $\gamma_{xy} = 0.4\%$.

Outline 5.18: $\varepsilon_x^\circ = 0 = \varepsilon_y^\circ = \gamma_{xy}^\circ = \kappa_x = \kappa_y$. $\kappa_{xy} = d_{66}M_{xy} = 19\,200/\text{MNmm} \times 1\,\text{N} = 19.2/\text{m}$.

Outline 5.19: The simultaneous equations to be solved are $\kappa_x = d_{11}M_x + d_{12}M_y$ and $\kappa_y = d_{12}M_x + d_{22}M_y$, where $M_x = 1$ N and $\kappa_y = 0$. κ_x and M_y are as yet unknown. We see $M_y = -(d_{12}/d_{22})M_x$, and hence $\kappa_x = (d_{11} - d_{12}^2/d_{22})M_x$. This leads to $M_y = +0.02553$ N; $\kappa_x = 2.736/$m.

Outline 5.20: (a) Under the boundary conditions $\varepsilon_x^\circ = 0$, $\varepsilon_y^\circ = 0.005\%$ we expect curvature to develop under the developed force resultants. Bearing in mind that $\varepsilon_x^\circ = B_{12} = 0$, we have to solve $N_x = A_{12}\varepsilon_y^\circ + B_{11}\kappa_x$; $N_y = A_{22}\varepsilon_y^\circ + B_{22}\kappa_y$; $(M_x =)0 = D_{11}\kappa_x + D_{12}\kappa_y$; and $(M_y =)0 = B_{22}\varepsilon_y^\circ + D_{12}\kappa_x + D_{22}\kappa_y$. Hence $\kappa_x = B_{22}D_{12}\varepsilon_y^\circ/(D_{11}D_{22} - D_{12}^2) = 7.013 \times 10^{-3}/$m; $\kappa_y = -(D_{11}/D_{12})\kappa_x = -0.2746/$m, $N_x = 0.0224$ N/mm, and $N_y = 0.8773$ N/mm. The stress profiles are in Figure A5.5(a).

(b) $N_x = A_{12}\varepsilon_y^\circ = 0.06024$ N/mm, $N_y = A_{22}\varepsilon_y^\circ = 2.359$ N/mm, $M_x = B_{12}\varepsilon_y^\circ = 0$; $M_y = B_{22}\varepsilon_y^\circ = 0.2698$ N. As curvatures have been suppressed, stresses through any ply are constant, as seen in Figure A5.5(b).

Outline 5.21: Most umbrellas rely on the fabric to bend the radial stays thus achieving a stiff membrane (of cloth) which will not invert in gusty winds. The fabric properly cut 'on the square' at the free edge will exert sufficient force on the stays to achieve a stable cover. Fabric cut to give a warp and weft at 45° to the free edge will stretch too much and will be unsatisfactory in gusty winds.

Outline 5.22: Let the fibre length between two nodal points of the braid be N, with longitudinal node spacing L and diametral node spacing D. The length of fibre between nodes is constant, according to $N^2 = L^2 + D^2$. The braid angle θ is given by $\tan\theta = D/L$.

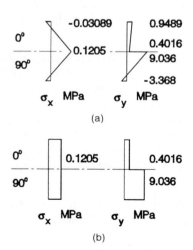

Figure A5.5 (a) Response of C + 4NS to $\varepsilon_x = 0$, $\varepsilon_y = 0.005\%$; (b) Response of C + 4NS to $\varepsilon_y = 0.005\%$, $\varepsilon_x = \kappa_x = \kappa_y = 0$.

Using the diameter as a subscript, for the given undeformed braid, $\tan\theta_{25} = 1$, so $L_{25} = D_{45}$, and $n = D_{25}\sqrt{2}$. For the 20 mm former we have $N^2 = L_{20}^2 + D_{20}^2 = 2D_{25}^2$, so $L_{20}^2 = 2 \times 625 - 400 = 850$, hence $L_{20} = 29.15$ mm, and $\theta_{20} = \tan^{-1}(D_{20}/L_{20}) = \tan^{-1}(20/29.15) = 34.45°$. Similarly for the 30 mm former: $N^2 = L_{30}^2 + D_{30}^2 = 2D_{25}^2$: $L_{30} = 18.7$ mm, $\theta_{30} = 58°$.

Outline 5.23: Maximum shear stiffness obtains when $\theta = \pm 45°$. For an inextensible braid of negligible thickness compared with the diameter we have for a 25 mm tube: $\tan\theta_{25} = D_{25}/L_{25}$, where $\theta_{25} = 30°$, hence $L_{25} = 25/\tan 30 = 43.33$ mm, and the distance between nodes is $N^2 = 25^2 + 43.33^2$. For $\theta = 45°$, $L = D$, so $N^2 = 2D^2$: $D^2 = N^2/2 = (625 + 1877)/2$, so $D = 35.37$ mm.

Outline 5.24: For a unidirectional lamina stressed at θ to its principal direction, or for a symmetrical angleply laminate stressed at θ to its principal directions, we can write

$$g = 1/G_{xy} = A\cos^2\theta\sin^2\theta + B(\sin^4\theta + \cos^4\theta)$$
$$= B + (A - 2B)\sin^2\theta - (A - 2B)\sin^4\theta$$

i.e. $dg/d\theta = 2(A - 2B)\sin\theta\cos\theta(1 - 2\sin 2\theta) = 0$ for a minimum, so *one* root is $\sin\theta = 1/\sqrt{2}$.

Outline 5.25: Using the data for a_{ij} in Table 5.6 we find $E_x = 1/(a_{11}h_L) = 1/(3225 \times 10^{-6} \times 4) = 77.52$ MPa. $v_{12} = -a_{12}/a_{11} = -(-8817)/3225 = 2.734$. This is an exceptionally high value, even for composite materials. $E_y = 1/(a_{22}h_L) = 9.56$ MPa. $G_{xy} = 1/(a_{66}h_L) = 327.4$ MPa. Note that although $G_{12} = 2.5$ MPa when stressed in the principal directions, the application of τ_{xy} in global co-ordinates gives a much larger value of $G_{xy}(+30°)$.

Outline 5.26: $N_x = A_{11}\varepsilon_x^\circ = 11.84$ N/mm ($\sigma_x = 2.96$ MPa), $N_y = A_{12}\varepsilon_x^\circ = 3.991$ N/mm ($\sigma_y = 0.9978$ MPa). Using the transformation matrix, $\sigma_1 = 3.93$ MPa, $\sigma_2 = 0.0278$ MPa. The shear stress and strain profiles are shown in Figure A5.6.

Outline 5.27: If we assume that the sheet is free to twist, we therefore must solve the three equations $M_x = D_{12}\kappa_y + D_{16}\kappa_{xy}$; $M_y = D_{22}\kappa_y + D_{26}\kappa_{xy}$; $M_{xy} = 0 = D_{26}\kappa_y + D_{66}\kappa_{xy}$. Hence $\kappa_{xy} = -(D_{26}/D_{66})\kappa_y = -0.2142$/m, $M_y = 0.1642$ N and $M_x = 0.4053$ N. The stress profiles are in Figure A5.7.

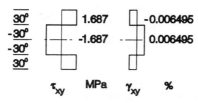

Figure A5.6 Shear response of RRX30S to $\varepsilon_x = 0.3\%$, $\varepsilon_y = 0$.

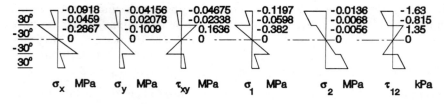

Figure A5.7 Responses of RRX30S to $\kappa_x = 0$, $\kappa_y = 0.5/\text{m}$.

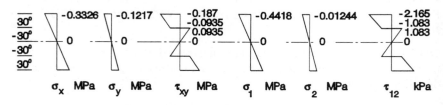

Figure A5.8 Responses of RRX30S to $\kappa_x = \kappa_{xy} = 0$, $\kappa_y = 0.5/\text{m}$.

Outline 5.28: (a) $\kappa_x = 0$, $\kappa_y = 0.5/\text{m}$, $\kappa_{xy} = 0$. We assume that the curvature of all the plies is κ_y. Hence $M_x = D_{12}\kappa_y = 0.8869\,\text{N}$, $M_y = D_{22}\kappa_y = 0.3244\,\text{N}$, $M_{xy} = 0.374\,\text{N}$. The stress profiles are in Figure A5.8.

(b) $\kappa_x = 0.5/\text{m}$, $\kappa_y = 0$, $\kappa_{xy} = 0$. $M_x = D_{11}\kappa_x = 2.631\,\text{N}$, $M_y = D_{12}\kappa_x = 0.8869\,\text{N}$, $M_{xy} = D_{16}\kappa_x = 1.124\,\text{N}$.

Outline 5.29: The laminate is regular, with a total thickness h, so the general co-ordinate of the fth lamina lower boundary will be $h_f = -(h/2) + fh/n$, where n is the total (even) number of laminae, and $h_{f-1} = -h/2 + (f-1)h/n$. Using the definition of [B], we have for any element

$$B = (\tfrac{1}{2}) \sum_{f=1}^{n} Q^*(h_f^2 - h_{f-1}^2) = (\tfrac{1}{2}) \sum_{f=1}^{n} Q^*(h_f + h_{f-1})(h_f - h_{f-1})$$

and

$$(h_f + h_{f-1})(h_f - h_{f-1}) = -h^2/n\{1 - (2f-1)/n\}$$

For B_{11}, B_{12}, B_{22} and $B_{66}, Q^*(+\theta) = Q^*(-\theta)$, and these $B_{ij} = 0$. For B_{16} and $B_{26}, Q^*(-\theta) = -Q^*(+\theta)$, so it is necessary to write

$$B_{16} = -(h^2 Q_{16}^*/2n) \sum_{f=1}^{n} \{(-1)^f [1 - (2f-1)/n]\} = h^2 Q_{16}^*/2n$$

i.e. B_{16} becomes small as n becomes large, so that the B matrix tends to zero and the laminate becomes sensibly symmetric.

Outline 5.30: The balanced symmetrical laminate of thickness h is wound as $h/4$ at $+\theta$, $h/2$ at $-\theta$, and $h/4$ at $+\theta$, where θ is the angle to the axial x direction. All plies have the same stiffness properties. The hoop stress is

$\sigma_y = pD_m/2h$, and the axial stress in the casing is $\sigma_x = pD_m/4h$. For a symmetric laminate $[\varepsilon] = [A]^{-1}[N]$, where $A_{11} = hQ_{11}^*$, $A_{12} = hQ_{12}^*$, $A_{22} = hQ_{22}^*$, $A_{66} = hQ_{66}^*$ and $A_{16} = A_{26} = 0$. Denoting $J = (Q_{11}^* \ Q_{22}^* - Q_{12}^{*2})$ for convenience,

$$\begin{pmatrix} \varepsilon_x \\ \varepsilon_y \\ \gamma_{xy} \end{pmatrix} = \begin{pmatrix} Q_{22}^*/J & -Q_{12}^*/J & 0 \\ -Q_{12}^*/J & Q_{11}^*/J & 0 \\ 0 & 0 & Q_{66}^* \end{pmatrix} \begin{pmatrix} \sigma_x \\ \sigma_y \\ \tau_{xy} \end{pmatrix}$$

Hence the required condition for zero hoop strain is $2Q_{11}^* = Q_{12}^*$, provided $J \neq 0$. Substituting for Q_{ij}^* in terms of Q_{ij} gives a quadratic of the form $a \sin^4 \theta + b \sin^2 \theta + c = 0$. The condition is satisfied is $b^2 > 4ac$. Substituting numerical values, suitable winding angles would be 69° or 65.27°. This is not a complete solution to the problem, of course; for example the strength also has to be checked, especially in the axial direction.

Outline 5.31: Using the data in Table 5.7 for 70°, $v_{yx} = -\varepsilon_x^\circ/\varepsilon_y^\circ = 4.988$.

Outline 5.33: See Figure A5.9.

Outline 5.34: (a), (c), and (e) are antisymmetric; (b) is symmetric; (d) is only 'antisymmetric with respect to θ_1', so is not truly antisymmetric. In (c) there is no distinction between $+0$ and -0, so the 0° ply (notionally divided) causes no problem.

Outline 5.35: The geometrical boundary conditions are not completely specified. One solution (a) is to assume that no twisting moment is applied, and an alternative (b) is to specify that $\kappa_{xy} = 0$, the choice in practice depending on the real local conditions for the design study.

(a) We solve $N_x = A_{11}\varepsilon_x^\circ + A_{12}\varepsilon_y^\circ + B_{16}\kappa_{xy}$; $N_y = A_{12}\varepsilon_x^\circ + A_{22}\varepsilon_y^\circ + B_{26}\kappa_{xy}$; $M_{xy} = 0 = B_{16}\varepsilon_x^\circ + B_{26}\varepsilon_y^\circ + D_{66}\kappa_{xy}$, whence $\kappa_{xy} = -(B_{16}\varepsilon_x^\circ + B_{26}\varepsilon_y^\circ)/D_{66} = 8.582/m$, $N_x = 7.091 \, N/mm$, and $N_y = 2.667 \, N/mm$. The stress profiles in global coordinates are shown in Figure A5.10.

(b) Preventing the development of κ_{xy} eliminates bending stresses. We find $N_x = A_{11}\varepsilon_x^\circ + A_{12}\varepsilon_y^\circ = 26.39 \, N/mm$ ($\sigma_x = 6.597 \, MPa$), $N_y = A_{12}\varepsilon_x^\circ + A_{22}\varepsilon_y^\circ =$

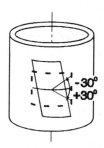

Figure A5.9 Deformation of element of (30/− 30/30) thin-walled tube under internal pressure.

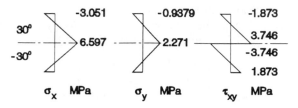

Figure A5.10 Responses in global co-ordinates of RRX30NS2 to $\varepsilon_x = \varepsilon_y = 0.5\%$.

9.085 N/mm ($\sigma_y = 2.271$ MPa), and $M_{xy} = B_{16}\varepsilon_x^\circ + B_{26}\varepsilon_y^\circ = -14.98$ N. The shear stress profile is $\tau_{xy}(+30^\circ) = 3.746$ MPa and $\tau_{xy}(-30^\circ) = -3.746$ MPa. The stresses in the principal directions are $\sigma_1 = 8.759$ MPa, $\sigma_2 = 0.1086$ MPa, $\tau_{12} = 0$.

Outline 5.36: Considering only blade forces N_x caused by rotation of the blades about one end, the greatest amount of twist per unit axial load would obtain by maximizing B_{16}, using a non-symmetric flat laminate, based on two plies laid at $(\theta/-\theta)_T$.

From the definition of [B], and recalling that $Q_{16}^*(-\theta) = -Q_{16}^*(+\theta)$, we have for this regular two ply laminate $B_{16} = -(h^2/2)Q_{16}^*(+\theta)$. Thus the simplest problem reduces to choice of angle to maximize B_{16}. From Chapter 3, $Q_{16}^* = (Q_{11} - Q_{12} - 2Q_{66})\cos^3\theta\,\sin\theta - (Q_{22} - Q_{12} - 2Q_{66})\,\cos\theta\,\sin^3\theta$, which we can write more conveniently as $Q_{16}^* = A\cos^3\theta\sin\theta + B\cos\theta\sin^3\theta$. Expressing in terms of $\cos^2\theta$ we find $dQ_{16}^*/d\theta = 4(A - B)\cos^4\theta + (5B - 3A)\cos^2\theta - B = 0$ for max or min; two roots are given by $\cos^2\theta = \{(3A - 5B) \pm \sqrt{((5B - 3A)^2 + 16B(A - B))}\}/(8(A - B))$.

Clearly the angle which maximizes $Q_{16}^*(\theta)$ depends on the stiffnesses of the material. For the ply HM-CARB having the reduced stiffnesses $Q_{11} = 180.7$ GPa, $Q_{12} = 2.41$ GPa, $Q_{22} = 8.032$ GPa, $Q_{66} = 5$ GPa, the maximum value of Q_{16}^* occurs at about 30°. (For many types of unidirectional composite, $Q_{11} \gg Q_{12}, Q_{22}, Q_{66}$, so 30° is a typical angle.)

This is the simplest possible approach. The blade has no aerodynamic profile, and no account has been taken of the bending stiffnesses of the blade. A fuller treatment based on a sensible practical arrangement of balanced woven glass fibre cloth on a aerofoil shaped former, with one principal axis of the cloth at $+20^\circ$ on the upper surface and -20° on the lower surface, is given by Karaolis *et al.* (1988).

Outline 5.37: $v_{xy} = v_{yx} = -a_{12}/a_{11} = 8.013/29.07 = 0.2756$.

Outline 5.38: The [A] and [a] matrices are the same, confirming that the laminate is indeed isotropic under in-plane loading. Under rotation, there is some decrease in bending stiffnesses D_{11} and D_{22}, and an increase in both D_{12} and D_{66}; D_{16} and D_{26} have changed sign, and the magnitude of D_{16} is considerably increased.

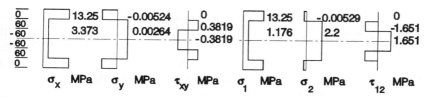

Figure A5.11 Responses of QI60S to $N_x = 10$ N/mm.

Outline 5.39: QI60S responds with a curvature $\kappa_{xy} = -4.519 \times 10^{-3} \times 20 = -0.09038$/m. QI60Sa shows much larger curvature of opposite sign $\kappa_{xy} = 52.25 \times 10^{-3} \times 20 = +1.045$/m. The isotropic sheet does not twist because $d_{16} = 0$.

Outline 5.40: The strains are $\varepsilon_x^\circ = a_{11}N_x = 29.07 \times 10^{-6} \times 10 = 2.907 \times 10^{-4}$ $\varepsilon_y^\circ = a_{12}N_x = -8.013 \times 10^{-5}$. The stresses in the fth ply are calculated using $[\sigma]_f = [Q^*]_f[\varepsilon^\circ]$, using the appropriate value of angle for the fth ply, and this leads to the profiles in Figure A5.11.

Outline 5.41: Because $\kappa_y = \kappa_{xy} = 0$, the bending generates moments M_y and M_{xy} as well as M_x. Solving the equations $\kappa_x = d_{11}M_x + d_{12}M_y + d_{16}M_{xy}$, $\kappa_y = 0 = d_{12}M_x + d_{22}M_y + d_{26}M_{xy}$, and $\kappa_{xy} = 0 = d_{16}M_x + d_{26}M_y + d_{66}M_{xy}$, we find $M_x = 2.589$ N, $M_y = 0.3307$ N and $M_{xy} = 0.06799$ N. The stress profiles are in Figure A5.12.

Outline 5.42: Laminates a, c, e and g can be excluded because they are non-symmetric. Laminate (b) only uses positive angles and will therefore include shear-extension coupling not present in a truly random-in-plane laminate. Laminate (d) is unsatisfactory because it contains two plies in the 0° and two in the 90° directions. Laminate (f) is a satisfactory representation, and it can be shown that it is quasi-isotropic, because it has plies oriented at angular intervals of $2\pi/n$, where $n = 6$.

Outline 5.43: The laminate is 1.2 mm thick. For N_x acting alone we have $\varepsilon_x = a_{11}N_x = a_{11}h\sigma_x$, hence $E_x = (a_{11}h)^{-1} = (10.99 \times 10^{-3} \times 1.2)^{-1}$ kN/mm^2 $= 75.83$ kN/mm$^2 = 75.83$ GPa. $E_y = (a_{22}h)^{-1} = E_x$. $G_{xy} = (a_{66}h)^{-1} = (28.87$

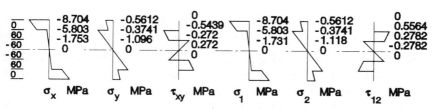

Figure A5.12 Response of QI60S to $\kappa_x = 0.25$/m, $\kappa_y = \kappa_{xy} = 0$.

$\times 10^{-3} \times 1.2)^{-1} = 28.86$ GPa. For N_x acting alone $\varepsilon_y = a_{12}N_x$, and $\varepsilon_x = a_{11}N_x$, hence $v_{xy} = -\varepsilon_y/\varepsilon_x = -a_{12}/a_{11} = +3.444/10.99 = 0.3134$. Note that for an isotropic material $G = E/(2(1 + v)) = 75.83/(2 \times 1.3134) = 28.87$ GPa. Thus under in-plane loading the laminate CRNDM7 is a good representation of an isotropic-in-the-plane random laminate. But in bending the representation is less convincing. $D_{16} \neq D_{26} \neq 0$, so there is twist-bend coupling, and v_{xy} taken as $-d_{12}/d_{11} = 0.34$, which is a little higher than the value of v_{xy} for in-plane loading. Bending behaviour is discussed further in Section 5.8.3.

Outline 5.44: Consider a unidirectional lamina stressed uniaxially in the x direction at θ to its principal direction: $\sigma_x = Q_{11}^*(\theta)\varepsilon_x$. For three identical laminae a, b, and c, inclined at $\theta_a, \theta_b, \theta_c$, the total stress will be $\sigma_x = (\frac{1}{3})[Q_{11}^*(\theta_a) + Q_{11}^*(\theta_b) + Q_{11}^*(\theta_c)]\varepsilon_x$. For a very large number of microlaminae randomly oriented in the plane, the total stress is given by

$$\sigma_x = \left(\frac{1}{\pi}\right)\int_{-\pi/2}^{+\pi/2} Q_{11}^*(\theta) d\theta \varepsilon_x$$

Substituting gives

$$Q_{11}^* = Q_{11}\cos^4\theta + 2(Q_{12} + 2Q_{66})\sin^2\theta\cos^2\theta + Q_{22}\sin^3\theta$$

$$E_r = \sigma_x/\varepsilon_x = [3Q_{11} + 2(Q_{12} + Q_{66}) + 3Q_{22}]/8$$
$$= 3E_1/8J + 3E_2/8J + v_{12}E_2/4J + G_{12}/2,$$

where

$$J = 1 - v_{12}v_{21}$$

Using $v_{12} \approx 0.3$ and assuming $E_f \gg E_p$, it follows that $J \approx 1$; micromechanics relations $1/E_2 = V_f/E_f + V_p/E_p$, and $1/G_{12} = V_f/G_f + V_p/G_p$ lead to $G_{12} \approx E_2/2.6$ and $v_{12}E_2/ = 0.075$. Thus the terms in E_2 amount to $0.267E_2$, i.e. about $E_2/4$.

Outline 5.45: To avoid in-plane direct shear coupling, use a balanced crossply or angleply laminate (Sections 5.3, 5.4, 5.6).

To eliminate all direct-bending coupling in flat plates use a symmetric laminate (Sections 5.3.2, 5.3.3, 5.5.1, 5.5.2, 5.5.5, 5.6.1).

To avoid bending-twisting use any crossply laminate (Section 5.3), or an antisymmetric laminate (5.5.3, also Table 5.19).

To avoid extension-bending, use any symmetric laminate, or an antisymmetric angleply laminate.

To eliminate extension-twisting use any crossply laminate or any symmetric angleply laminate.

Outline 5.46: To achieve balance ($A_{16} = A_{26} = 0$) we must have the same number of plies in the $+\theta$ and $-\theta$ directions. Two plies $(+\theta/-\theta)_T$ gives [B] $\neq 0$. Four plies as $(+\theta/-\theta)_s$ retains $D_{16} = D_{26} \neq 0$ and $(+\theta/-\theta/+\theta/-\theta)$ ensures [B] $\neq 0$. Six plies can achieve balance but not symmetry. Eight plies laid up as $(+\theta/-\theta/-\theta/+\theta/-\theta/+\theta/+\theta/-\theta)_T$ satisfies the requirements.

Outline 5.47: SL1 has a thickness $h_L/3$, with its ply co-ordinates (in units of $h_L/12$) $-2, -1, 0, +1, +2$. From the definition of $[B]$ we have $[B] = (h_L^2/288)\{[Q(\theta_1)]_1((-1)^2 - (-2)^2 + (+2)^2 - (+1)^2 + [Q(\theta_2)]_2(0^2 - (-1)^2 + (+1)^2 - 0^2\} = 0$. The approach for SL2 has the same format. For the laminate-as-a-whole the ply co-ordinates (in multiples of $h_L/12$) are $-6, -5, -4, -3, -2, 0, +2, +4, +6$; so $[B] = (h_L^2/288)\{[Q(\theta_1)]_1((-5)^2 - (-6)^2 + (-2)^2 - (-3)^2 + (2)^2 - 0 + (+4)^2 - (+2)^2) + [Q(\theta_2)]_2((-4)^2 - (-5)^2 + (-3)^2 - (-4)^2 + 0 - (-2)^2 + (+6)^2 - (+4)^2\} = 0$.

Outline 5.48: Consider the sublaminate $(\theta_1/\theta_2)_T$ having co-ordinates from the top $-h/2, 0$ and $+h/2$. We find terms such as

$$A_{11} = Q_{11}(\theta_1) \cdot (h/2) + Q_{11}(\theta_2) \cdot (+h/2) = (h/2)[Q_{11}(\theta_1) + Q_{11}(\theta_2)]$$
$$D_{11} = (\tfrac{1}{3})[Q_{11}(\theta_1)(0^3 - (-h/2)^3) + Q_{11}(\theta_2)((+h/2)^3 - 0^3)]$$

Hence

$$A_{11}/h = D_{11}/(h^3/12) = (\tfrac{1}{2})[Q_{11}(\theta_1) + Q_{11}(\theta_2)]$$

CHAPTER 6

Outline 6.1: Applying the stresses gives the individual results:

Uniaxial tensile stress	$\sigma_1 = +20\,\text{MPa}$	$F = 62$
Uniaxial compressive stress	$\sigma_1 = -20\,\text{MPa}$	$F = 11.5$
Uniaxial tensile stress	$\sigma_2 = +20\,\text{MPa}$	$F = 1.5$
Uniaxial compressive stress	$\sigma_2 = -20\,\text{MPa}$	$F = 7$
Shear stress	$\tau_{12} = 20\,\text{MPa}$	$F = 3$

As expected, it can be seen that the values of load factor vary widely with the character and direction of the applied stress.

Outline 6.2: Applying the Tsai–Hill criterion to the Kevlar-epoxy lamina we obtain the results shown in Table A6.1.

The values of Tsai–Hill load factors are mainly as expected. Note in particular the values of F for the stresses $(\sigma_1 = \sigma_2 = \tau_{12} = +20\,\text{MPa})$ is the same as that for $(\sigma_1 = 0, \sigma_2, = \tau_{12} = +20\,\text{MPa})$ because of the negative term in the Tsai–Hill equation. Indications of the most likely mode of failure under the loading conditions shown can, with due experimental care, be gauged from the stress/strength ratios, with the dominant term in bold type. For $\sigma_1 = \sigma_2 = -20\,\text{MPa}$ the two ratios are similar in value.

Outline 6.3: Calculating the limiting ratios using data for ply EG leads to the load factors shown in Table A6.2 using the maximum stress and strain criteria.

In each loading situation but one, the value of load factor obtained from the maximum stress or maximum strain criterion is less than that using the Tsai–Hill failure criterion, because interactions are ignored. For

Table A6.1 Failure data for 1 mm thick Kelvar/epoxy (see also data of Problem 6.1)

σ_1	σ_2	τ_{12}	F	σ_1/σ_{1max}	σ_2/σ_{2max}	τ_{12}/τ_{12max}
+20	+20	—	1.5	0.016	**0.667**	—
−20	+20	—	1.475	0.087	**0.667**	—
+20	−20	—	6.912	0.016	**0.143**	—
−20	−20	—	7	0.087	0.143	—
+20	—	20	2.996	0.016	—	**0.333**
−20	—	20	2.903	0.087	—	**0.333**
+20	+20	20	1.342	0.016	**0.667**	0.333
−20	+20	20	1.323	0.087	**0.667**	0.333
+20	−20	20	2.752	0.016	0.143	**0.333**
−20	−20	20	2.757	0.087	0.143	**0.333**
—	+20	20	1.342	—	**0.667**	0.333
—	−20	20	2.757	—	0.143	**0.333**

Table A6.2 Load factors for lamina 30 G

Load, N	σ_{1max}/σ_1	σ_{2max}/σ_2	τ_{12max}/τ_{12}	$\varepsilon_{1max}/\varepsilon_1$	$\varepsilon_{2max}/\varepsilon_2$	$\gamma_{1max}/\gamma_{12}$
N_x 10	267	80	**37**	295	99	**37**
N_y 10	800	**27**	37	4490	**27**	37
N_{xy} 10	231	62	**32**	180	52	**32**
N_x 10 $\left.\right\}$ N_y 10	200	20	—	275	21	—
N_x 10 $\left.\right\}$ N_y 10 N_{xy} 10	107	149	**32**	109	1460	**32**

$N_x = N_y = 10$ N/mm, the first two terms in the full Tsai–Hill criterion cancel out so the result of that calculation degenerates to the maximum stress criterion.

Outline 6.4: Referring to Section 3.3.2 and substituting $\sigma_1 = 2\tau_{xy} \sin \theta \cos \theta$, $\sigma_2 = -2\tau_{xy} \sin \theta \cos \theta$, $\tau_{12} = \tau_{xy}(\cos^2 \theta - \sin^2 \theta)$ into the modified Tsai–Hill criterion gives

$$\tau_{xy} = [4 \sin^2 \theta \cos^2 \theta/(\sigma_{1max})^2 + 4 \sin^2 \theta \cos^2 \theta/(\sigma_{2max})^2 + (\cos^2 \theta - \sin^2 \theta)^2/(\tau_{12max})^2]^{-0.5}$$

Note that whereas $\sigma_{xmax}(\theta)$ varies widely (from σ_{1max} to σ_{2max}), τ_{xymax} varies but little over the range $0° < \theta < 90°$.

Outline 6.5: The hoop stress $\sigma_H(= \sigma_x) = 2\sigma_A$. Transforming into stresses in the principal directions we find $\sigma_1 = \sigma_A(1 + \cos^2 \theta)$, $\sigma_2 = \sigma_A(1 + \sin^2 \theta)$ and

$\tau_{12} = -\sigma_A \sin \theta \cos \theta$. Using $\sigma_A = pR/2h$, the shortened Tsai–Hill criterion gives:

$$p_{max} = (2\sigma_A h/R)[(1 + \cos^2 \theta)^2/(\sigma_{1max})^2 + (1 + \sin^2 \theta)^2/(\sigma_{2max})^2 + \sin^2 \theta \cos^2 \theta/(\tau_{12max})^2]^{-0.5}$$

from which it can be seen that the failure pressure varies by a factor of about 2 over the range $0° < \theta < 90°$.

Outline 6.6: The responses of lamina θK to the compressive load $N_x = -10$ N/mm may be inferred from Table 6.11 by changing the sign of stresses and strains. Using appropriate values of compressive strength, the values shown in Table A6.3 of load factor are obtained.

The first conclusion must be that the load factors under compressive load are much more uniform (over this range of angle) than under tensile load, and that there is a minimum in the region of 45°, caused not least by a change in the likely dominant mode of failure. Using stress/strength ratios, we infer compressive longitudinal (yielding) failure at $\theta = 0°$, shear for $\theta = 30°$ and 45°, and transverse compressive failure for $\theta = 75°$ and 90°. There are competing claims for $\theta = 15°$ (shear, longitudinal compression) and $\theta = 45°$ (mainly shear but with growing influence from transverse compression).

Outline 6.7: The shear strength will depend on whether the applied shear stress is positive or negative, which is important for composites, because the resolved stresses in the principal directions are limited by differing tensile and compressive strengths, which need to be taken into account when applying the Tsai–Hill criterion.

Values of stresses in the principal directions and the factor F by which the applied force has to be multiplied to cause failure are given in Table A6.4.

These data confirm that for a single ply a negative shear stress is more damaging than a positive shear stress, because the transverse stresses in the principal direction are tensile, thus providing a splitting force in a weak direction, and hence a lower load factor to failure.

On the basis of stress/strength ratios, at small and large values of θ the failure of Lamina θG is dominated by shear. Around $\theta = 45°$ the failure is transverse dominated (zero shear stress at 45°), but the influence of transverse tensile failure is broader (and more damaging – below 30° and above 60° – for $N_{xy} = -1$ N/mm, than for $N_{xy} = +1$ N/mm where the transverse stress is compressive.

Table A6.3 Tsai–Hill load factor for Lamina θK under $N_x = -10$ N/mm

Angle $\theta°$	0	15	30	45	60	75	90
Factor	23	12.84	11.29	11.03	11.29	12.4	14

Table A6.4 Load factors for shear stress applied to lamina θG

N_{xy} N/mm	$\theta°$	σ_1 MPa	σ_2 MPa	τ_{12} MPa	F
+1	0	0	0	1	80
+1	15	0.5	−0.5	0.866	90.87
+1	30	0.866	−0.866	0.5	140.2
+1	45	1	−1	0	252.2
+1	60	0.866	−0.866	−0.5	140.2
+1	75	0.5	−0.5	−0.866	90.87
+1	90	0	0	−1	80
−1	0	0	0	−1	80
−1	15	−0.5	0.5	−0.866	83.81
−1	30	−0.866	0.866	−0.5	93.39
−1	45	−1	1	0	99.61
−1	60	−0.866	0.866	0.5	93.39
−1	75	−0.5	0.5	0.866	83.81
−1	90	0	0	1	80

Table A6.5 Most likely modes of failure for lamina θG

$\theta°$	$M_x = 5\,N + M_{xy} = 10\,N$	$M_x = 2\,N + M_y = 10\,N$	$M_x = 2\,N + M_y = -10\,N$
0	shear	transverse tension	transverse tension
15	shear	transverse tension	transverse tension
30	transverse tension	transverse tension	transverse t/shear
45	transverse tension	transverse t/shear	shear
60	shear	shear/transverse t	shear
75	shear	shear/transverse t	shear
90	shear	transverse tension	transverse tension

Outline 6.8: Using the stress/strength ratios as a *rough* guide, the dominant failure mechanisms are likely to be as shown in Table A6.5; where two modes are given, the one likely to be most influential is quoted first, even though this situation is difficult to interpret.

Comments:
Table 6.20: σ_1^t and σ_2^t have the same sign through the range of θ, whereas τ_{12}^t changes sign part way through the range. The location of first failure is on the bottom surface.

Table 6.21: Curvature κ_x changes sign but all stresses in principal directions maintain their signs over a range of θ; all first failures occur on the bottom surface.

Table 6.22: Curvature κ_{xy} and the direct stresses in the principal directions change their signs over the range of θ; the location of first failure varies with θ.

Outline 6.9: Lamina θK has a low value of compressive strength (in practice it is a yield stress for this material), and the first obvious conclusion is that for low values of θ the top surface fails in compression, then as θ increases beyond about 30° the lower surface fails in transverse tension. In the region of $\theta = 15°$ to 30° there are roughly equal contributions from two stresses: at 15° longitudinal compression competes with shear, and at 30° transverse tension competes with shear.

In view of the yield and likely non-linear stress/strain curves in longitudinal compression and shear, the predictions of first failure at low values of θ are likely to be rather approximate for Lamina θK.

Under moment $M_x = 1 \, N$, θG is consistently stronger than θK. The material θK has an extremely low value of longitudinal compressive strength (which induces premature fibre kinking), so there is a difference in failure mode at 0° (compressive for θK, tensile for θG). At 15° θG is dominated by transverse tension on the bottom surface whereas θK fails on top under comparable amounts of longitudinal compression and shear. From 60° to 90° failure is dominated by transverse tension in both materials.

Outline 6.10: We take the stresses in the 1 direction ply-by-ply. For $N_x = 100 \, N/mm$, ply 1 is at 0°, so $\sigma_x = \sigma_1$; ply 2 is at 90° so $\sigma_y = \sigma_1$ and so on. The plots in Figure A6.1 may seem unfamiliar until you remember that the real direction of the stress changes through the laminate at each interface where a change of fibre angle is met. There is no problem of discontinuities in stress profiles in either global or principal directions for a single ply.

Using the stress profiles in Figure A6.1 we can test each ply surface and interface and find which combination of stresses will give the lowest load factor when the Tsai–Hill failure criterion is applied under a loading $M_y = 10 \, N$. By inspection and before going into detailed calculations, the bottom of ply 4 at 0° carries a high transverse stress and the stress/strength ratio (0.1061) suggests that this will be much more damaging than the effect of the higher longitudinal stress at the bottom of ply 3 (90°); we ought to keep an eye on the top surface 0° where $\sigma_2/\sigma_{2max} = -10.61/-270 = 0.0393$, and on the top of ply 4 (0°) where $\sigma_2/\sigma_{2max} = 5.303/100 = 0.053$. The formal calculation for failure at the bottom surface is

$$(-0.3133/-1600)^2 - (-0.3133 \times 10.61)/(-1600)^2 + (10.61/100)^2 = 1/F^2$$

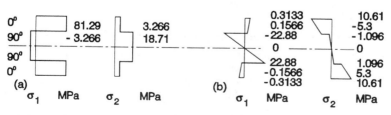

(a) σ_1 MPa σ_2 MPa (b) σ_1 MPa σ_2 MPa

Figure A6.1 Stress profiles in principal directions for 0/90SG: (a) $N_x = 100 \, N/mm$; (b) $M_y = 10 \, N$.

Table A6.6 Failure sequence for 0/90 SG under $M_y = 10$ N

Moment (N)	κ_x (/m)	κ_y (/m)	κ_{xy} (/m)	Tsai–Hill factor	Place	Angle degrees
M_y 10	−0.0706	0.9903	0	9.428	4b	0
				18.86	4t	0
				25.43	1t	0

Thus we arrive at the sequence for the first three failure events (Table A6.6):

Comparing these results with those for $M_x = 10$ N in Table A6.6, we see that, as expected, the midplane curvatures and failure sequences are quite different when bending moments are applied in the two different principal directions.

Outline 6.11: For $N_{xy} = +100$ N/mm the stress profiles are (Table A6.7(a)): First ply failure occurs in plies 2 and 3 at $-45°$ with $F = 1.517$ (and the $+45°$ plies fail at $F = 2.309$).

For $N_{xy} = 100$ N/mm we find the following stress profiles (uniform in each ply) (Table A6.7(b)):

Applying the Tsai–Hill criterion we incur first ply failure (in shear) in the $0°$ and $90°$ plies at $F = 1.4$ (then $-45°$ at $F = 2.125$, and $+45°$ at $F = 3.233$). But modes for later failures are not easy to determine from these combinations of direct stresses.

The conclusion is that the thicker tube (b) fails at a rather lower torque than the thinner tube (a); if for other reasons it is necessary to use tube (b) then it would be desirable to protect the $0°$ and $90°$ plies by putting them inside the $45°$ plies in an arrangement such as $(+45°/-45°/0°/90°)_s$.

Table A6.7 Stress profiles
(a) in $(+45°/-45°)_s$, $N_{xy} = 100$ N/mm

Ply No.	Angle °	σ_1 (MPa)	σ_2 (MPa)	τ_{12} (MPa)
1	+45	338.2	−61.76	0
2	−45	−338.2	61.76	0

(b) in $(0°/90°/+45°/-45°)_s$, $N_{xy} = 100$ N/mm

Ply No.	Angle °	τ_{xy} (MPa)	τ_{12} (MPa)	σ_1 (MPa)	σ_2 (MPa)
1	0	57.18	57.18	0	0
2	90	57.18	−57.18	0	0
3	+45	142.8	0	241.5	−44.11
4 midplane	−45	142.8	0	−241.5	44.11

Table A6.8 Tsai–Hill factors for $(\theta/-\theta)$ SG under $N_x = -10$ N/mm

$\theta°$	0	15	30	45	60	75	90
Factor	320	149	57.01	31.66	30.42	41.96	54

Outline 6.12: $\theta = 0°$ longitudinal tension, $\theta = 15°$ shear with substantial transverse tension, $\theta = 30°$, $45°$ shear, $\theta = 60°$ almost equal shear and transverse tension, $\theta = 75° - 90°$ transverse tension.

Outline 6.13: See Table A6.8. The conclusion for this laminate is that the load factor in uniaxial compression in the x direction is always greater than that for uniaxial tension.

Outline 6.14: The reference direction x in the angle ply laminate is taken as the hoop direction. The response for $(\theta/-\theta)$ SG under $N_x = 10$ N/mm, $N_y = 5$ N/mm are as shown in Table A6.9. From the tabulated data we can see the following:

- The maximum value of Tsai–Hill load factor occurs at about $33°$.
- The shear stress τ_{12} is zero at about $26.5°$.
- The transverse stress σ_2 has a minimum value at about $35°$.
- The longitudinal stress σ_1 has a maximum value at about $35°$.
- The global transverse strain ε_y is zero at about $45°$ and $75°$.

Working on the basis of σ_1/σ_{1max}, σ_2/σ_{2max}, τ_{12}/τ_{12max}, we deduce the following as the most likely modes of failure: $\theta = 0°$ transverse tension, $\theta = 15° - 20°$ shear but much transverse tension, $\theta = 30° - 40°$ transverse tension, $\theta = 45° - 55°$ transverse tension but much shear, $\theta = 60° - 90°$ transverse tension.

Table A6.9 Load factors for laminates $(\theta/-\theta)$ SG under $N_x = 10$ N/mm, $N_y = 5$ N/mm

θ °	ε_x (%)	ε_y (%)	γ_{xy} (%)	σ_1^t (MPa)	σ_2^t (MPa)	τ_{12} (MPa)	Tsai–Hill factor	Place
0	0.009463	0.02029	0	5	2.5	0	39.6	all
15	0.0098	0.01774	0	5.311	2.189	0.204	44.62	all
20	0.01032	0.01566	0	5.53	1.97	0.1766	49.23	all
25	0.01133	0.01294	0	5.77	1.73	0.06354	55.63	all
30	0.01307	0.009654	0	5.986	1.514	−0.1522	62.08	all
35	0.01577	0.006046	0	6.114	1.386	−0.4699	62.55	all
40	0.01949	0.002545	0	6.083	1.417	−0.8578	53.91	all
45	0.024	−0.000317	0	5.852	1.648	−1.25	43.02	all
55	0.03333	−0.002734	0	4.822	2.618	−1.742	29.23	all
60	0.03715	−0.002382	0	4.298	3.202	−1.759	25.7	all
65	0.04008	−0.001435	0	3.753	3.747	−1.635	23.43	all
75	0.04359	0.000847	0	2.941	4.559	−1.098	21.02	all
90	0.0451	0.002478	0	2.5	5	0	20.03	all

Outline 6.15: (a) Once global strains in the laminate have been calculated using $\varepsilon_{xL} = d_{11}N_x + d_{12}N_y + d_{16}N_{xy}$, it is necessary to calculate the stresses within each fth ply using the appropriate values of Q_{ij}^*. From Section 3.3.5 we noted that for $Q_{11}^*, Q_{12}^*, Q_{22}^*$ and $Q_{66}^*, Q^*(-\theta) = Q(+\theta)$, but $Q_{16}^*(-\theta) = -Q_{16}^*(+\theta)$ and $Q_{26}^*(-\theta) = -Q_{26}^*(+\theta)$. Hence in the fth ply we find different values of stress such as $\sigma_{xf} = Q_{11f}^*\varepsilon_{xL} + Q_{12f}^*\varepsilon_{yL} + Q_{16f}^*\gamma_{xyL}$, depending on whether the ply is laid up at $+\theta$ or $-\theta$ to the fibre (θ) direction. These differences in global stresses are then translated into differences in the signs of σ_1 and σ_2 for $+\theta$ and $-\theta$, and hence different strengths. (b) No change of strength for $\theta = 0°$ or $90°$. In the tabulated range $15°$ to $75°$ it will be the outer plies (1,4) which will fail first.

Outline 6.16: The response of $(45°/-45°)$SG to $N_x = 10\,\text{N/mm}$ and $N_{xy} = 10\,\text{N/mm}$ is as given in Table A6.10. Note that under a combination of direct and shear loads the direct stresses have quite different values for the $+45°$ and $-45°$ plies.

Outline 6.17: By calculating the stress/strength ratios, we see that at most values of θ except close to $45°$ the likely mode of failure is shear. At $45°$ the longitudinal tensile stress/strength ratio is 0.0085 and the transverse compressive ratio is 0.0057, which would suggest that the failure mode will probably be longitudinal tension.

Outline 6.18: All first ply failures are at the bottom of the lowest ply, for which all stresses in the principal directions have opposite signs to those shown in Table 6.36. The stress/strength ratios and likely dominant failure mechanism for $(\theta/-\theta)$SG under $M_x = 10\,\text{N}$ are as shown in Table A6.11.

Table A6.10

ε_x (%)	ε_y (%)	γ_{xy} (%)	σ_1 (MPa)	σ_2 (MPa)	τ_{12} (MPa)	Tsai–Hill factor	Failure place
0.03221	−0.01642	0.03894	−4.555	2.643	−2.5	24.34	2, 3
			12.36	−0.4454	+2.5	29.65	1, 4

Table A6.11 Dominant modes of failure for $(\theta/-\theta)$ SG under $M_x = 10\,\text{N}$

$\theta°$	$\sigma_1/\sigma_{1\max}$	$\sigma_2/\sigma_{2\max}$	$\tau_{12}/\tau_{12\max}$	Dominant failure mode
0	**0.015**	0	0	longitudinal tension
15	0.014	0.005	**0.037**	shear
30	0.012	0.019	**0.068**	shear
45	0.08	0.055	**0.094**	shear
60	0.004	**0.106**	0.086	transverse compression + shear
75	0.001	**0.139**	0.048	transverse compression
90	0	**0.15**	0	transverse compression

Outline 6.19: Using the concept of stress/strength ratios as a rough guide, shear failure is dominant in the range $\theta = 0°$ to $25°$ and $65°$ to $90°$. In the region $30°$ to $60°$ transverse tensile failure is dominant; at $45°$ no shear stress is present in the principal directions.

Outline 6.20: The directions of transverse stresses in the laminate $(90°/0°)_s$ are reversed compare with those in $(0°/90°)_s$, and so the free edges between $90°$ and $0°$ plies are in compression and therefore not prone to splitting. This is therefore an advantageous arrangement where narrow laminates, or wide laminates with holes, are to be used.

CHAPTER 7

Outline 7.1: At T_0 the block has dimensions $L_1 L_2 L_3$, and at T_1 the new dimensions are $L_1(1 + \alpha\Delta T) \times L_2(1 + \alpha\Delta T) \times L_3(1 + \alpha\Delta T)$. The fractional increase in volume of the block, i.e. the volume strain, is given by $\varepsilon_v = \Delta V/V_0 = (V - V_0)/V = (1 + \alpha\Delta T)^3 - 1$. Neglecting second order terms of small quantities, $(1 + \alpha\Delta T)^3 \approx 1 + 3\alpha\Delta T$, hence $\varepsilon_v \approx 3\alpha\Delta T$.

Outline 7.2: The new major axes in the principal directions are $D_1 = D_0(1 + \alpha_1\Delta T) = 40(1 + 6 \times 10^{-6}(60 - 15)) = 40.011$ mm. $D_2 = D_0(1 + \alpha_2\Delta T) = 40.063$ mm. The circular hole at 15 °C has become slightly elliptical at 60 °C.

Outline 7.3: Let the co-ordinate of the midplane of the sheet be $z = 0$, so that fluid x is in contact with the surface $z = -h/2$ and y in contact with $z = +h/2$. The midplane strain in the sheet is given by $\varepsilon° = (\varepsilon^x + \varepsilon^y)/2 = (\alpha(T_1 - T_0) - \alpha(T_2 - T_0))/2 = (\alpha/2)(T_1 + T_2 - 2T_0)$. The midplane curvature can be found from Equation (3.11), $[\varepsilon] = [\varepsilon°] + z[\kappa]$, hence $\varepsilon^x = \varepsilon° + (-h/2)\kappa$ and $\varepsilon^y = \varepsilon° + (h/2)\kappa$, giving $\kappa = \alpha(T_2 - T_1)/h$.

Outline 7.4: $\kappa = 0.4438$ m, $\sigma_a^i = 42.33$ MPa, $\sigma_b^i = -19.38$ MPa, $\varepsilon^i = 6.435 \times 10^{-4}$. (For interest $F_a = 57.85$ N.)

Outline 7.5: The analysis must now be reworked for different thicknesses of the component strips. Let the thickness of a be h_a, and of b be h_b. The force equilibrium $F_a = -F_a$ still holds, but the moment equilibrium is (Figure A7.1)

$$M_a + M_b = F_a(h_a + h_b)/2 \tag{A7.1}$$

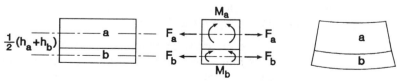

Figure A7.1 Non-symmetrical two-ply laminate.

The strains at the interface now become

$$F_a/E_aA_a + M_ah_a/2E_aA_a + \alpha_a\Delta T = F_b/E_bA_b + M_b(-h_b/2)/E_bI_b + \alpha_b\Delta T \quad (A7.2)$$

Using

$$M_a + M_b = M_a(1 + E_bI_b/E_aI_a) = F_a(h_a + h_b)/2 \quad (A7.3)$$

we find, after some manipulation, the curvature:

$$\kappa = \{6(\alpha_b - \alpha_a)E_bh_b(h_a + h_b)\Delta T\}/\{E_ah_a^3(1 + E_bh_b^3/E_ah_a^3)(1 + E_bh_b/E_ah_a)$$
$$+ 3E_bh_b(h_a + h_b)^2\}$$

Having found the internal forces and moments, we can use Equations (7.5) and (7.4) to calculate the stress and strain profiles through the thickness of the strip. The interfacial stresses are:

$$\sigma_a^i = [E_ah_a\kappa/6]\{(h_a(1 + E_bh_b^3/E_ah_a^3)/(h_a + h_b) + 3\}$$
$$\sigma_b^i = -[E_a\kappa/6h_b]\{(h_a^3(1 + E_bh_b^3/E_ah_a^3)/(h_a + h_b) + 3h_b^2E_b/E_a\}$$

Each of these equations reduces to the case of equal thickness components as appropriate.

Substituting numerical values we find $\kappa = 0.4775$/m, $\sigma_a^i = 35.03$ MPa, $\sigma_b^i = -21.82$ MPa.

Outline 7.6: Setting $T^b = 2\,°C$ and $T^t = 12\,°C$, with room temperature $T^o = 20\,°C$ we have $(\Delta T^t + \Delta T^b)/2 = -13$ K. The thermal force resultants can be calculated from $N_x^T = (Q_{11}\alpha_1 + Q_{12}\alpha_2)(\Delta T^t + \Delta T^b)h/2 = -28.04$ N/mm. Because the material is isotropic, $N_y^T = N_x^T$. The thermal strains can now be calculated as $\varepsilon_x^T = a_{11}N_x^T + a_{12}N_y^T = -0.286\%$. $\varepsilon_x^T = \varepsilon_y^T$.

The thermal moments can be found from $[M]^T = [Q][\alpha](T^b - T^t)h^2/12$ to be $M_x^T = M_y^T = -0.00897$ N. The curvatures are $\kappa_x^T = d_{11}M_x^T + d_{12}M_y^T = -0.4392$/m, and $\kappa_y^T = \kappa_x^T$.

Outline 7.7: We calculated the thermal force resultants $N_x^T = N_y^T = -0.02804$ N/mm in the previous problem. The total strains are therefore given by $[\varepsilon^{otot}] = [a][N^a + N^T]$, so that $\varepsilon_x^{otot} = a_{11}(N_x^a + N_x^T) + a_{12}(N_y^a + N_y^T) = 0.469\%$ and $\varepsilon_x^{otot} = a_{12}(N_x^a + N_x^T) + a_{22}(N_y^a + N_y^T) = -0.276\%$.

The thermal moments are the same as in the previous problem and so the curvatures are $\kappa_x^{tot} = \kappa^T = -0.4392$/m $= \kappa_y^{tot} = \kappa^T$.

The stress profiles arise solely from the application of external mechanical loads, so $\sigma_x^{tot} = \sigma_x^m = 0.01$ MPa, and $\sigma_y^{tot} = 0.005$ MPa $= \sigma_y^m$. In particular $\sigma_x^T = \sigma_y^T = 0$. The total strain profiles are the sums of the thermal strains and the mechanical strains, shown in Figure A7.2.

Outline 7.8: Using x to denote the hoop direction, we have $N_x = pD_m/2 = 0.05$ N/mm, and $N_y = 0.025$ N/mm. The total force resultants per unit width are therefore $N_x^{tot} = 0.05 - 0.02804 = +0.02196$ N/mm, and $N_y^{tot} = -0.00304$ N/mm, the same as for the previous problem. The net result is that

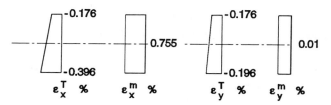

Figure A7.2 Response of NR5 to $T^b = 2\,^\circ C$, $T^t = 12\,^\circ C$, $N_x^a = 0.05$ N/mm, $N_y^a = 0.025$ N/mm.

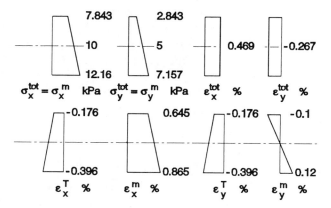

Figure A7.3 Response of NR5 to $T^b = 2\,^\circ C$, $T^t = 12\,^\circ C$, $N_x^a = 0.05$ N/mm, $N_y^a = 0.025$ N/mm, $\kappa_x^{tot} = \kappa_y^{tot} = 0$.

the midplane strains are $\varepsilon_x^o = 0.469\%$ and $\varepsilon_y^o = -0.276\%$. However the axisymmetric pipe resists the thermal moments M_x^T and M_y^T, such that $\kappa_x^{tot} = \kappa_y^{tot} = 0$. We therefore find that the pipe applies its own internal moments $M_x = M_y = 0.008987$ N, which induce total and mechanical stress profiles but not thermal stress. The stress and strain profiles are shown in Figure A7.3.

Outline 7.9: Assume fibres are aligned in the 1 direction and that $\alpha_3 = \alpha_2$. $V = L_1(1 + \alpha_1 \Delta T)L_2(1 + \alpha_2 \Delta T)L_3(1 + \alpha_3 \Delta T) \sim [1 + (\alpha_1 + 2\alpha_2)\Delta T]L_1 L_2 L_3$. The volume strain is $\varepsilon_v = (V - V_o)/V_o \sim (\alpha_1 + 2\alpha_2)\Delta T$. If the material were isotropic, $\alpha_1 = \alpha_2 = \alpha$, so $\varepsilon_v = 3\alpha\Delta T$ as seen in Problem 7.1.

Outline 7.10: $\varepsilon_1 = \alpha_1 \Delta T = 7 \times 10^{-6} \times (65 - 20) = 3.15 \times 10^{-4}$. $L_1(65^\circ) = L_1(20^\circ)(1 + \alpha_1\Delta T) = 2.00063$ m. $\varepsilon_2 = \alpha_2\Delta T = 3 \times 10^{-5} \times (65 - 20) = 1.35 \times 10^{-3}$, $D_2(65^\circ) = D_2(20^\circ)(1 + \alpha_2\Delta T) = 10.0135$ mm.

Outline 7.11: Using the above equations and solving for $\varepsilon_1 = \varepsilon_2 = 0$, we find terms such as $\sigma_2 = -Q_{12}\alpha_1\Delta T - Q_{22}\alpha_2\Delta T$, leading to $\sigma_2 = -[E_2(\alpha_2 +$

$v_{12}\alpha_1)\Delta T]/(1 + v_{12}v_{21})$. The negative sign confirms that a compressive stress had to be applied to counter the free thermal strain caused by an increase in temperature.

Outline 7.12: We assume that $\varepsilon_x = 0$, and that the rod is free to change its transverse dimensions, hence

$$
\begin{pmatrix} \varepsilon_1 \\ \varepsilon_2 \\ \gamma_{12} \end{pmatrix} = \begin{pmatrix} S_{11} & S_{12} & 0 \\ S_{12} & S_{22} & 0 \\ 0 & 0 & S_{66} \end{pmatrix} \begin{pmatrix} \sigma_1 \\ 0 \\ 0 \end{pmatrix} + \begin{pmatrix} \alpha_1 \\ \alpha_2 \\ 0 \end{pmatrix} \Delta T
$$

Thus we have $\varepsilon_1 = 0 = \sigma_1/E_1 + \alpha_1\Delta T$, i.e. $\sigma_1 = -E_1\alpha_1\Delta T = -E_1\alpha_1(T - T_o) = -45.6 \times 10^9 \times 7 \times 10^{-6} \times (-25 - 20) = +14.364$ MPa. $\varepsilon_2 = \sigma_1(-v_{12}/E_1) + \alpha_2\Delta T = 14.364 \times 10^6 \times (-0.274/45.6 \times 10^9) + 3 \times 10^{-5} \times -45 = -0.1436\%$.

It is worth studying the more formal approach, using the thermal force resultants. We have $N_1^T = \Delta T \cdot h(Q_{11}\alpha_1 + Q_{12}\alpha_2)$, $N_2^T = \Delta T \cdot h(Q_{12}\alpha_1 + Q_{22}\alpha_2)$. The problem states that $\varepsilon_1 = 0 = a_{11}[N_1^m + N_1^T] + a_{12}N_2^T$. Recalling that $a_{12}/a_{11} = -v_{12}$, we find after some manipulation that $N_1^m = -\Delta T h E_1\alpha_1$.

Outline 7.13: The inverse strain transformation matrix for $30°$ is

$$
[T]^{-1} = \begin{pmatrix} 0.75 & 0.25 & -0.866 \\ 0.25 & 0.75 & 0.866 \\ 0.433 & -0.433 & 0.5 \end{pmatrix}
$$

$$\alpha_x = 0.75 \times 1 \times 10^{-5} + 0.25 \times 2 \times 10^{-4} = 5.75 \times 10^{-5}/\text{K}$$

$$\alpha_y = 0.25 \times 1 \times 10^{-5} + 0.75 \times 2 \times 10^{-4} = 1.525 \times 10^{-4}/\text{K}$$

$$\alpha_{xy}/2 = 0.433 \times 1 \times 10^{-5} - 0.433 \times 2 \times 10^{-4} = -8.227 \times 10^{-5} = \alpha_{xy}/2$$

hence $\alpha_{xy} = -1.6454 \times 10^{-4}/\text{K}$. The sheet dimensions become

$$L_x = 350(1 + 5.75 \times 10^{-5} \times 35) = 350.70 \text{ mm}$$

$$L_y = 230(1 + 1.525 \times 10^{-4} \times 35) = 231.23 \text{ mm}$$

$$\gamma_{xy} = \alpha_{xy}\Delta T = -1.6454 \times 10^{-4} \times 35 = -5.759 \times 10^{-3} = -0.5759\%$$

Outline 7.14: Because the sheet is orientated at $30°$ to the principal directions, we must now use the transformed reduced stiffnesses $[Q^*]$ to find the stresses from the strains. The thermal force resultant is given by

$$
\begin{pmatrix} N_x^T \\ N_y^T \\ N_{xy}^T \end{pmatrix} = \begin{pmatrix} Q_{11}^* & Q_{12}^* & Q_{16}^* \\ Q_{12}^* & Q_{22}^* & Q_{26}^* \\ Q_{16}^* & Q_{26}^* & Q_{66}^* \end{pmatrix} \begin{pmatrix} \alpha_x \\ \alpha_y \\ \alpha_{xy} \end{pmatrix} \Delta T \cdot h
$$

Using data from Table 3.7, we calculate $N_x^T = 0.531$ N/mm, $N_y^T = 0.246$ N/mm, and $N_{xy}^T = (+)0.248$ N/mm. It must be emphasized that these forces are only needed when all three constraints to thermal change are imposed.

Outline 7.15: (a) We have already found the expansion coefficients to be $(\alpha_x, \alpha_y, \alpha_{xy}) = (0.575, 1.525, -1.6453) \times 10^{-4}/\text{K}$. Hence $(N_x^T, N_y^T, N_{xy}^T) = [Q^*][\alpha]\Delta T \cdot$

$h = (0.4555, 0.2109, 0.2119)$ N/mm. Using the [a] matrix in Table 3.8 the thermal strains are $[\varepsilon^{oT}] = [a][N^T]$: $(\varepsilon_x^{oT}, \varepsilon_y^{oT}, \gamma_{xy}^{oT}) = (0.1725, 0.4575, -0.4936)\%$.

(b) The thermal force resultants given by $[N^T] = [Q^*][\alpha](\Delta T^b + \Delta T^t)h/2$ are $(N_x^T, N_y^T, N_{xy}^T) = (-0.2809, -0.1299, -0.1307)$ N/mm and hence $(\varepsilon_x^{oT}, \varepsilon_y^{oT}, \gamma_{xy}^{oT}) = [a][N^T] = (-0.1064\%, -0.2821\%, +0.3044\%)$.

The thermal moment resultants given by $[M^T] = [Q^*][\alpha](T^b - T^t)h^2/12$ are $(M_x^T, M_y^T, M_{xy}^T) = (-0.05821, -0.02693, -0.02708)$ N. Using the [d] matrix from Table 3.8 we can now calculate the curvatures for three situations: (i) for a flat plate where all curvatures are free to develop; (ii) for the thin-walled tube where κ_x and κ_y are prevented but the tube is free to twist; and (iii) for a thin-walled tube where all three curvatures are prevented.

(i) $[\kappa^T] = [d][M^T]$: $(\kappa_x, \kappa_y, \kappa_{xy}) = (-0.6612, -1.754, +1.892)$/m.

(ii) This tube problem is rather more tricky. The tube form prevents the development of curvatures in the hoop and longitudinal directions. We therefore need to solve simultaneously the equations $\kappa_x^T = 0 = d_{11}M_x + d_{12}M_y + d_{16}M_{xy}^T$; $\kappa_x^T = 0 = d_{12}M_x + d_{22}M_y + d_{26}M_{xy}^T$; $\kappa_{xy}^T = d_{16}M_x + d_{26}M_y + d_{66}M_{xy}^T$, where M_x and M_y include the internal mechanical moment needed to suppress the development of thermal curvatures κ_x^T and κ_y^T. Hence $M_x = (d_{12}d_{26} - d_{16}d_{22})M_{xy}^T/(d_{11}d_{22} - d_{12}^2) = 1.712 M_{xy}^T$, and $M_y = (d_{11}d_{26} - d_{12}d_{16})M_{xy}^T/(d_{12}^2 - d_{11}d_{22}) = 0.5692 M_{xy}^T$. Hence $\kappa_x^T = \kappa_y^T = 0$ as required, and $\kappa_{xy}^T = -0.2366$/m.

It is worth looking more carefully at the values of M_x and M_y: they each consist of the thermal moment M_i^T (which here is negative), and the additional internal mechanical moment M_i^m (here positive) which counterbalances the thermally-induced curvature, i.e. $M_x = M_x^T + M_x^m$. The thermal moments do not induce stress in a single ply, but the internal mechanical moments *do* induce stresses which are shown in Figure A7.4.

(iii) To achieve $\kappa_x = \kappa_y = \kappa_{xy}$ in this problem we can take the commonsense approach of applying internal mechanical moments to overcome the thermal moments: $M_i = -M_i^T$, so that $(M_x^m, M_y^m, M_{xy}^m) = (+0.05821, +0.02694, +0.02707)$ N. These internal mechanical moments induce the total and mechanical stress profiles shown in Figure A7.5.

(c) $N_x^a = pd_m/2 = 0.04 \times 30/2 = 0.6$ N/mm, and $N_y^a = 0.3$ N/mm. The midplane strains now correspond to the effective force resultant $N^e = N^a + N^T$.

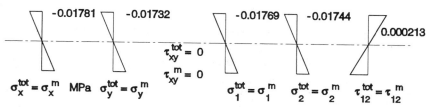

Figure A7.4 Response of NE30 tube to $\kappa_x = \kappa_y = 0$, free to develop κ_{xy}. Stresses in MPa.

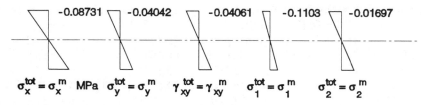

Figure A7.5 Response of NE30 tube to $\kappa_x = \kappa_y = \kappa_{xy} = 0$. Stresses in MPa.

Thus $N_x^e = 0.6 - 0.2809 = 0.3191 \, \text{N/mm}$, $N_y^e = 0.3 - 0.1299 = 0.1701 \, \text{N/mm}$, and $N_{xy}^e = N_{xy}^T = -0.1307 \, \text{N/mm}$. Hence $\varepsilon_x^{\text{tot}} = a_{11}N_x^e + a_{12}N_y^e + a_{16}N_{xy}^e = 2.117\%$, $\varepsilon_y^{\text{tot}} = 0.09528\%$ and $\gamma_{xy}^{\text{tot}} = -3.715\%$.

Stresses are calculated from the mechanical strains and curvatures. Thus from $[\varepsilon^{\text{om}}] = [a][N^a]$ and $[\kappa_m] = [d][M^m]$, $[\sigma^m(z)] = [Q^*][\varepsilon^{\text{om}} + z\kappa^m]$.

For the midplane we have $\varepsilon_x^{\text{om}} = a_{11}N_x^m + a_{12}N_y^m = 0.02223$, $\varepsilon_y^{\text{om}} = a_{12}N_x^m + a_{22}N_y^m = 0.003771$, and $\gamma_{xy}^{\text{om}} = a_{16}N_x^m + a_{26}N_y^m = -0.040194$. Hence $\sigma_x^m = Q_{11}^*\varepsilon_x^{\text{om}} + Q_{12}^*\varepsilon_y^{\text{om}} + Q_{16}^*\gamma_{xy}^{\text{om}} = 0.298 \, \text{MPa}$. Alternatively because the in-plane load is uniform across the wall thickness, we simply find $\sigma_x^m = N_x^a/h = 0.6/2 = 0.3 \, \text{MPa}$, with a minor difference attributable to rounding off error between computed and manual calculations.

For stresses away from the midplane we first need to calculate the curvatures induced by the mechanical internal moments. $\kappa_x^m = d_{11}M_x^m + d_{12}M_y^m + d_{16}M_{xy}^m = 0.6612/\text{m}$, $\kappa_y^m = +1.754/\text{m}$, and $\kappa_{xy} = -1.892/\text{m}$. Hence at $z = -0.001 \, \text{m}$ we obtain $\varepsilon_x^{\text{om}} + z\kappa_x^m = 0.02157$, $\varepsilon_y^{\text{om}} + z\kappa_y^m = 0.002017$, and $\gamma_{xy}^{\text{om}} + z\kappa_{xy}^m = -0.0383$, leading to $\sigma_x^m(-0.001 \, \text{m}) = 0.2127 \, \text{MPa}$, $\sigma_y^m(-0.001 \, \text{m}) = 0.1096 \, \text{MPa}$, and $\tau_{xy}^m(-0.001 \, \text{m}) = -0.04061 \, \text{MPa}$. We then obtain the

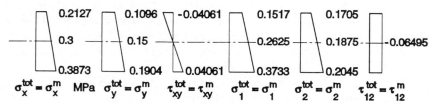

Figure A7.6 Response of NE30 tube to $\kappa_x = \kappa_y = \kappa_{xy} = 0$. $p = 40 \, \text{kPa}$. Stresses in MPa.

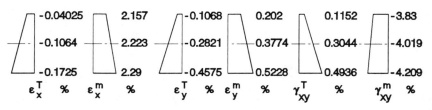

Figure A7.7 Strain response of NE30 tube to internal pressure and temperature gradient.

mechanical and total stress profiles shown in Figure A7.6, based on the strain profiles given in Figure A7.7.

Outline 7.16: The thermal strains in the laminate are related to the laminate thermal expansion coefficients by $[\varepsilon]_L^T = [\alpha]_L \Delta T$, and hence $[\alpha]_L = [\varepsilon]_L^T / \Delta T = [a][N^T]/\Delta T = \{[a]\Sigma\int[Q]_f[\alpha]_f \Delta T dz\}/\Delta T$. For the special case of uniform temperature change in a symmetric laminate we have $[\alpha]_L = [a]\Sigma[Q]_f[\alpha]_f(h_f - h_{f-1})$.

For SAL4S, $N_x^T = N_y^T = 83.709$ N/mm: $\varepsilon_x^T = \varepsilon_y^T = 0.042558\%$, and hence $\alpha_1 = \varepsilon_1^T/\Delta T = 4.2558 \times 10^{-4}/30 = 1.4186 \times 10^{-5}$/K. As expected, the thermal expansion coefficient for the laminate falls within the values for the individual isotropic plies. In general we have a thickness- and stiffness-weighted thermal expansion coefficient for the laminate.

Outline 7.17: The stress distribution is shown in Figure A7.8. The calculations are as follows:

$$N_x^c = \Delta T\{(Q_{11S}\alpha_S + Q_{12S}\alpha_S)h/2 + (Q_{11A}\alpha_A + Q_{12A}\alpha_A)h/2\} = -334.85 \text{ N/mm} = N_y^c$$
$$\varepsilon_x^{oc} = a_{11}N_x^c + a_{12}N_y^c = -1.7024 \times 10^{-3} = \varepsilon_y^{oc}$$

At the midplane

$$\varepsilon_{xA}^e = -1.7024 \times 10^{-3} - 2.3 \times 10^{-5} \times -120 = +1.058 \times 10^{-3}$$
$$\sigma_{xA}^c = (Q_{11A} + Q_{12A})(\varepsilon_{xA}^e) = 110.5 \text{ MPa}.$$

Outline 7.18:

(a) $T^o = 20\,°C$; $T^t = 35\,°C$; $T^b = 10\,°C$; $T^c = 20\,°C$
(b) $T^o = 20\,°C$; $T^t = 10\,°C$; $T^b = 35\,°C$; $T^c = 20\,°C$

Outline 7.19: As a matter of revision we need to calculate $[Q]$ and $[\alpha]$ for each orientation: $v_{21} = v_{12}E_2/E_1 = 0.01957$, thus $(1 - v_{12}v_{21}) = 0.994$. Hence

$$[Q(0°)] \text{ kN/mm}^2 \qquad\qquad [Q(90°)] \text{ kN/mm}^2$$

$$\begin{pmatrix} 138.8 & 2.716 & 0 \\ 2.716 & 9.05 & 0 \\ 0 & 0 & 7.1 \end{pmatrix} \qquad \begin{pmatrix} 9.05 & 2.716 & 0 \\ 2.716 & 138.8 & 0 \\ 0 & 0 & 7.1 \end{pmatrix}$$

$$\alpha_1(0°) = -0.3 \times 10^{-6}/K \quad \alpha_1(90°) = 28.1 \times 10^{-6}/K$$
$$\alpha_2(0°) = 28.1 \times 10^{-6}/K \quad \alpha_2(90°) = -0.3 \times 10^{-6}/K$$

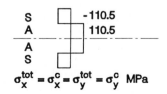

$$\sigma_x^{tot} = \sigma_x^c = \sigma_y^{tot} = \sigma_y^c \text{ MPa}$$

Figure A7.8 Response of SAL4S to assembly at 140 °C. Other $T = 20$ °C.

We can now find the thermal force resultants:

$$N_1^T = \Delta T \Sigma Q_f \alpha_f (h_f - h_{f-1})$$
$$= (-120)[(2 \times 0.125)(Q_{11}(0)\alpha_1(0) + Q_{12}(0)\alpha_2(0))$$
$$+ (4 \times 0.125)(Q_{11}(90)\alpha_1(90) + Q_{12}(90)\alpha_2(90))]$$
$$= (-120)[0.25(138.8 \times 10^3 \times -0.3 \times 10^{-6} + 2.716 \times 10^3 \times 28.1 \times 10^{-6})$$
$$+ 0.5(9.05 \times 10^3 \times 28.1 \times 10^{-6} + 2.716 \times 10^3 \times -0.3 \times 10^{-6})]$$
$$= -16.25 \text{ N/mm}$$

$$N_2^T = (-120)[(2 \times 0.125)(Q_{12}(0)\alpha_1(0) + Q_{22}(0)\alpha_2(0)) + (4 \times 0.125)(Q_{12}(90)\alpha_1(90)$$
$$+ Q_{22}(90)\alpha_2(90))] = -9.685 \text{ N/mm}$$

The thermal strains in the laminate are now:

$$\varepsilon_x^{oc} = a_{11}N_1^T + a_{12}N_2^T = 25.53 \times 10^{-6} \times -16.25 - 0.7255 \times 10^{-6} \times -9.986$$
$$= -4.079 \times 10^{-4}$$

$$\varepsilon_y^{oc} = a_{12}N_1^T + a_{22}N_2^T = -0.7255 \times 10^{-6} \times 16.25 + 13.97 \times 10^{-6} \times -9.685$$
$$= +1.235 \times 10^{-4}$$

The stress-inducing strains are the next step:

$$\varepsilon_x^e(0) = \varepsilon_x^{oc} - \alpha_1(0)\Delta T = -4.079 \times 10^{-4} - (-0.3 \times 10^{-6})(-120) = -4.439 \times 10^{-4}$$
$$\varepsilon_y^e(0) = \varepsilon_y^{oc} - \alpha_2(0)\Delta T = -1.235 \times 10^{-4} - (28.1 \times 10^{-6})(-120) = +3.249 \times 10^{-3}$$
$$\varepsilon_x^e(90) = -4.079 \times 10^{-4} - (28.1 \times 10^{-6})(-120) = +2.964 \times 10^{-3}$$
$$\varepsilon_y^e(90) = -1.235 \times 10^{-4} - (-0.3 \times 10^{-6})(-120) = -1.595 \times 10^{-4}$$

Finally the thermal stresses are

$$\sigma_x(0) = 138.8 \times 10^3 \times -4.439 \times 10^{-4} + 2.716 \times 10^3 \times 3.249 \times 10^{-3}$$
$$= -52.79 \text{ MPa}$$
$$\sigma_y(0) = 2.716 \times 10^3 \times -4.439 \times 10^{-4} + 9.05 \times 10^3 \times 3.249 \times 10^{-3}$$
$$= 28.2 \text{ MPa}$$
$$\sigma_x(90) = 9.05 \times 10^3 \times 2.964 \times 10^{-3} + 2.716 \times 10^3 \times -1.595 \times 10^{-4}$$
$$= 26.39 \text{ MPa}$$
$$\sigma_y(90) = 2.716 \times 10^3 \times 2.964 \times 10^{-3} + 138.8 \times 10^3 \times -1.595 \times 10^{-4}$$
$$= -14.09 \text{ MPa}$$

Outline 7.20: From Section 7.5.1, $\alpha_{11L} = \alpha_{22L} = \varepsilon_1^\circ/\Delta T = 2.269 \times 10^{-4}/30 = 7.563 \times 10^{-6}/\text{K}$.

Outline 7.21: Midplane strains depend on $a_{11} = a_{22}$, and are the same in this temperature field, but the curvatures κ_x and κ_y depend on d_{11} which is different from d_{22}. Because the curvatures are different, the slopes of the strain profiles are different, and hence the stress profiles are also different remembering also that terms in 11 and 22 in $[Q(0°)]$ and $[Q(90°)]$ are different:

$$\sigma_x(0°) = Q_{11}(0°)\{\varepsilon_x° + z\kappa_x\} + Q_{12}(0°)\{\varepsilon_y° + z\kappa_y\}$$
$$\sigma_x(90°) = Q_{11}(90°)\{\varepsilon_x° + z\kappa_x\} + Q_{12}(90°)\{\varepsilon_y° + z\kappa_y\}$$
$$\sigma_y(0°) = Q_{12}(0°)\{\varepsilon_x° + z\kappa_x\} + Q_{22}(0°)\{\varepsilon_y° + z\kappa_y\}$$
$$\sigma_y(90°) = Q_{12}(90°)\{\varepsilon_x° + z\kappa_x\} + Q_{22}(90°)\{\varepsilon_y° + z\kappa_y\}$$

Outline 7.22: To suppress κ_2 we need to apply a moment M_y^m which alone gives $\kappa_2^m = -\kappa_2^T$, where κ_2^T is the thermally-induced curvature from example 1.3, i.e. $\kappa_2^m = -0.6569/m$. Hence $M_y^m = -\kappa_2^T/d_{22} = -0.6569/3.245 = -0.2024$ N. $\kappa_1^m = d_{12}M_y^m = -41.94 \times 10^{-3} \times -0.2024 = 0.008489/m$. The total curvatures are now $\kappa_2 = \kappa_2^T + \kappa_2^m = 0.6569 + (-0.6569) = 0$, as required. The total midplane strains are the same as for Examples 3 and 4. The total stress profiles are given in Figure 7.20.

Outline 7.23: The temperature gradient through the thickness induces a bending strain linear through the thickness. The discontinuity in stress $\sigma_y^{tot}(z)$ arises from the change of expansion coefficient at the interface between dissimilar plies and from the change of stiffness. At the upper interface the 0° ply if unbonded would expand in the y direction by $\alpha_2\Delta T^i$ and the 90° ply if unbonded would expand freely by the smaller amount $\alpha_1\Delta T^i$. To achieve the same strain ε_y^i at the interface, we have to apply an internal compressive force to the 0° ply and a tensile force to the 90° ply, as shown in Figure 7.20.

Outline 7.24: In order to suppress the bending caused by the temperature gradient, it is necessary to apply moments $M_x^m = 0.5518$ N and $M_y^m = 0.2108$ N. The midplane strains arising from the temperature gradient are $\varepsilon_x° = \varepsilon_y° = 0.005673\%$. The total and thermal stress profiles are shown in Figure A7.9, and the mechanical and thermal strains are shown in Figure A7.10.

Outline 7.25: By inspection the total stresses profile is that given in Figure A7.11. The mechanical strain profile can be constructed graphically from the thermal strain profile $\varepsilon_x^T(z)$ in the figure, and hence the mechanical stress $\sigma_x^m(z)$ calculated from $\sigma_x^m(0°) = Q_{11}\varepsilon_x^m(z) + Q_{12}\varepsilon_y^m(z)$.

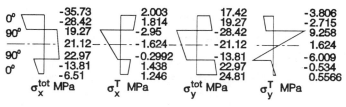

Figure A7.9 Stress response of C + 4S to $T° = 20°C$. $T^t = 40°C$, $T^b = 15°C$, $T^c = 125°C$.

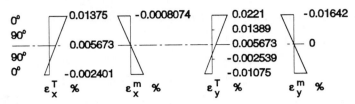

Figure A7.10 Strain response of $C+4S$ to $T° = 20 °C$, $T^t = 40 °C$, $T^b = 15 °C$, $T^c = 125 °C$.

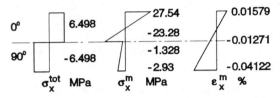

Figure A7.11 Response of $C + 4S$ to $T° = T^c = 20 °C$, $T^t = T^b = 50 °C$, $\kappa_1 = \kappa_2 = 0$.

Outline 7.26: $\alpha_1 = 7 \times 10^{-6}/K$, $\alpha_2 = 3 \times 10^{-5}/K$; using $[T(30°)]^{-1}$:

$$\alpha_x(30°) = 0.75 \times 7 \times 10^{-6} + 0.25 \times 3 \times 10^{-5} = 1.275 \times 10^{-5}/K$$
$$\alpha_y(30°) = 0.25 \times 7 \times 10^{-6} + 0.75 \times 3 \times 10^{-5} = 2.425 \times 10^{-5}/K$$
$$\alpha_{xy}/2 = 0.433 \times 7 \times 10^{-6} - 0.433 \times 3 \times 10^{-5} = -9.959 \times 10^{-6}/K$$

Hence

$$\alpha_{xy}(30°) = -1.992 \times 10^{-5}/K$$

To evaluate thermal expansion coefficients for the $-30°$ plies we use:

$$[T(-30°)]^{-1} = \begin{pmatrix} 0.75 & 0.25 & 0.866 \\ 0.25 & 0.75 & -0.866 \\ -0.433 & 0.433 & 0.5 \end{pmatrix}$$

Hence for $-30°$: $(\alpha_x, \alpha_y, \alpha_{xy}) = (1.275, 2.425, +1.992) \times 10^{-5}/K$.

We can now calculate the thermal force resultants. In general we have

$$[N^T] = \Delta T \Sigma [Q^*]_f [\alpha]_f (h_f - h_{f-1})$$
$$= \Delta T\{[Q^*(+30°)]_f [\alpha(+30°)]_f (h/2) + [Q^*(-30°)]_f [\alpha(-30°)]_f (h/2)\}$$

Unpacking this into the fine detail we have terms such as

$$N_x^T = \Delta T(h/2) \{(Q_{11}^*(30°)\alpha_x(30°) + Q_{12}^*(30°)\alpha_y(30°) + Q_{16}^*(30°)\alpha_{xy}(30°))$$
$$+ Q_{11}^*(-30°)\alpha_x(-30°) + Q_{12}^*(-30°)\alpha_y(-30°) + Q_{16}^*(-30°)\alpha_{xy}(-30°)\}$$

Substituting numerical values we find

$$(N_x^T, N_y^T, N_{xy}^T) = \Delta T(h/2)(79.62977, 73.0185, 0) \times 10^4 \text{ SI units}$$

We note that $N_{xy}^T = 0$. This is confirmed by common sense for a symmetric laminate, and more formally from the relevant full equation for N_{xy}^T, where $-Q_{16}^*(+30°) = Q_{16}^*(-30°)$ and $-\alpha_{xy}(+30°) = \alpha_{xy}(-30°)$.

Because $(\varepsilon^T)_L = [a][N^T] = [\alpha]_L \Delta T$, we have $[\alpha]_L = [a][N^T]/\Delta T$. Using values of $[a]$ from Table 7.2, and $\Delta T = 1K$, we find

$$(\alpha_{xL}, \alpha_{yL}, \alpha_{xyL}) = (6.463, 21.99, 0) \times 10^{-6}/K$$

Outline 7.27: By addition of terms in Section 7.6.2 we have $\varepsilon_x^\circ = 0.0555\%$, $\varepsilon_y^\circ = 0.06022\%$, $\kappa_{xy} = 1.083/m$. To suppress twist we need $M_{xy} = -5.728$ N, so that $\varepsilon_x^\circ = 0.03841\%$, $\varepsilon_y^\circ = 0.05411\%$.

Outline 7.28: 1. Apply twist. 2. Reduce the uniform temperature.

CHAPTER 8

Outline 8.1: For the symmetric laminate A: $\varepsilon_x = a_{11}N_x = 7.19\,(mm/MN) \times 1\,N/mm = 7.19 \times 10^{-6} = 0.000719\%$; $\varepsilon_y = -0.0002106\%$.

For the unsymmetric laminate B, we must note that the curvatures κ_x and κ_y are suppressed by the application of internal moments M_x and M_y. We need to calculate these values of moments using

$$\binom{\varepsilon}{\kappa} = \binom{a \quad b}{h \quad d}\binom{N}{M}$$

so we find $\kappa_x = 0 = h_{11}N_x + d_{11}M_x + d_{12}M_y$; $\kappa_y = 0 = h_{12}N_x + d_{12}M_x + d_{22}M_y$ and hence we find the moments in terms of the applied load N_x: $M_x = N_x(h_{11}d_{22} - h_{12}d_{12})/(d_{12}^2 - d_{11}d_{22})$; $M_y = N_x(h_{11}d_{12} - h_{12}d_{11})/(d_{11}d_{22} - d_{22}^2)$. Substituting numerical values, $M_x = -0.1224\,N_x$, and $M_y = 5.263 \times 10^{-3}\,N_x$, so we are now able to calculate the axial strain response as $\varepsilon_x = a_{11}N_x + b_{11}M_x + b_{12}M_y = 7.192 \times 10^{-4}\%$.

The reduced stiffness matrices [Q] for steel and aluminium are given in Tables 4.1 and 4.2. The stresses can now be found: $\sigma_{xs} = Q_{11s}\varepsilon_x + Q_{12s}\varepsilon_y = 225.7 \times 7.19 \times 10^{-6} + 63.19 \times -2.106 \times 10^{-6} = 1.489 \times 10^{-3}\,GPa = 1.489$ MPa; $\sigma_{ys} = -0.02095$ MPa; $\sigma_{xA1} = 0.5102$ MPa, and $\sigma_{yA1} = +0.02095$ MPa. The stress profiles are shown in Figure A8.1.

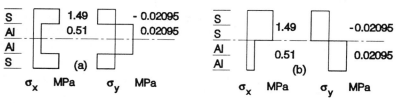

Figure A8.1 Stress profiles in: (a) SAL4S; (b) SAL2NS.

Outline 8.2: Webs in a channel section are not wide, so edge effects need to be assessed. Using $(0°/90°)_S$ will put the interface between $90°$ and $0°$ plies in compression, which is safer.

Outline 8.3: The cross-sectional area $A = 8 \times 10^{-6}\,m^2$ and $I = 4 \times 2^3 \times 10^{-12}/12 = 2.666 \times 10^{-12}\,m^4$. The radius of gyration is $r = \sqrt{(I/A)} = 0.577\,mm$, so the slenderness ratio is $L/r = 87$, so Euler buckling is likely. $E_x = (a_{11}h)^{-1} = (10.96 \times 2\,mm^2/MN)^{-1} = 45.6\,GPa$, and hence the critical buckling load is $P_c = 4\pi^2 \times 45.6 \times 10^9 \times 2.666 \times 10^{-12}/0.01 = 478\,N$. The stress $240\,MPa$ corresponds to a strain 0.526%. Note that the compressive strength of this material is much higher than the buckling strength.

Outline 8.4: Under axial tension we can see from Table 6.32 that failure would occur at $N_x = 556.6\,N$. For a mean circumference $2\pi R_m = 471\,mm$, the failure load in tension is expected to be $262.2\,kN$.

Under axial compression we saw from Problem 6.13 that failure would occur at $N_x = -570.1\,N$ provided buckling did not occur, i.e. a compressive load of $-268.5\,kN$.

For this thin-walled tube the radius of gyration is $r = \sqrt{(I/A)} = \sqrt{(R^2_m/4)} = 37.5\,mm$, and hence the slenderness ratio is $L/r = 0.1/0.375 = 2.7$, so Euler buckling is not expected.

For local buckling, we estimate from Section 2.8.5 the critical buckling stress as $\sigma_c \approx 0.5E_x h/D_m$. From Table 6.31, $a_{11} = 19.02\,mm/MN$, so $E_x = (a_{11}h)^{-1} = 26\,GPa$, and hence $\sigma_c = -0.5 \times 26 \times 10^9 \times 2/50 = -173\,MPa$, or $P_c = -163\,kN$. The local buckling stress formula is for an isotropic material. We should note that $a_{22} = 43.09\,mm/MN$, so there is more 'give' round the circumference which is likely to reduce the buckling load a little. The conclusion remains however that under axial compression the tube would fail by buckling rather than by reaching its axial compressive strength.

Outline 8.5: For 0/90SG $a_{11} = 21.21\,mm/MN$ and for $45/-45$SG $a_{11} = 110.3\,mm/MN$. Buckling load is proportional to modulus, so the $+45/-45$SG buckles at only about one-fifth the load for 0/90SG. Men's neckties are normally cut from woven cloth cut 'on the bias', because the low modulus E_x helps to make a neat knot with a neat gather (a buckling effect). It is much more difficult to tie an elegant knot in a necktie cut 'on the square' from woven cloth.

Outline 8.6: The [A] matrix (but not its inverse) is the same for both laminates (a) and (b), as expected. The non-symmetric laminate (b) has a non-zero [B] matrix, also as expected. The compliance matrices for laminate (a) and (b) are clearly quite different, and hence the superficial (but incorrect) conclusion must be that the strains in the two tubes under the same pressure are quite different.

For a flat element, the strain responses in the symmetric laminate can be calculated from such general equations as $\varepsilon_x = a_{11}N_x + a_{12}N_y$; $\varepsilon_y = a_{12}N_x + a_{22}N_y$. Setting $N_A = N_x = 5\,N/mm$ and $N_H = N_y = 10\,N/mm$ for the tube we find $\varepsilon_x = (38.04 \times 5 - 23.78 \times 10) \times 10^{-6} = -47.6 \times 10^{-6}$, and $\varepsilon_y = +742.9 \times 10^{-6}$. For a thin-walled tube these strains also apply.

For the flat non-symmetric laminate we have the response $k_{xy} = b_{16}N_x + b_{26}N_y$. But when the non-symmetric laminate takes the form of a tube, the wall is not free to take up twisting curvature caused by internal pressure. It therefore follows that this curvature must be suppressed by applying an internal twisting moment M_{xy}, which with the direct force resultants makes $k_{xy} = 0$. Thus for the non-symmetric laminate the simultaneous equations to be solved are $\varepsilon_x = a_{11}N_x + a_{12}N_y + b_{16}M_{xy}$; $\varepsilon_y = a_{12}N_x + a_{22}N_y + b_{26}M_{xy}$; $\kappa_{xy} = 0 = b_{16}N_x + b_{26}N_y + d_{66}M_{xy}$. Hence $M_{xy} = -(b_{16}N_x + b_{26}N_y)/d_{66}$. Letting $N_x = 5$ N/mm, and $N_y = 10$ N/mm, we have $M_{xy} = -(119.3 \times 5 + 42.69 \times 10)/1512 = -0.677$ N. The hoop strain ε_x is found to be $\varepsilon_x = (47.45 \times 5 - 20.41 \times 10 + 119.3 \times (-0.677)) \times 10^{-6} = -47.62 \times 10^{-6}$. Similarly $\varepsilon_y = 742.95 \times 10^{-6}$.

Thus we see that the strain responses to tubes made from both laminates (a) and (b) are identical (within rounding errors), provided the boundary conditions are correctly identified.

Outline 8.7: Note that the angle has been specified with reference to the circumferential direction, so the natural co-ordinate system to use to solve the problem is x for the hoop direction and y for the axial direction. This is not the normal convention, but provides a good opportunity to adapt the theory to this circumstance.

The wall thickness $h = 3$ mm, diameter $D = 120$ mm, pressure $p = 0.4$ MPa $= 0.4$ N/mm^2. The hoop stresses may be expressed in terms of forces per unit width: $N_H = \sigma_H h = pD/2 = 24$ N/mm, $N_A = 12$ N/mm.

We may now calculate the strains using $[\varepsilon^\circ] = [a][N]$:

$$\varepsilon_H^\circ = a_{11}N_H + a_{12}N_A = (7.383 \times 24 - 19.87 \times 12) \times 10^{-3} = -0.0612$$

$$\varepsilon_A^\circ = a_{12}N_H + a_{22}N_A = (-19.87 \times 24 + 59.31 \times 12) \times 10^{-3} = +0.2348$$

$$\gamma_{AH}^\circ = a_{16}N_H + a_{26}N_A = (-0.444 \times 24 + 0.091 \times 12) \times 10^{-3} = -0.00956$$

Hence under air pressure (and ignoring localized end effects) the new diameter is $D(1 + \varepsilon_H) = 112.66$ mm, new length $= L(1 + \varepsilon_A) = 12.348$ m, and there is a radial twist $\theta = L\gamma_{AH}/R = 10 \times -0.00956/0.06 = -1.59^c = -91.29^\circ$, as shown in Figure A5.9.

The mass of water is not negligible. At the top end of the hose the axial force per unit width is $N_{Aw} = D\rho g L/4 = 0.12 \times 1000 \times 9.81 \times 10/4 = 0.03 \times 10^5$ N/m $= 3$ N/mm. Under an internal water pressure of 0.4 MPa we have the hoop force $N_H = 24$ N/mm as before, but the axial force is now increased to $N_A = 15$ N/mm. This leads to $\varepsilon_H = -0.121$, $\varepsilon_A = +0.4128$ and $\gamma_{AH} = -0.0093$. Compared with air pressure using water in this test nearly doubles the hoop strain, gives over half as much axial strain but hardly changes the twist at all. At the lower end of the pipe the response to air and water are the same.

The change in length of the pipe under water alone can be found by establishing the force acting on the underside of an element of tube of cross-section A solely due to the weight of water underneath it. At a distance z from the lower end, the force on the lower side of an element of length dz is $F(z) = A\rho gz$. The extension of the element, $d\Delta = (F/AE)dz$. The increase in length Δ of the pipe is the sum of all the elements, i.e.:

$$\Delta = \int_0^L (A\rho g/AE)z\,dz = A\rho gL^2/2AE = WL/2AE$$

where W is the weight of the water in the pipe, and E is the effective modulus of the material. It can therefore be seen that the total change in length of the pipe under the water is the same as half the total weight of water acting at the lower end.

Outline 8.8: Although a flat sheet made from C + 4NS would develop curvature under a direct stress, this tube cannot bend when direct stresses are applied to it, so where appropriate, it is necessary to include the boundary conditions $\kappa_x = \kappa_y = 0$, in addition to the applied force resultant(s).

(a) and (b): Noting that $b_{12} = 0$, we must solve the four simultaneous equations: $\varepsilon_x = a_{11}N_x + a_{12}N_y + b_{11}M_x$; $\varepsilon_y = a_{12}N_x + a_{22}N_y + b_{22}M_y$; $\kappa_x = 0 = b_{11}N_x + d_{11}M_x + d_{12}M_y$; $\kappa_y = 0 = b_{22}N_y + d_{12}M_x + d_{22}M_y$. The moments M_x and M_y can be re-expressed in terms of N_x, N_y, and the various coefficients: $M_x = (d_{12}b_{22}N_y - d_{22}b_{11}N_x)/(d_{11}d_{22} - d_{12}^2)$ and $M_y = (d_{11}b_{22}N_y - d_{12}b_{11}N_x)/(d_{12}^2 - d_{11}d_{22})$. These values of moments may now be used in the expressions above to calculate ε_x and ε_y under externally applied N_x and N_y.

(c) When the tube is subject to a couple, the external load applied, N_{xy}, induces a shear strain $\gamma_{xy} = a_{66}N_{xy}$. There is no shear-bending coupling, so no moments need to be applied to suppress curvature.

Outline 8.9: For the symmetric laminate, the shear stress is given by $\tau_{xy} = T/(2\pi R^2h) = 30/(2\pi(0.025)^2 \times 5 \times 10^{-4}) = 15.28$ MPa. The force resultant per unit width is $N_{xy} = \tau_{xy}h = 7639$ N/m $= 7.639$ N/mm. This leads to $\gamma_{xy} = a_{66}N_{xy} = 43.49 \times 10^{-6} \times 7.639 = 3.322 \times 10^{-4}$. The angle of twist is $\theta = \gamma L/R = 0.0106^c = 0.607°$.

In tube form the bending curvatures are suppressed by applying internal moments M_x and M_y which can be related to the force N_{xy} using $\kappa_x = 0 = b_{16}N_{xy} + d_{11}M_x + d_{12}M_y$; $\kappa_y = 0 = b_{26}N_{xy} + d_{12}M_x + d_{22}M_y$. This gives $M_x = M_y = N_{xy}(b_{26}d_{11} - b_{16}d_{12})/(d_{12}^2 - d_{11}d_{22})$. Substituting numerical values we find $M_x = 0.8923$ N, and hence $\gamma_{xy} = a_{66}N_{xy} + b_{16}M_x + b_{26}M_y = 331.6 \times 10^{-6}$, which agrees within rounding errors.

The stress profiles are shown in Figure A8.2.

Outline 8.10: $d_{11} = 32.89 \times 10^{-3}$/Nm, hence $w_{max} = F_zL^3d_{11}/3b = 1 \times 0.001 \times 32.89 \times 10^{-3}/3 \times 0.01 = 1.1$ mm.

Outline 8.11: Let the thickness of the composite beam be h_c. Its stiffness is proportional to $E_1h_c^3$, and $E_1 = V_fE_f + V_pE_p$. The mass of beam per unit cross-sectional area is

$$m = \rho_ch_c = \rho_fh_f + \rho_ph_p$$

Figure A8.2 Stress profiles in tube under torque.

i.e. $\rho_c = \alpha V_f + \rho_p$, where $\alpha = \rho_f - \rho_p$. Hence

$$E_1 h_c^3 = [V_f E_f + (1 - V_f)E_p]m^3/(\alpha V_f + \rho_p)^3$$

and for a given mass per unit area, the maximum stiffness is $d(E_1 h_c^3)/dV_f = 0$
$= -3\alpha[V_f E_f + (1 - V_f)E_p]/(\alpha V_f + \rho_p)^4 + (E_f - E_p)/(\alpha V_f + \rho_p)^3$. If $E_p \ll E_f$,
then $E_f(2\alpha V_f - \rho_p) = 0$, so $\rho_p = 2\alpha V_f$. Taking $\rho_f \sim 2500 \text{ kg/m}^3$ and $\rho_p \sim 1000$
kg/m^3, $V_{f\,opt} \sim 0.33$.

Outline 8.12: (a) The principle of the bending deflection calculation is the
same is that for Problem 8.10. $d_{11} = 66.74$ /MNmm, so $w_{bmax} = 2.23$ mm. The
beam also twists in accordance with the bending moment, and the bend-twist
compliance $d_{16} = -40.91$/MNmm. Unlike the plate problems in Chapter 5,
the moment in the beam varies along the length. Recalling the discussion in
Section 2.5.9, the local angle of twist per unit length may be expressed as
$(d\phi/dx) = \kappa_{xy}(x) = d_{16}M_x(x) = d_{16}M(x)/b$, where $M(x)$ is the local bending
moment in the beam at co-ordinate x from the fixed end, in (beam theory)
units of Nm. In a cantilever we know that the bending moment in $M(x) =$
$F_z(L - x)$. The local twist is $d\phi = d_{16}(F_z(L - x)/b)dx$, and integration over
the length of the beam gives the twist, ϕ, at the free end: $\phi = d_{16}PL^2/2b =$
$-40.91 \times 10^{-3} \times 1 \times 0.001/(2 \times 0.001) = -0.02046^c \approx -1.17°$. The principle behind this twisting calculation provides the explanation for the behaviour
of the corrugated cardboard shown in Figure 3.26.

(b) The load F_y will induce bending because of the in-plane longitudinal
modulus $E_x = (a_{11}h)^{-1}$. There will be no twist, but there will be a very small
shear (neglected in (a)) under the in-plane shear load F_y, as caused by the
in-plane shear modulus $G_{xy} = (a_{66}h)^{-1}$. $I_{zz} = hb^3/12 = 2 \times 1000 \times 10^{-12}/$
$12 = 1.667 \times 10^{-10} \text{ m}^4$. $a_{11} = 19.02 \text{ mm/MN} = 19.02 \times 10^{-9} \text{ m/N}$. $v_{bmax} =$
$F_y L^3 h a_{11}/3I_{zz} = 1 \times 10^{-3} \times 2 \times 10^{-3} \times 19.02 \times 10^9/3 \times 1.667 \times 10^{-10} =$
0.076 mm. $N_{xy} = F_y/b = 0.01 \text{ N/mm}$, so $\gamma_{xy} = a_{66}N_{xy} = 45.81 \times 0.1 = 4.581 \times$
10^{-6}, which gives the negligible $v_s = L\gamma_{xy} = 0.00046$ mm.

Outline 8.13: A balanced symmetric laminate is desirable to eliminate unwanted coupling effects. The simplest approach is to assume that $0°$ plies will
take much (but not all) of the bending load, and that $\pm 45°$ plies will cope with
much of the shear.

For simplicity we assume that a tube of the same mean diameter may be used. Thus we seek a composite tube in which $G_s J_s = G_c J_c$ and $E_s I_s = E_c I_c$, where subscripts s and c denote steel and composite. From Chapter 2 we know that for a thin-walled tube $I = \pi R_m^3 h$ and $J = 2I$, so the stiffness criteria reduce to a pair of simultaneous equations: $G_o h_o + G_{45} h_{45} = G_s h_s$, and $E_o h_o + E_{45} h_{45} = E_s h_s$, in which o refers to properties in the principal directions and 45 to the 45° directions. $h_c = h_o + h_{45}$.

The values of E_{45} and G_{45} must relate to a balanced symmetric angleply laminate $(45°/-45°)_s$, and not just to a single ply, in order to include the constraint of suppressing shear coupling. Values of density derive from E_1 and E_f using the rule of mixtures, $\rho_c = \rho_f V_f + \rho_m V_m$.

The data in the tables lead to the following:

Property	Steel	EG	HM–CARB
E_o GPa	208	45.6	180
G_o GPa	81.3	5.14	5
E_{45} GPa	–	15.5	22.7
G_{45} GPa	–	12.84	15.1
ρ_c kg/m^3	7830	2070	1522

We have two stiffness equations which can be solved simultaneously to obtain the total thickness needed of each ply orientation, and hence the total number of plies n needed. The results of these calculations, with an estimate of the ratio of the mass of the composite pipe to the steel pipe, are as follows:

Material	h_o	n_o	h_{45}	n_{45}^*	Relative mass
EG	2.79	22	5.2	42	2.1
HM-CARB	0.918	8	5.08	40	1.17

* Half the 45° plies will be set at $+45°$ and the other half at $-45°$.

Outline 8.14: The ratio E_x/G_{xy} is particularly large for unidirectional plies having a large anisotropy ratio, and is sensitive to changes in angle of the applied stress to the principal directions. See ply NE for $\theta = 0°$ and 30° in Chapter 3. The use of crossply structures may retain high anisotropy ratios (if the basic ply has a high anisotropy ratio), but the ratio E_x/G_{xy} will fall dramatically as the crossply laminate is rotated with respect to the stress direction; at 45° the peak in $G_{xy}(\theta)$ may give a fractional ratio of E_x/G_{xy}. See the sequence of laminates $C + 4S$, $C + 4S/-15$, $C + 4S/-30$, and $C + 4S/-45$ in Section 5.5.6.

Index